WITHDRAWN FROM LIBRARY

MONTGOMERY COLLEGE
ROCKVILLE CAMPUS LIBRARY
ROCKVILLE, MARYLAND

DIETARY PHYTOCHEMICALS IN CANCER PREVENTION AND TREATMENT

Edited under the auspices of the

American Institute for Cancer Research

Washington, D.C.

PLENUM PRESS • NEW YORK AND LONDON

DEC 04 2003

281617

Library of Congress Cataloging-in-Publication Data

Dietary phytochemicals in cancer prevention and treatment / edited under the auspices of the American Institute for Cancer Research.
p. cm. -- (Advances in experimental medicine and biology ; v. 401)
"Proceedings of the American Institute for Cancer Research's sixth annual research Conference on Dietary Phytochemicals in Cancer Prevention and Treatment, held August 31-September 1, 1995, in Washington, D.C."--T.p. verso.
Includes bibliographical references and index.
ISBN 0-306-45365-7
1. Cancer--Chemoprevention--Congresses. 2. Cancer--Nutritional aspects--Congresses. 3. Botanical chemistry--Congresses. 4. Cancer--Diet therapy--Congresses. I. American Institute for Cancer Research. II. Conference on Dietary Phytochemicals in Cancer Prevention and Treatment (1995 : Washington, D.C.) III. Series.
[DNLM: 1. Neoplasms--prevention & control--congresses. 2. Plants, Edible--chemistry--congresses. 3. Anticarcinogenic Agents--congresses. W1 AD559 v.401 1996 / QZ 200 D5655 1996]
RC268.15.D53 1996
616.99'4--dc20
DNLM/DLC
for Library of Congress 96-24704
CIP

Proceedings of the American Institute for Cancer Research's Sixth Annual Research Conference on Dietary Phytochemicals in Cancer Prevention and Treatment, held August 31 – September 1, 1995, in Washington, D.C.

ISBN 0-306-45365-7

© 1996 Plenum Press, New York
A Division of Plenum Publishing Corporation
233 Spring Street, New York, N. Y. 10013

10 9 8 7 6 5 4 3 2

All rights reserved

No part of this book may be reproduced, stored in a retrieval system, or transmitted in any form or by any means, electronic, mechanical, photocopying, microfilming, recording, or otherwise, without written permission from the Publisher

Printed in the United States of America

PREFACE

The sixth annual research conference of the American Institute for Cancer Research was held August 31 and September 1, 1995, at the Loews L'Enfant Plaza Hotel in Washington, DC. In view of the promising leads in the diet/nutrition and cancer research field, the conference was devoted to "Dietary Phytochemicals in Cancer Prevention and Treatment." The number of sessions was increased over that in previous conferences in order to accommodate the topics of interest. The conference overview, entitled "Plants and Cancer: Food, Fiber, and Phytochemicals," provided a framework for the following sessions. In addition, the attendees were reminded that for several decades epidemiologists have noted a lower risk of lung, esophageal, stomach, and colon cancer in populations consuming diets high in fruits and vegetables. However, isolation and ingestion of individual protective factors are not the preferred action since the complexity of the food and the matrix in which nutritional factors are embedded are important. The individual sessions then provided more insight as to why eating fruits and vegetables is associated with a lower risk of cancer.

The first of these sessions was on "Isothiocyanates" that induce both the Phase I and Phase II enzymes that increase detoxification and conjugation reactions, thus causing more rapid removal of any xenobiotic or carcinogen. Thus, less carcinogen is available for interaction with DNA or other critical cellular macromolecules. Natural isothiocyanates include phenethylisothiocyanate (PEITC) and benzylisothiocyanate (BITC) that occur as glucosinolates, bound in a complex fashion to the sugar glucose, in turnips, watercress, or cabbage. Chewing or mincing the vegetable causes the endogenous vegetable enzyme myrosinase to hydrolyze the glucosinolate; a rearrangement affords the free isothiocyanate that is the chemopreventive agent. In both mice and rats PEITC inhibited lung tumors from 4-(methylnitrosamino)-1-(3-pyridyl)-1-butanone (NNK), one of the major lung carcinogens in tobacco smoke, but it had no effect against lung tumors from benzo(a)pyrene (BP), another carcinogen in tobacco smoke. However, BITC inhibited the lung tumorigenicity of BP in mice. In addicted smokers who ate watercress at each meal for several days, there was higher urinary excretion of detoxified conjugates of NNK, indicating that the PEITC from the watercress blocked the metabolic activation pathway of NNK and increased the detoxification mode. The effect of isothiocyanates was explored further with a different animal model, namely esophageal cancer induced in rats by nitrosomethylbenzylamine (NMBA). Although PEITC was a fairly good inhibitor of NMBA, if given before and during NMBA administration, increasing the carbon chain from phenethyl to phenylpropyl (PPITC) afforded an even more potent compound. However further increases in chain length had little effect or even increased esophageal tumor response. Correspondingly, rats given PEITC or PPITC had lower levels of DNA-methylated products (O^6-methylguanine) in the esophagi than did control rats or those receiving other isothiocyanates.

Overall, some isothiocyanates inhibited esophageal carcinogenesis while others were more effective against lung carcinogens. Further investigation may delineate the separate mechanisms responsible and establish the role of isothiocyanates as chemopreventive substances.

Polyphenols, which are ubiquitous in plant products, were the subject of another session. Various precursors, including caffeic, ferulic, hydroxycinnamic acids, and others, are present in plants; when the plant material is ingested, the bacteria in the gastrointestinal tract convert these precursors to diols and lactones that have antioxidant properties. In model experiments plant phenolics showed antioxidant effects, inhibited nitrosation of secondary amines from endogenous nitrite, trapped electrophiles, inhibited the cyclooxgenase pathway of arachidonic acid metabolism, inhibited protein kinases, and showed antiestrogenic activity. More specifically, the polyphenols in green tea, epicatechin, epigallocatechin, and their gallate esters were noted as inhibiting the initiation, promotion, and progression stages of mouse skin carcinogenesis. Mechanistically, they depressed the activity of P450-dependent enzymes and decreased the levels of carcinogen–DNA adducts in the skin. In addition, extracts of green tea protected against tumor formation in lung, esophagus, gastrointestinal tract, liver, and other internal organs in model systems for both mice and rats. However, epidemiological studies of populations drinking tea versus those using other beverages are not conclusive as to whether green tea affords protection in humans.

Flavonoids, a subset of plant polyphenols, were the basis for presentations on somewhat diverse and complex investigations. An overview provided some insight into the interaction between estrogens and phytoestrogens, such as genistein and lignans, the relationship to breast cancer, and the factors involved in breast cancer risk, including age at menarche, pregnancy, and age at menopause. Estrogens act through estrogen receptors that may have different conformations, depending on whether the interacting substances are agonists, partial agonists, or antagonists. Some flavonoids can modulate the effects of estrogen by binding to the receptor or by directing the metabolism of estrogen toward the catechol estrogen path, yielding antiestrogens. Some natural products such as zearalenone act like estrogens and may represent a deleterious factor.

A study on genistein showed that although it reduced mammary tumors in female rats given a potent carcinogen, it did not appear to act via inhibition of a protein kinase. Labeled genistein was extensively recirculated in rats, and only a fifth of the label was excreted in a week. In humans consumption of a soy beverage led to plasma levels of genistein considered sufficient to regulate proliferation of epithelial cells in the breast and thus provide a chemopreventive effect.

Mechanistic studies on quercetin, a polyhydroxyflavone with wide distribution in fruits and plants, showed that it has multiple biochemical effects in mammalian cells, increasing cyclic adenosine monophosphate (cAMP), decreasing protein kinases, decreasing cAMP and cyclic guanosine monophosphate phosphodiesterases, and interacting with some estrogen binding sites. In H-*ras*-transformed mouse 3T3 fibroblastic cells, quercetin suppressed growth in a dose-dependent manner, perhaps by inactivating the p21 protein in the *ras*-transformed cells. In human mammary carcinoma cells carrying mutant p53, quercetin also decreased cell growth, probably by inhibiting translation of the p53 protein.

Other widely distributed plant products are the monoterpenes, *d*-limonene and perillyl alcohol, both of which have shown inhibitory action on carcinogenesis in animal models. These compounds affect isoprenylation of specific proteins that are *ras* gene products. *d*-limonene and mevinolin, which also affects protein isoprenylation, inhibited the growth of the CT-26 mouse colon tumor line in culture. Given in the diet, these agents decreased the growth of colonies of CT-26 cells that had been implanted in the spleen. Perillyl alcohol, a metabolite of *d*-limonene, was effective in chemoprevention of chemically induced rat mammary cancer, as well as in reducing significantly the growth of a transplantable pancreatic ductal adenocarcinoma in hamsters or even causing complete regression. Further mechanistic studies on the monoterpenes revealed that *d*-limonene impaired DNA synthesis in myeloid and lymphoid leukemia cell lines and did not cause accumulation of an unmodified RAS protein. Perillyl alcohol also decreased RAS protein levels, possibly by impairment of RAS transcription. However, the monoterpenes had no effect on hydroxymethyglutaryl coenzyme A reductase activity, an indication that they apparently do not inhibit farnesyl protein transferase.

Another diverse class of chemopreventive agents includes the organosulfides, present in *Allium* and *Brassica* vegetables. Many organosulfides are considered safe and are allowed as food additives. An overview of various epidemiological and laboratory studies showed stronger evidence for the preventive effects of garlic in animal carcinogenesis models than did the epidemiological reports. Although organosulfides from garlic or garlic extracts inhibited the growth of various tumor cell lines in culture or transplantable tumors in animals, extrapolation to reasonable levels of human consumption from the amounts found effective in the animal experiments is dubious. In *Allium* plants, the organosulfur compounds occur as S-allylcysteine S-oxides; when the plant is crushed or cut, C-S lyase enzymes (alliinases) convert the oxides to sulfenic acids that condense to form thiosulfinate esters. One of these had antitumor activity in animals. Selenium and selenoamino acids are also found in garlic, onions, broccoli, and cabbage. Separating the various sulfur and selenium constituents of these vegetables has been a challenge from the chemical aspect. Delineation of the dynamics of the metabolic disposition of these substances has also been demanding. Although animal experiments have shown inhibitory effects of the garlic constituents, in other situations they have enhanced or promoted tumor activity. Thus, extrapolation to humans must be approached cautiously. Nevertheless, S-allylcysteine, a water-soluble compound from garlic, decreased the incidence of dimethylhydrazine-induced colon tumors in mice and the nucleotoxicity from the carcinogen. The garlic agents are strong inducers of gluthathione-S-transferase, an enzyme involved in detoxification of many electrophiles. Furthermore, the garlic organosulfur compounds appear to suppress CYP2E1, the specific isoform of P450 that often governs the metabolic activation of several carcinogens.

A final paper in this session described combining the known chemopreventive effects of selenium, as selenite or selenomethionine, with the active constituents from garlic. Growing garlic bulbs in a greenhouse medium highly enriched with sodium selenite or selenate yielded garlic with selenium concentrations ranging up to 1300 ppm versus the usual 0.03 ppm in ordinary garlic. This high-selenium garlic was more effective than natural garlic in preventing mammary cancer in rats treated with 7,12-dimethylbenz (a)anthracene and was not toxic. Further trials led to the conclusion that the effect was due to the selenium rather than the garlic. The high-selenium garlic led to approximately a 50% decrease in carcinogen–DNA adducts, correlating reasonably well with the decrease in tumor incidence. This special garlic did not cause any large increase in selenium levels in tissues, affect selenoenzymes, or have an apparent toxic action.

The last session was devoted to the practical aspects of including phytochemicals in the U.S. diet, their role in cancer prevention, and whether dietary changes are needed. The following cautions were presented: Definitive data on the toxicity in animals and humans are not available, nor are data on bioavailability and levels of phytochemicals in diets; food composition tables are not refined enough to permit estimates of consumption of phytochemicals; phytochemical content of foods differs depending on many factors; many Americans fail to meet current recommendations for including fruits, vegetables, and grains in their diets; and there still are questions on any recommendations for including more or specific phytochemicals in diets. Appropriately, the National Cancer Institute had launched the National 5 A Day for Better Health Program, which encourages people to eat a combination of five servings of fruits and vegetables each day. Although it is not known which combinations of phytochemicals provide a chemopreventive effect, epidemiological studies indicate lower risks of various types of cancer in those eating more fruits and vegetables. The NCI is cooperating with the media, retail stores, the food industry, the research and evaluation segment, and the community through state health agencies in order to deliver the message about diet and cancer prevention to Americans.

The viewpoint of part of the food industry on phytochemicals was expressed in the remaining presentations. Official guidelines on dietary intake have urged Americans to consume less high-fat animal products and increase intake of fruits, vegetables, beans, and grains. The current USDA Food Guide Pyramid emphasizes grains as the foundation of the diet, followed by vegetables and fruits, then dairy and meat, with fats and sweets to be used sparingly. However, it still allows fairly generous portions of animal foods and includes beans, seeds, and nuts with meat and dairy foods. Since beans, seeds, and nuts have different nutrient, fiber, and phytochemical contents from those in meats, there should be revision of the guidelines. A transition from animal-based to plant-based diets should be emphasized. However, the food industry should not overpromise with phytochemicals, and since about half of the population does not eat sufficient fruits and vegetables or meet dietary guidelines, the recommended dietary levels for vitamins should be revised or reconsidered. In addition, there may be a need to develop hypernutritious foods to improve diets. As a sequel, improved diets would lead to disease prevention and lower healthcare costs, as well as new markets and new types of business for the food industry. However, some beneficial foods, such as high-fiber cereals and brans, especially wheat, are already available in grocery stores. In model animal experiments, dietary fiber has inhibited colon cancer, and it aids in removal of circulating estrogens, thus decreasing a risk factor for breast cancer.

At the poster session there were 72 presentations on various aspects of research in the phytochemical/cancer area. An additional evening session was devoted to a workshop, "Diet, Nutrition and Cancer Prevention: Research Opportunities, Approaches and Pitfalls." This session covered such topics as strengthening research on diet and cancer prevention, approaches to increase funding, improving grant applications, and opening lines of communication between funding agencies and research workers.

To summarize, many varied types of foods containing desirable phytochemicals with cancer preventive activities are readily available. Because of the concerns voiced during the last session of the conference, it is not prudent to isolate these substances and use them alone. Thus, educational motivation is needed to encourage everyone to eat a variety of fruits, vegetables, and grain products each day to ensure an approach to better health.

The Editors

CONTENTS

Chapter 3

Chapter 4

Chapter 5

1

CHEMOPREVENTION OF LUNG CANCER BY ISOTHIOCYANATES

Stephen S. Hecht

American Health Foundation
1 Dana Road
Valhalla, New York 10595

ABSTRACT

Naturally occurring and synthetic isothiocyanates are among the most effective chemopreventive agents known. A wide variety of isothiocyanates prevent cancer of various tissues including the rat lung, mammary gland, esophagus, liver, small intestine, colon, and bladder. Mechanistic studies have shown that the chemopreventive activity of isothiocyanates is due to favorable modification of Phase I and Phase II carcinogen metabolism, resulting in increased carcinogen excretion or detoxification and decreased carcinogen DNA interactions. In the majority of studies reported, the isothiocyanate must be present at the time of carcinogen exposure in order to observe inhibition of tumorigenesis. Our studies have focused on the naturally occurring isothiocyanates phenethyl isothiocyanate (PEITC) and benzyl isothiocyanate (BITC) as inhibitors of lung cancer. The carcinogens employed in these studies have been the major lung carcinogens in tobacco smoke – 4-(methylnitrosamino)-1-(3-pyridyl)-1-butanone (NNK) and benzo[a]pyrene (BaP). Combinations of chemopreventive agents that inhibit tumorigenesis by NNK and BaP in rodents may be effective in addicted smokers. PEITC is an effective inhibitor of lung tumor induction by NNK in F-344 rats and A/J mice. BITC but not PEITC inhibits BaP induced lung tumorigenesis in A/J mice. PEITC is a selective inhibitor of the metabolic activation of NNK in the rodent lung, and studies in smokers who consumed watercress, a source of PEITC, indicate that the metabolic activation of NNK is also inhibited by PEITC in humans. Combinations of chemopreventive agents active against different carcinogens in tobacco smoke may be useful in the chemoprevention of lung cancer.

INTRODUCTION

Isothiocyanates (R–N=C=S) occur as their glucosinolate conjugates in a wide variety of cruciferous vegetables.[1] When vegetable cells are damaged, the enzyme myrosinase is released which catalyzes the hydrolysis of glucosinolates. Isothiocyanates are then formed

Dietary Phytochemicals in Cancer Prevention and Treatment
Edited under the auspices of the American Institute for Cancer Research, Plenum Press, New York, 1996

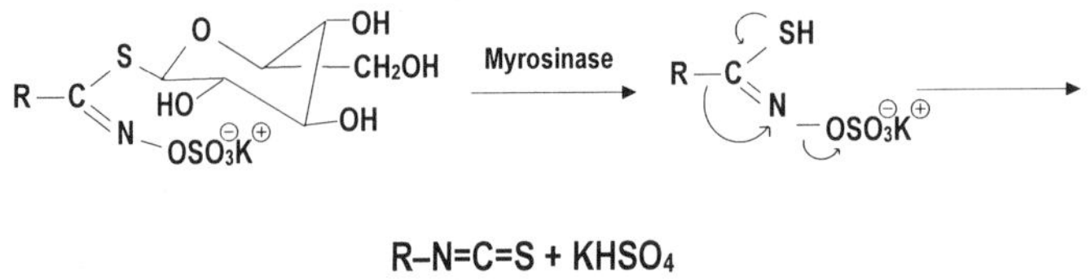

$R-N=C=S + KHSO_4$

Figure 1. Formation of isothiocyanates from glucosinolates.

by a Lossen type rearrangement, as illustrated in Figure 1. Isothiocyanates are responsible in part for the sharp taste associated with certain cruciferous vegetables. Consumption of normal amounts of vegetables such as watercress or cabbage releases milligram amounts of isothiocyanates.[1,2]

Naturally occurring and synthetic isothiocyanates have been tested as chemopreventive agents, and over twenty compounds with a variety of structural features have been assessed.[3-25] Isothiocyanates prevent tumors in numerous target tissues including rat lung, mammary gland, esophagus, liver, small intestine, colon, and bladder. A high degree of selectivity with respect to target tissue and isothiocyanate structure has been observed in some studies. For example, phenethyl isothiocyanate (PEITC) inhibits lung cancer in rats treated with 4(methylnitrosamino)-1-(3pyridyl)-1butanone (NNK), but has no effect on tumor induction in the liver or nasal cavity of these animals.[17] In mice, various isothiocyanates are effective inhibitors of lung and forestomach tumors, but none have shown efficacy against skin carcinogenesis.[13] Studies by Chung and co-workers demonstrated that the lipophilicity of the isothiocyanate increases chemopreventive efficacy against NNK in mouse lung while reactivity with glutathione decreases efficacy.[19,23] The rich variety of isothiocyanate inhibitors already discovered suggests that mechanism based structure activity studies can lead to the design of highly effective isothiocyanates for inhibition of carcinogenesis in a variety of systems.

In studies carried out to date, most isothiocyanates have shown chemopreventive activity in protocols involving administration of the isothiocyanate either before or during exposure to the carcinogen. One exception is benzyl isothiocyanate (BITC) which inhibited the induction of mammary tumors in rats by 7,12dimethylbenz[*a*]anthracene (DMBA) when given after the carcinogen.[11] Mechanistic studies have clearly shown that isothiocyanates are effective inhibitors of cytochrome P450 enzymes which metabolize carcinogens as well as being enhancers of certain Phase II enzymes such as NAD(P)H:quinone reductase, glutathione-S-transferase and UDP-glucuronyl transferase involved in carcinogen detoxification, consistent with their activity as blocking agents. Enzymatic mechanisms involved in inhibition of carcinogenesis by isothiocyanates have been extensively studied.[26,27] There is a significant degree of structural specificity in the effects of isothiocyanates on enzymes involved in carcinogen activation and detoxification. For example, treatment of mice with PEITC had no effect on ethoxyresorufin Odealkylase (EROD) activity, associated with P450 1A, while BITC was a strong inhibitor of this enzyme in lung.[28] As discussed below, these observations correlate with the specific chemopreventive activities of PEITC and BITC against mouse lung tumorigenesis by NNK or BaP.

Inhibition of NNK-Induced Lung Tumorigenesis by PEITC

Chemoprevention may be a way to decrease the risk for lung cancer in smokers who cannot give up the tobacco habit, even after having undergone smoking cessation programs. Smokers who have quit, but are still at higher risk for lung cancer, may also be helped by

Table 1. Incidence of lung, liver, and nasal cavity tumors after treatment with NNK, NNK + PEITC, and PEITC

		Number of rats with tumors (%)								
		Lung			Liver			Nasal Cavity		
Treatment	# of rats	Adenoma	Carcinoma	Total	Adenoma	Hepatocellular carcinoma	Total	Benign[a]	Malignant[b]	Total
NNK	40	8	24	32 (80)	12	3	15 (38)	8	3	11 (28)
NNK + PEITC	40	5	12[c]	17 (43)[d]	9	5	14 (35)	6	1	7 (18)
PEITC	20	0	0	0 (0)	4	2	6 (30)	0	1	1 (5)
Control	20	1	0	1 (5)	3	1	4 (20)	0	1	1 (5)

a. Squamous cell papillomas, transitional-cell papillomas, polyps
b. Squamous cell carcinoma
c. One squamous cell carcinoma, 11 adenocarcinoma
d. $P<0.05$ compared to NNK group

chemoprevention approaches. The American Cancer Society estimated that there would be 157,000 deaths from lung cancer in the United States in 1995.[29] At least 80% of these deaths will have been caused by cigarette smoking.[30] If the use of chemopreventive agents could delay or prevent the development of lung cancer in even a small percentage of addicted smokers, a large number of these deaths could be avoided.

Our approach to the development of effective chemopreventive agents for lung cancer has been to focus on inhibiting the activity of lung carcinogens in tobacco smoke. Among the approximately 50 known carcinogens present in tobacco smoke, polynuclear aromatic hydrocarbons (PAH), typified by BaP, and the tobacco-specific nitrosamine NNK are the most likely causes of lung cancer in smokers. This evaluation is based on tumor induction studies in laboratory animals, biochemical studies with lung tissue and cells from laboratory animals and humans, and detection of DNA adducts in the lungs of smokers. In smokers, exposure to PAH and NNK is chronic, resulting in a steady state level of DNA adducts of various types which are plausibly the cause of the multiple genetic alterations in oncogenes and tumor suppressor genes associated with the carcinogenic process. Since these compounds require metabolic activation, agents which decrease the extent of formation of the resulting electrophilic DNA binding intermediates should decrease the extent of DNA damage and thereby inhibit carcinogenesis. Isothiocyanates, both naturally occurring and synthetic, are able to inhibit the metabolic activation and carcinogenicity of PAH and NNK and they have been our main target for development as chemopreventive agents. Tobacco smoke also contains tumor promoters and, in the classical initiation-promotion paradigm, inhibitors of tumor promotion should also be effective chemopreventive agents. Therefore, we believe that it is likely that combinations of chemopreventive agents including those that inhibit metabolic activation as well as those that inhibit oxidative damage and tumor promotion will eventually be necessary for maximum effectiveness in the prevention of lung cancer in smokers.

PEITC has received the most attention in our studies to date. PEITC is a naturally occurring isothiocyanate, being found as its glucosinolate conjugate gluconasturtiin in several vegetables including watercress. PEITC is released from watercress upon chewing by the action of myrosinase. Consumption of approximately 50 g of watercress releases 1015 mg of PEITC.[2]

When PEITC was added to NIH-07 diet at a concentration of 498 ppm (3 mol/g diet) before and during treatment of male F-344 rats with NNK, it caused a significant and selective 50% reduction in the incidence of adenocarcinoma of the lung[17] (Table 1). There were no toxic effects of PEITC at this dose. A single dose of 5 mol of PEITC administered to A/J mice 2h prior to treatment with 10 mol of NNK resulted in a significant 62% reduction in lung tumor multiplicity.[20] Other studies using multiple doses of PEITC have shown similar results in A/J mice.[18,19] Thus, PEITC has been firmly established as an effective inhibitor of lung tumorigenesis induced by NNK in both rats and mice.

An overview of the major metabolic activation and detoxification pathways of NNK is illustrated in Figure 2.[31] In laboratory animals and humans, NNK is rapidly converted to 4(methylnitrosamino)-1-(3-pyridyl)-1-butanol (NNAL) by carbonyl reductase enzymes. NNAL is also a potent pulmonary carcinogen. NNAL is partially converted to its diastereomeric glucuronides, NNAL-Gluc. These glucuronides are believed to be detoxification products of NNK. Pyridine N-oxidation of NNK and NNAL gives the corresponding N-oxides which are detoxification products. Metabolic activation of NNK proceeds by α-hydroxylation of the methylene and methyl carbons producing unstable intermediates **1** and **2**. These spontaneously decompose with formation of aldehydes and the electrophilic diazohydroxides **4** and **5**. Diazohydroxide **4** methylates DNA of NNK target tissues producing permanent mutations, mainly of the G to A type. Diazohydroxide **5** alkylates DNA producing both G to A and G to T mutations. It also reacts with hemoglobin to form ester

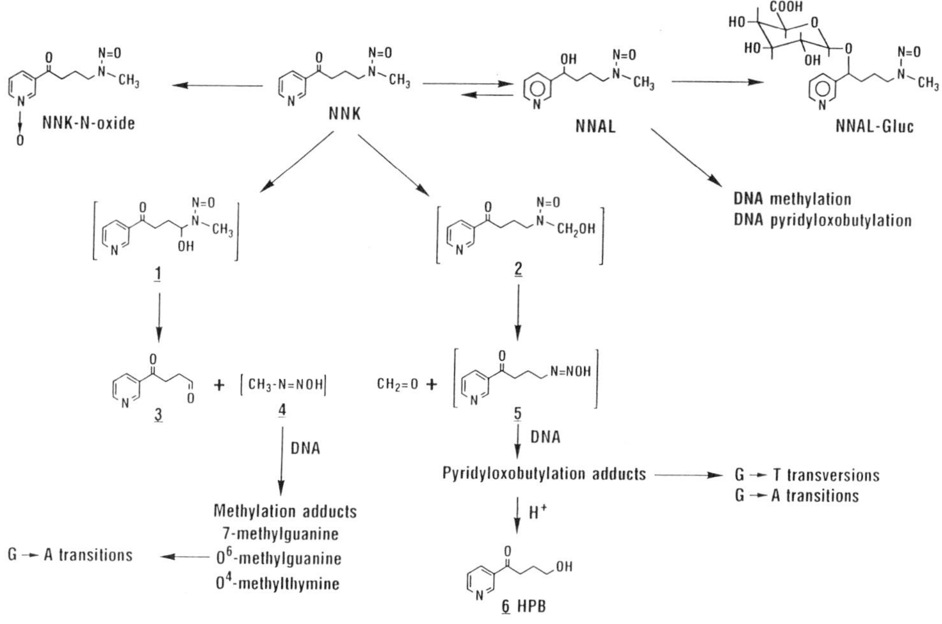

Figure 2. Metabolism of NNK: an overview

adducts. Hydrolysis of DNA or hemoglobin obtained from animals treated with NNK or from smokers produces HPB (**6**), which is a biomarker of the metabolic activation of NNK.[32] Smokers' urine contains quantifiable amounts of NNAL and NNAL-Gluc as biomarkers.

The mechanism of inhibition of NNK carcinogenesis by PEITC has been examined. Initial studies demonstrated that PEITC inhibited the metabolic activation of NNK to electrophiles which methylate and pyridyloxobutylate pulmonary DNA in rats; inhibition of hemoglobin adduct formation was also observed.[17] Subsequently, detailed investigations of the effects of PEITC on NNK metabolism in mouse and rat liver and lung, as well as studies of other enzyme activities have clearly demonstrated that the inhibitory effect of PEITC on NNK carcinogenesis is due to inhibition of NNK metabolic activation to methylating and pyridyloxobutylating electrophiles and enhancement of its detoxification.[26,28] In rats treated with PEITC by gavage or by addition to the diet, a persistent inhibition of metabolic activation of NNK is observed in lung microsomes, resulting from inhibition of cytochrome P450 enzymes. In contrast, a persistent inhibition in liver microsomes is not observed; rather, there is induction after initial inhibition. Experiments in vitro have shown that PEITC is a competitive inhibitor of NNK metabolic activation in rat lung microsomes and in explants of rat lung, with IC_{50} ranging from 150-210 nM.[28,33]

The effects of PEITC on NNK metabolism have also been examined *in vivo*. In these experiments, the goal was to determine whether the observed inhibition of tumorigenesis was due to specific inhibition of metabolic activation of NNK, or whether treatment with PEITC might have caused a change in distribution of NNK resulting in diminished amounts of the carcinogen reaching extrahepatic tissues. In experiments carried out using a protocol essentially identical to that employed in the carcinogenicity study described above, it was shown that the levels of NNK and its primary metabolite NNAL were not markedly different in tissues of PEITC treated and control rats. However, the data clearly indicated a decrease

Table 2. Effect of PEITC on excretion of NNAL and NNALGluc in rats treated with NNK

A. 68 week data	Metabolites in Urine (nmol/24 h)[b]			
Group[a]	NNAL	NNAL-Gluc	Total	Fold Increase in Total
NNK	4.1 ± 1.0	9.4 ± 3.8	13.5 ± 4.8	–
NNK + PEITC	11.6 ± 2.5	44.6 ± 15.8	56.1 ± 17.9	4.2*
B. 79 week data	Metabolites in Urine (nmol/24 h)[c]			
Group[a]	NNAL	NNAL-Gluc	Total	Fold Increase in Total
NNK	3.1	8.8	11.9	–
NNK + PEITC	14.3 ± 5.7	59.6 ± 36.8	73.9 ± 42.4	6.2*

[a]NNK in drinking water (2 ppm): PEITC in diet (3 μmol/g) for 68 or 79 weeks

[b]Mean ± S.D., n=3

[c]Mean of 2 rats (NNK group); mean ± S.D. (n=3), PEITC group

* $P<1x10^{-4}$

S. Carmella, A. Borukhova, N. Trushin, and S.S. Hecht, unpublished data.

in the levels of NNK metabolic activation in almost all tissues examined of the PEITC treated rats.[34]

The effects of chronic PEITC treatment on hemoglobin adducts and urinary metabolites of NNK have been examined in rats. Results of the urinary metabolite analyses are summarized in Table 2. Chronic PEITC treatment caused significant 4-6 fold increases in the levels of NNAL and NNAL-Gluc in urine; this most likely results from a decrease in metabolic activation of NNK since hemoglobin adducts of NNK also decreased (data not shown). The ratio of NNAL-Gluc to NNAL, a potential biomarker of NNK detoxification, increased upon PEITC treatment. Collectively, the results of these studies clearly show that PEITC exerts a specific inhibitory effect on the metabolic activation of NNK while enhancing detoxification, without causing any apparent toxic effects in rats.

Inhibition of BaP-Induced Lung Tumorigenesis by BITC

While PEITC is an effective inhibitor of lung carcinogenesis by NNK, studies to date have not demonstrated efficacy with respect to BaP. In one study in A/J mice, PEITC was administered by gavage prior to i.p. injection of BaP. No inhibition of BaP induced lung tumorigenesis was observed over a range of PEITC doses.[22] In a second study, a single dose of 6.7 mol PEITC was given by gavage to A/J mice, 15 min prior to gavage of 7.9 mol of BaP. No inhibition of lung tumorigenesis was observed, although PEITC did inhibit forestomach tumor induction by BaP. In contrast, a 7.9 mol dose of BITC given by the same protocol did result in a statistically significant 50% reduction of BaP-induced lung tumor multiplicity in the A/J mouse model (Table 3).[13] These results are in agreement with previously reported data on inhibition of BaP induced lung tumorigenesis by BITC.[12] The contrasting effects of PEITC and BITC on lung tumorigenesis by BaP in A/J mice are consistent with mechanistic studies discussed above which have shown that BITC but not PEITC significantly inhibited ethoxyresorufin O-dealkylase activity in A/J mouse lung microsomes, indicative of inhibition of P450 1A which may be involved in the metabolic activation of BaP.[28] In ongoing studies, we are examining the effects of BITC and PEITC on the metabolic activation and DNA binding of BaP in A/J mouse lung and liver.

Table 3. Effects of BITC and PEITC on BaP-induced lung and forestomach tumorigenesis in A/J mice[a,b]

	% Mice with Tumors		Tumors per Mouse	
Group	Lung	Forestomach	Lung	Forestomach
BaP only	95	95	4.8	4.8
BITC/BaP	80	95	2.6*	4.9
PEITC/BaP	90	90	4.0	2.5*

*P<0.001
[a]Female A/J mice given 6.7 mol isothiocyanate i.g. 15 min prior to 7.9 mol BaP, 3 times at 2 week intervals and sacrificed 26 weeks after first dose
[b]Reference 13

Effects of Watercress Consumption on NNK Metabolism in Smokers

The studies described above demonstrate that PEITC inhibits NNK induced lung tumorigenesis in rats and mice by inhibiting its metabolic activation. We wanted to determine whether similar effects would occur in smokers. The source of PEITC used in this study was watercress (*nasturtium officinale*), which contains substantial amounts of the gluconasturtin, the glucosinolate precursor of PEITC.[2]

Eleven smokers maintained constant smoking habits and avoided cruciferous vegetables and other sources of isothiocyanates throughout the study. The protocol is summarized in Figure 3. They donated 24h urine samples on 3 consecutive days (baseline period). One to three days later, they consumed 2 oz (56.8 g) of watercress at each meal for three days and donated 24h urine samples on each of these days (watercress consumption period). One and two weeks later they again donated 24h urine samples on 2-3 consecutive days (follow up periods). The samples were analyzed for two metabolites of NNK; NNAL and NNAL-Gluc, as well as PEITC-NAC, a metabolite of PEITC. Minimum exposure to PEITC during the watercress consumption period averaged 19-38 mg per day. Seven of the 11 subjects had increased levels of urinary NNAL plus NNAL-Gluc on Days 2 and 3 of the watercress consumption period compared to the baseline period. Overall, the increase in urinary NNAL plus NNAL-Gluc in this period was significant [mean S.D., 0.924 1.12 nmol/24 h (33.5%), P<0.01]. Urinary levels of NNAL plus NNAL-Gluc returned to near baseline levels in the follow up periods. These data are summarized in Figure 4. The percent increase in urinary NNAL plus NNAL-Gluc during Days 2 and 3 of the watercress consumption period correlated with intake of PEITC during this period as measured by total urinary PEITC-NAC (r=0.62, P=0.04) as shown in Figure 5. The results of this study support our hypothesis that PEITC inhibits the oxidative metabolism of NNK in humans, as seen in rodents, and support further development of PEITC as a chemopreventive agent against lung cancer.

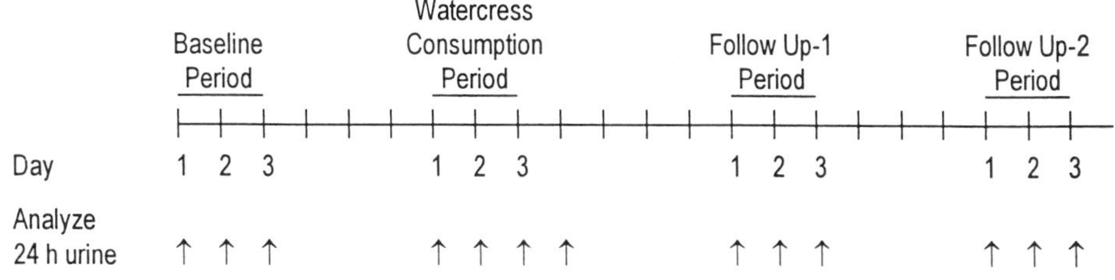

Figure 3. Protocol for examining the effects of watercress consumption on NNK metabolism in smokers.

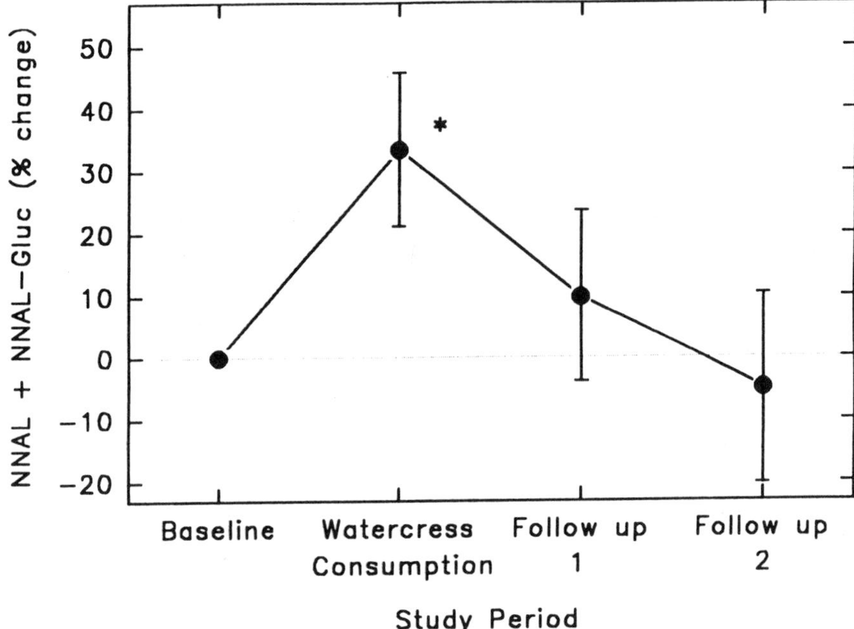

Figure 4. Overall percentage change (meanS.E.) in urinary NNAL plus NNAL-Gluc on Days 2 and 3 of the watercress consumption period and in the follow up periods compared to the baseline period; n=11 for baseline and watercress consumption period, n=8 for follow up1 period, n=5 for follow up2 period. *The difference between the watercress consumption period and baseline period was significant, P<0.01. The differences between the follow up periods and baseline period were not significant.

SUMMARY

Naturally occurring and synthetic isothiocyanates are effective chemopreventive agents in a variety of animal models. We have focused on the development of isothiocyanates as inhibitors of lung cancer. PEITC inhibits lung cancer induction by the tobacco-specific carcinogen NNK in rats and mice, mainly by blocking its metabolic activation to intermediates that cause permanent mutations in DNA. Decreased metabolic activation of NNK results in increased urinary excretion of two metabolites: NNAL and NNAL-Gluc. The latter is believed to be a detoxification product of NNK. One good dietary source of PEITC is watercress. In smokers who consumed watercress, levels of NNAL and NNAL-Gluc increased in urine during the watercress consumption period, indicating that inhibition of NNK metabolic activation is occurring in smokers, as observed in rodents. While PEITC may be an effective chemopreventive agent against NNK, it does not inhibit tumorigenesis by BaP, another important tobacco smoke lung carcinogen. BITC inhibits lung tumorigenesis induced by BaP in mice. It is likely that combinations of agents, including blocking agents such as PEITC and BITC, as well as suppressing agents, will be necessary for chemoprevention of lung cancer in smokers. The best strategy for preventing lung cancer is avoiding tobacco products and chemoprevention should never be considered as an alternative to quitting. However, motivated individuals who are addicted to nicotine may benefit from chemoprevention.

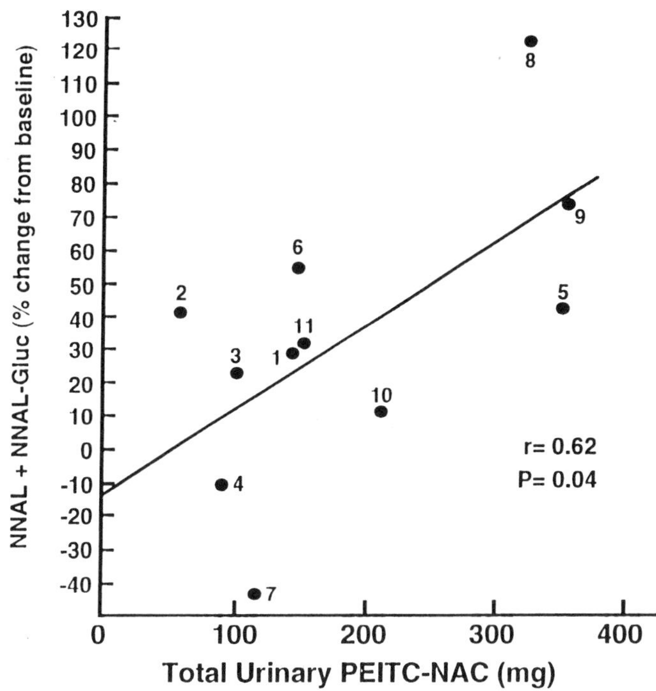

Figure 5. Relationship between percent change from baseline of urinary NNAL plus NNAL-Gluc on Days 2 and 3 of the watercress consumption period and total urinary PEITC-NAC on Days 1-3 of the watercress consumption period for the 11 smokers.

ACKNOWLEDGMENTS

Our studies on chemoprevention are supported by Grant No. CA46535 from the National Cancer Institute.

REFERENCES

1. Sones, K., R.K. Heaney, and G.R. Fenwick, An estimate of the mean daily intake of glucosinolates from cruciferous vegetables in the U.K, *J. Sci. Food Agric.* 35:712 (1984).
2. Chung, F.-L., M.A. Morse, K.I. Eklind, and J. Lewis, Quantitation of human uptake of the anticarcinogen phenethyl isothiocyanate after a watercress meal, *Cancer Epidemiol., Biomarkers Prev.* 1:383 (1992).
3. Sasaki, S., Inhibitory effects by a-naphthylisothiocyanate on liver tumorigenesis in rats treated with 3'-methyl-4-dimethyl-aminoazobenzene, *J. Nara Med. Assoc.* 14:101 (1963).
4. Sidransky, H., N. Ito, and E. Verney, Influence of *a*-naphthyl-isothiocyanate on liver tumorigenesis in rats ingesting ethionine and N-2-fluorenylacetamide, *J. Natl. Cancer Inst.* 37:677 (1966).
5. Lacassagne, A., L. Hurst, and M.D. Xuong, Inhibition, par deux naphthylisothiocyanates, de l'hepato-cancérogenèse produit, chez le rat, par le p-diméthylaminoazobenzène (DAB), *C.R. Séances Soc. Biol. Fil.* 164:230 (1970).
6. Ito, N., Y. Hiasa, Y. Konishi, and M. Marugami, The development of carcinoma in liver of rats treated with m-toluylenediamine and the synergistic and antagonistic effects with other chemicals, *Cancer Res.* 29:1137 (1969).
7. Makiura, S., Y. Kamamoto, S. Sugihara, K. Hirao, Y. Hiasa, M. Arai, and N. Ito, Effect of 1-naphthyl isothiocyanate and 3-methylcholanthrene on hepatocarcinogenesis in rats treated with diethylnitrosoamine, *Jpn. J. Cancer Res.* 64:101 (1973).

8. Ito, N., K. Matayoshi, K. Matsumura, A. Denda, T. Kani, M. Arai, and S. Makiura, Effect of various carcinogenic and non-carcinogenic substances on development of bladder tumors in rats induced by N-butyl-N-(4-hydroxybutyl)nitrosoamine, *Jpn. J. Cancer Res.* 65:123 (1974).
9. Wattenberg, L.W., Inhibition of carcinogenic effects of polycyclic hydrocarbons by benzyl isothiocyanate and related compounds, *J. Natl. Cancer Inst.* 58:395 (1977).
10. Morse, M.A., S.G. Amin, S.S. Hecht, and F.-L. Chung, Effects of aromatic isothiocyanates on tumorigenicity, O^6methylguanine formation, and metabolism of the tobacco-specific nitrosamine 4-(methylnitrosamino)-1-(3-pyridyl)-1-butanone in A/J mouse lung, *Cancer Res.* 49:2894 (1989).
11. Wattenberg, L.W., Inhibition of carcinogen-induced neoplasia by sodium cyanate, tert-butyl isocyanate, and benzyl isothiocyanate administered subsequent to carcinogen exposure, *Cancer Res.* 41:2991 (1981).
12. Wattenberg, L.W., Inhibitory effects of benzyl isothiocyanate administered shortly before diethylnitrosamine or benzo[a]pyrene on pulmonary and forestomach neoplasia in A/J mice, *Carcinogenesis* 8:1971 (1987).
13. Lin, J.-M., S. Amin, N. Trushin, and S.S. Hecht, Effects of isothiocyanates on tumorigenesis by benzo[a]pyrene in murine tumor models, *Cancer Lett.* 74:151 (1993).
14. Morse, M.A., J.C. Reinhardt, S.G. Amin, S.S. Hecht, G.D. Stoner, and F.-L. Chung, Effect of dietary aromatic isothiocyanates fed subsequent to the administration of 4(methylnitrosamino)-1-(3-pyridyl)-1-butanone on lung tumorigenicity in mice, *Cancer Lett.* 49:225 (1990).
15. Sugie, S., A. Okumura, T. Tanaka, and H. Mori, Inhibitory effects of benzyl isothiocyanate and benzyl thiocyanate on diethylnitrosamine-induced hepatocarcinogenesis in rats, *Jpn. J. Cancer Res.* 84:865 (1993).
16. Sugie, S., K. Okamoto, A. Okumura, T. Tanaka, and H. Mori, Inhibitory effects of benzyl thiocyanate and benzyl isothiocyanate on methylazoxymethanol acetate-induced intestinal carcinogenesis in rats, *Carcinogenesis* 15:1555 (1994).
17. Morse, M.A., C.-X. Wang, G.D. Stoner, S. Mandal, P.B. Conran, S.G. Amin, S.S. Hecht, and F.-L. Chung, Inhibition of 4-(methylnitrosamino)-1-(3-pyridyl)-1-butanone-induced DNA adduct formation and tumorigenicity in lung of F344 rats by dietary phenethyl isothiocyanate, *Cancer Res.* 49:549 (1989).
18. Morse, M.A., K.I. Eklind, S.G. Amin, S.S. Hecht, and F.-L. Chung, Effects of alkyl chain length on the inhibition of NNKinduced lung neoplasia in A/J mice by arylalkyl isothiocyanates, *Carcinogenesis* 10:1757 (1989).
19. Morse, M.A., K.I. Eklind, S.S. Hecht, K.G. Jordan, C.-I. Choi, D.H. Desai, S.G. Amin, and F.-L. Chung, Structure-activity relationships for inhibition of 4(methylnitrosamino)-1-(3-pyridyl)-1-butanone lung tumorigenesis by arylalkyl isothiocyanates in A/J mice, *Cancer Res.* 51:1846 (1991).
20. Morse, M.A., K.E. Eklind, S.G. Amin, and F.L. Chung, Effect of frequency of isothiocyanate administration on inhibition of 4-(methylnitrosamino)-1-(3-pyridyl)-1-butanone-induced pulmonary ademona formation in A/J mice, *Cancer Lett.* 62:77 (1992).
21. Stoner, G.D., D. Morrissey, Y.-H. Heur, E. Daniel, A. Galati, and S.A. Wagner, Inhibitory effects of phenethyl isothiocyanate on *N*-nitrosobenzylmethylamine carcinogenesis in the rat esophagus, *Cancer Res.* 51:2063 (1991).
22. Adam-Rodwell, G., M.A. Morse, and G.D. Stoner, The effects of phenethyl isothiocyanate on benzo[*a*]pyrene-induced tumors and DNA adducts in A/J mouse lung, *Cancer Lett.* 71:35 (1993).
23. Jiao, D., K.I. Eklind, C.I. Choi, D.H. Desai, S.G. Amin, and F.L. Chung, Structure-activity relationships of isothiocyanates as mechanism-based inhibitors of 4-(methylnitrosamino)-1-(3-pyridyl)-1-butanone-induced lung tumorigenesis in A/J mice, *Cancer Res.* 54:4327 (1994).
24. Zhang, Y., T.W. Kensler, C.-G. Cho, G.H. Posner, and P. Talalay, Anticarcinogenic activities of sulforaphane and structurally related synthetic norbornyl isothiocyanates, *Proc. Natl. Acad. Sci. USA* 91:3147 (1994).
25. Hecht, S.S., Chemoprevention by isothiocyanates, *J. Cell. Biochem.* [*Suppl.*] 22:195 (1995).
26. Yang, C.S., T.J. Smith, and J.-Y. Hong, Cytochrome P-450 enzymes as targets for chemoprevention against chemical carcinogenesis and toxicity: opportunities and limitations, *Cancer Res.* 54:1982s (1994).
27. Zhang, Y. and P. Talalay, Anticarcinogenic activities of organic isothiocyanates: chemistry and mechanisms, *Cancer Res.* 54:1976s (1994).
28. Guo, Z., T.J. Smith, E. Wang, K.I. Eklind, F.-L. Chung, and C.S. Yang, Structure-activity relationships of arylalkyl isothiocyanates for the inhibition of 4-(methylnitrosamino)-1-(3-pyridyl)-1-butanone metabolism and the modulation of xenobiotic-metabolizing enzymes in rats and mice, *Carcinogenesis* 14:1167 (1993).
29. Wingo, P.A., T. Tong, and S. Bolden, Cancer statistics, 1995, *CA* 45:8 (1995).

30. Shopland, D.R., H.J. Eyre, and T.F. Pechachek, Smoking-attributable cancer mortality in 1991: Is lung cancer now the leading cause of death among smokers in the United States?, *J. Natl. Cancer Inst.* 83:1142 (1991).
31. Hecht, S.S., Metabolic activation and detoxification of tobacco-specific nitrosamines-a model for cancer prevention strategies, *Drug Metab. Rev.* 26:373 (1994).
32. Hecht, S.S., S.G. Carmella, P.G. Foiles, and S.E. Murphy, Biomarkers for human uptake and metabolic activation of tobacco-specific nitrosamines, *Cancer Res.* 54:1912s (1994).
33. Doerr-O'Rourke, K., N. Trushin, S.S. Hecht, and G.D. Stoner, Effect of phenethyl isothiocyanate on the metabolism of the tobacco-specific nitrosamine 4(methylnitrosamino)-1-(3-pyridyl)-1-butanone by cultured rat lung tissue, *Carcinogenesis* 12:1029 (1991).
34. Staretz, M.E. and S.S. Hecht, Effects of phenethyl isothiocyanate on the tissue distribution of 4-(methylnitrosamino)-1-(3-pyridyl)-1-butanone and metabolites in F344 rats, submitted (1995).

2

ISOTHIOCYANATES AS INHIBITORS OF ESOPHAGEAL CANCER

Gary D. Stoner and Mark A. Morse

Division of Environmental Health
The Ohio State University School of Public Health
Arthur G. James Cancer Hospital and Research Institute
Columbus, Ohio 43210

INTRODUCTION

Esophageal cancer in humans occurs worldwide with a variable geographic distribution. A recent estimate has placed it seventh in order of cancer occurrence for both sexes combined[1]. There is a higher incidence in males, with a male to female ratio of 2:1 or higher. The highest incidence rates are found in China, Iran, parts of central Asia, and in the Transkei region of South Africa.[2-4] The disease occurs consistently among the poor in most areas of the world where the diet is often restricted and nutritional imbalance is common. Worldwide, more than 90% of esophageal cancers are squamous cell carcinomas. The characteristics of this neoplasm include keratinization with occasional keratin "pearl" formation and intracellular bridges. About 5% of esophageal cancers are adenocarcinomas. Adenocarcinomas usually arise in Barrett's esophagus, a condition in which the normal squamous epithelium of the esophagus is replaced by glandular epithelium. The remaining 5% of esophageal neoplasms represent metastases from other organs.

Esophageal cancer has a complex etiology. In high incidence areas such as the Caspian littoral of Iran and Linxian County, China, major risk factors are thought to include micronutrient deficiencies, low levels of protective factors which exist naturally in fresh fruits and vegetables, and consumption of foods containing mycotoxins and nitrosamines.[4-6] Nitrosamines and alcoholic beverages have been implicated in the high incidence of esophageal cancer in the Transkei region in South Africa, while in Northwest France and the United States, alcohol consumption and tobacco use are suspected to be etiologic factors.[5-7] Indeed, tobacco and alcohol use are assuming a more important role in the occurrence of this disease worldwide. There is an emerging literature suggesting a role for human papilloma virus (HPV) in the etiology of squamous cell carcinoma of the esophagus. Recent studies using sensitive molecular techniques have detected HPV-16 or HPV-18 in about 15% of esophageal tumor samples.[8]

The rat has been used almost exclusively as an animal model for studies in esophageal cancer. In rats, nitrosamines are the most powerful and versatile esophageal carcinogens

Dietary Phytochemicals in Cancer Prevention and Treatment
Edited under the auspices of the American Institute for Cancer Research, Plenum Press, New York, 1996

known. Many asymmetric nitrosamines act as fairly specific inducers of tumors in the rat esophagus including, N-nitrosonornicotine, N-nitrosomethylamylamine, and N-nitrosomethylbenzylamine (NMBA). NMBA is by far the most potent inducer of esophageal tumors in rats; tumors can be induced in 15 weeks or less.[9,10] A number of readily discernible preneoplastic lesions are produced as well, including simple hyperplasia, leukoplakia, and dysplastic lesions.[11] The majority of tumors induced by NMBA are squamous cell papillomas, although a small incidence of basal or squamous cell carcinomas may also be detected.[11] Due to occlusion of the esophagus and/or induction of respiratory distress, a large esophageal papilloma can be life-threatening. Thus, many NMBA-treated animals are unable to survive long enough to develop carcinomas.

All presently available data indicate that α-hydroxylation is the major metabolic activation pathway of nitrosamines which are carcinogenic to the esophagus. This is illustrated for NMBA in Figure 1. Microsomes isolated from the epithelium of the esophagus contain cytochrome P-450 enzymes which can catalyze the α-hydroxylation of NMBA.[12] Hydroxylation of the methylene carbon gives the α-hydroxy derivative [1] which spontaneously decomposes to the electrophile methane diazohydroxide [3] and benzaldehyde [4].

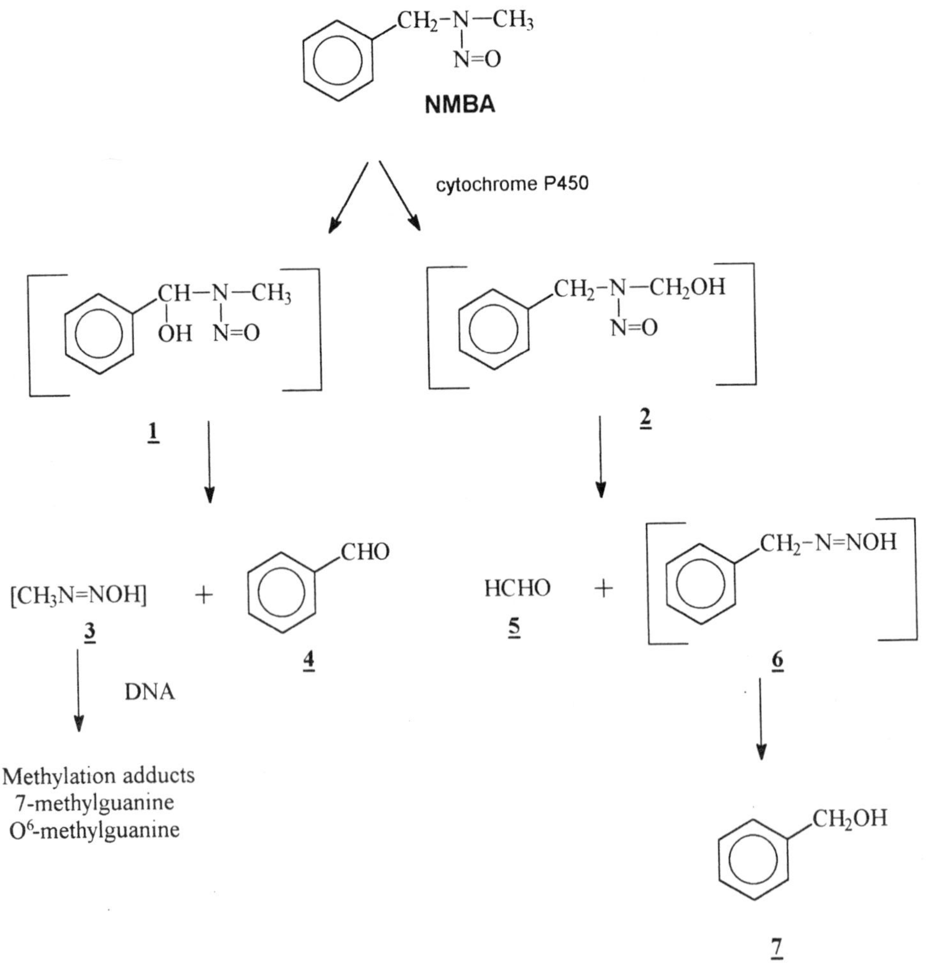

CH_2–N–CH_3, N=O

NMBA

cytochrome P450

CH(OH)–N(N=O)–CH_3 **1**

CH_2–N(N=O)–CH_2OH **2**

[CH_3N=NOH] **3** + CHO **4**

HCHO **5** + CH_2–N=NOH **6**

DNA

Methylation adducts
7-methylguanine
O^6-methylguanine

CH_2OH **7**

Figure 1. Preferred pathway of NMBA activation in rat esophagus.

Methane diazohydroxide methylates DNA to give the adducts 7-methylguanine and O^6-methylguanine. Hydroxylation of the methyl carbon of NMBA yields formaldehyde [5] and benzyldiazohydroxide [6]. Benzylation of DNA has recently been reported.[13] Both pathways of NMBA metabolism have been detected in esophageal epithelial microsomes. The methylene hydroxylation pathway far exceeds methyl hydroxylation.

The development of NMBA-induced tumors in the rat esophagus is associated with a series of molecular events. Among these is mutational activation of the H-*ras* oncogene in 80-100% of papillomas. The principal mutation is a G → A transition, a mutation that can be attributed to the formation of O^6-methylguanine and the subsequent mispairing of this adducted base with thymine.[14] Other molecular events include mutational inactivation of the p53 tumor suppressor gene in a minority of tumors[15] and overexpression of transforming growth factor-α (TGFα) and epidermal growth factor receptor (EGFR).[16]

In recent years, our laboratory has investigated the ability of a series of isothiocyanate compounds to inhibit NMBA-tumorigenesis in the rat esophagus. Initial studies were conducted with phenethyl isothiocyanate (PEITC).[17] PEITC is a primary product of thioglucosidase-catalyzed hydrolysis of gluconasturtiin, a glucosinolate found in many cruciferous vegetables such as cabbage, Brussels sprouts, cauliflower, turnips, etc.[18,19] Previous studies in other laboratories have demonstrated the ability of PEITC to inhibit the metabolism and DNA methylation of a series of nitrosamine carcinogens both *in vivo* and *in vitro*. PEITC has also been shown to inhibit lung tumor induction in rats[20] and in mice[21] by the tobacco specific nitrosamine 4-(methylnitrosamino)-1-(3-pyridyl)-1-butanone. In the present report, the results of testing PEITC at various dietary concentrations for its ability to inhibit NMBA tumorigenesis in the rat esophagus are discussed. In addition, since isothiocyanates of longer alkyl chain length were found to be more effective inhibitors of NNK-tumorigenesis in strain A/J mouse lung, we examined the effect of alkyl chain length on NMBA tumorigenesis in the rat esophagus.[22] Isothiocyanates that were compared for their ability to inhibit NMBA tumorigenesis included benzyl(BITC), phenethyl (PEITC), 3-phenylpropyl (PPITC), 4-phenylbutyl (PBITC), and 6-phenylhexyl (PHITC) isothiocyanates (Figure 2). The ability of these compounds to inhibit NMBA-induced esophageal tumorigenicity was compared with their inhibitory effects on NMBA-induced DNA methylation in rat esophageal tissues.

METHODS

Chemicals

BITC, PEITC and DMSO were purchased from Aldrich Chemical Company (Milwaukee, WI). PPITC and PBITC were purchased from LKT Laboratories (Minneapolis, MN). PHITC was kindly provided by Dr. Shantu G. Amin of the American Health Foundation, Valhalla, NY. NMBA was synthesized as described previously.[23] All chemicals were analyzed for purity by reversed-phase HPLC and were found to be approximately 99% pure.

Animals

Male F-344 rats (age 5-6 weeks) were purchased from Harlan Sprague-Dawley, Inc. (Indianapolis, IN) and were viral antibody free. The rats were housed three per cage and maintained under standard conditions (20 ± 2°C, 50 ± 10% relative humidity, 12 h light/dark cycle). Modified AIN-76A diet, Dyets (Bethlehem, PA) and fresh tap water were provided *ad libitum*. The modified AIN-76A diet contained 20% casein, 0.3% D,L- methionine, 52% cornstarch, 13% dextrose, 5% cellulose, 5% corn oil, 3.5% AIN salt mix, 1% AIN vitamin mix and 0.2% choline bitartrate.

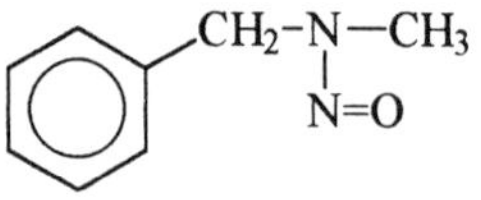

CH_2-N—CH_3
N=O

N-nitrosomethylbenzylamine (NMBA)

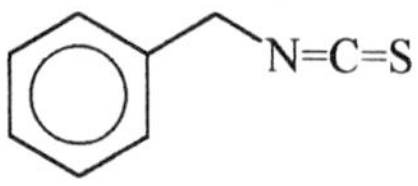

N=C=S

benzyl isothiocyanate (BITC)

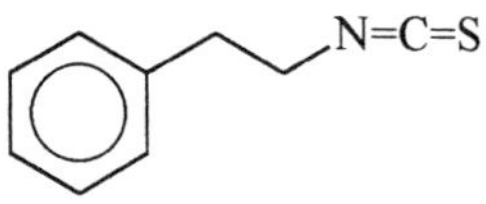

N=C=S

phenethyl isothiocyanate (PEITC)

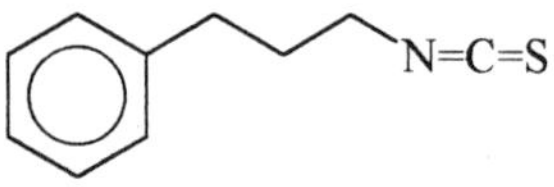

N=C=S

3-phenylpropyl isothiocyanate (PPITC)

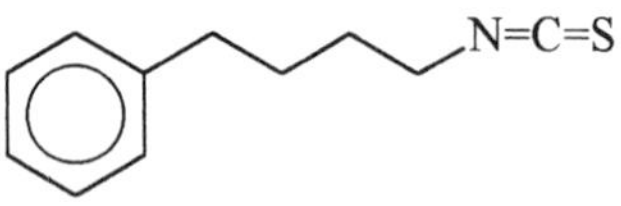

N=C=S

4-phenylbutyl isothiocyanate (PBITC)

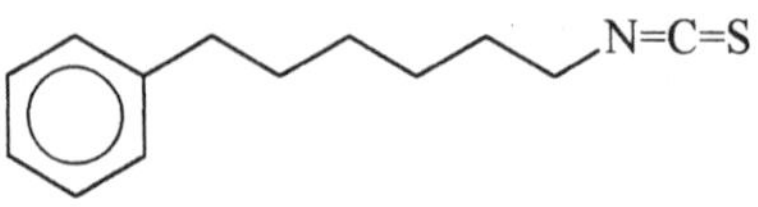

N=C=S

6-phenylhexyl isothiocyanate (PHITC)

Figure 2. Structures of NMBA and test compounds.

Experimental Diets

Experimental diets were prepared every two weeks with the aid of a Hobart D-300 compact mixer (Troy, OH). AIN-76A diet was mixed with the appropriate amount of each isothiocyanate and blended at a setting of "2" for 15 min. Diets were stored at 4°C before feeding. Preliminary studies indicated that all isothiocyanates were stable in the diet for a period of two weeks when stored at 4°C.[22]

Tumorigenesis Bioassays

Dose Response Bioassay of PEITC. In separate experiments, groups of 13 - 27 rats (6-7 weeks of age) were placed on control AIN-76A or experimental diets containing 0.325, 0.75, 1.5, or 3.0 µmol PEITC per gram diet for two weeks before NMBA dosing. N-nitrosomethylbenzylamine was administered subcutaneously once each week at a dose of 0.5 mg/kg (in 0.1 ml of 10% DMSO in water) for 15 weeks. Three additional groups of 15 rats were fed PEITC at concentrations of 0.0, 0.75, and 3.0 µmol/g and administered 10% DMSO subcutaneously for 15 weeks. All animals were maintained on control or experimental diets during NMBA dosing and for an additional eight weeks. Body weights were measured weekly until week 19 and every two weeks thereafter. Food consumption was monitored weekly throughout the course of the experiment. At the end of this 25-week period rats were killed. The esophagi were excised, opened longitudinally, placed on flat white index cards, and examined for tumors under a dissecting microscope. Tumors > 0.5 mm in diameter were quantitated. Each esophagus was then divided into three parts: upper, middle and lower, and fixed in 10% neutral buffered formalin. The esophagi were later embedded in paraffin blocks, sectioned, and stained for histopathological evaluation of preneoplastic lesions (acanthoses and hyperkeratoses, leukoplakias and leukokeratoses,) as well as papilloma, and carcinoma.[11]

Effects of Isothiocyanates of Different Alkyl Chain Length. Male F-344 rats (age 6-7 weeks) were randomized into groups consisting of 15 animals each and fed modified AIN-76 diet or modified AIN-76A diet containing the five different isothiocyanates (BITC, PEITC, PPITC, PBITC, and PHITC) at concentrations of 2.5, 1.0 and 0.4 µmol/g diet. Rats were maintained on their respective diets throughout the 25-week bioassay. Two weeks after initiation of the respective diets, rats in the carcinogen control group and in isothiocyanate-treated groups received NMBA subcutaneously, at a concentration of 0.5 mg/kg of body weight in 20% DMSO (vehicle) based on a dosing volume of 1 ml/kg. NMBA was administered once weekly for fifteen weeks. Control animals received either the vehicle or they were untreated. The isothiocyanate control groups received only the isothiocyanates at 2.5 µmol/g diet. Food consumption and body weights were recorded weekly throughout the bioassay and at the end of the bioassay animals were sacrificed and tumors were quantitated as described above for the PEITC bioassays.

Analysis of DNA Adducts

For isolation and quantitation of DNA adducts, groups of 20 F-344 rats were placed on control diets or diets containing 2.5 µmol/g of BITC, PEITC, PPITC, PBITC, or PHITC. Two weeks after assignments to diets, rats were administered NMBA at a dose of 0.5 mg/kg by subcutaneous injection. Groups of rats were killed by CO_2 asphyxiation at 24 and 72 h after NMBA dosing. At necropsy, the esophagi of all animals were excised. The esophagi of four rats were pooled to yield a single sample for a total of five samples per group. DNA was isolated and purified as described previously.[24] Quantitation of O^6-methylguanine levels in esophageal DNA samples was performed after acidic hydrolysis by strong cation exchange HPLC coupled with fluorescence detection (excitation wavelength = 290 nm; emission wavelength = 360 nm).

Statistical Analysis

Analysis of variance followed by Newman-Keuls' ranges test was used to statistically compare tumor body weights, food consumption data, and DNA methylation data. Tumor multiplicities were compared by analysis of variance followed by either Tukey's ranges test or Newman-Keuls' ranges test. Comparison of tumor incidence among groups was performed using Fisher's exact probability test.

RESULTS

Effects of PEITC on NMBA-Induced Esophageal Tumorigenesis

Esophageal tumor incidence and multiplicity data are summarized in Table 1. Dietary administration of PEITC at 0.75 and 3 μmol/g (groups 2 and 3) had no adverse impact in survival, growth or food consumption of the rats (data not shown). In rats treated with NMBA alone (group 4), mean body weight and food consumption were significantly decreased ($p < 0.05$) during the final weeks of the experiment as compared with the vehicle control group. Rats treated with NMBA and fed PEITC at 1.5 or 3 μmol/g had body weights similar to vehicle controls (data not shown).

Rats fed PEITC at concentrations of 0.0, 0.75, and 3.0 μmol/g and administered vehicle developed no esophageal tumors, while rats fed the control diet and administered NMBA had a tumor multiplicity of 9.3 ± 0.9 (mean ± S.E.) tumors per rat. Dietary PEITC inhibited NMBA-induced esophageal tumors in a dose-related manner. Statistically significant reductions in tumor incidence were evident at 1.5 and 3.0 μmol/g diet, and in tumor multiplicity at 0.75, 1.5, and 3.0 μmol/g diet. The 0.325 μmol/g diet had no significant inhibitory activity.

Table 1. Effect of dietary PEITC on NMBA-induced esophageal tumorigenicity in F344 rats

Group	Treatment	Tumor Incidence (% inhibition)[a]	Tumor Multiplicity (% inhibition)[b]
1	Vehicle Control	0^1	0.0^1
2	0.75 μmol/g PEITC	0^1	0.0^1
3	3.0 μmol/g PEITC	0^1	0.0^1
4	NMBA Control	100^2	9.3 ± 0.9^2
5	0.325 μmol/g PEITC + NMBA	100^2 (0)	10.7 ± 1.1^2 (0)
6	0.75 μmol/g PEITC + NMBA	100^2 (0)	5.7 ± 1.2^3 (39)
7	1.5 μmol/g PEITC + NMBA	60^3 (40)	0.9 ± 0.2^1 (90)
8	3.0 μmol/g PEITC + NMBA	0^1 (100)	0.0^1 (100)

[a] Values that bear different superscripts are statistically different from one another as determined by Fisher's exact probability test ($p < 0.01$).

[b] Mean ± standard error. Values that bear different superscripts are statistically different from one another as determined by analysis of variance and Tukey's ranges test ($p < 0.05$).

Effects of Various Arylalkyl Isothiocyanates on NMBA-Induced Esophageal Tumorigenesis

Body weight gain throughout the 25-week bioassay was not severely decreased in the NMBA-treated groups, but there was a slight decrease towards the end of the study (data not shown). There were no significant differences in food consumption among the groups throughout the study. As indicated in Table 2, the tumor incidence in the NMBA-treated control rats (group 2) was 100%. A similar response was found in the groups treated with BITC + NMBA, PBITC + NMBA and PHITC + NMBA. Tumor multiplicities for the NMBA-treated control rats were 6.7 ± 0.8 (Table 2, experiment 1) and 7.2 ± 0.7 (Table 2, experiment 2); these results are in close agreement with previous studies using this same model system. Significant inhibition of tumor multiplicity was found in the groups treated with NMBA and either 1.0 μmol BITC/g diet (group 4), 2.5 and 0.4 μmol PBITC/g diet (groups 12 and 14), and in all treatments of the PEITC and PPITC groups (groups 6-11) (experiment 1). The remaining NMBA- treated groups showed slight inhibition of tumor multiplicity, but this inhibition was not statistically significant. In rats that received PHITC + NMBA (experiment 2), tumor multiplicities were increased in a dose-related manner with 8.7 ± 3.2 tumors per rat in the 0.4 μmol/g group, 11.6 ± 3.7 tumors per rat in the 1.0 μmol/g group, and 12.2 ± 3.6 tumors per rat in the 2.5 μmol/g group. However, only the tumor multiplicities of the 1.0 and 2.5 μmol/g groups were significantly different from the

Table 2. Effects of arylalkyl isothiocyanates on NMBA-Induced esophageal tumorigenicity in F344 rats

Group	Treatment	Tumor Incidence (% inhibition)[a]	Tumor Multiplicity (% inhibition)[b]
Exp. 1			
1	Vehicle Control	0^{1}	0.0 ^{1}
2	NMBA Control	100^{4}	6.7 ± 0.8^{3}
3	2.5 μmol/g BITC + NMBA	100^{4} (0)	6.5 ± 0.6^{3} (3)
4	1.0 μmol/g BITC + NMBA	100^{4} (0)	4.1 ± 0.6^{2} (38)
5	0.4 μmol/g BITC + NMBA	100^{4} (0)	5.6 ± 0.7^{2,3} (17)
6	2.5 μmol/g PEITC + NMBA	7^{1,2} (93)	0.1 ± 0.1^{1} (99)
7	1.0 μmol/g PEITC + NMBA	40^{1,2,3} (60)	0.4 ± 0.1^{1} (94)
8	0.4 μmol/g PEITC + NMBA	57^{2,3,4} (43)	1.1 ± 0.5^{1} (83)
9	2.5 μmol/g PPITC + NMBA	0^{1} (100)	0.0 ± 0.0^{1} (100)
10	1.0 μmol/g PPITC + NMBA	7^{1,2} (93)	0.1 ± 0.1^{1} (99)
11	0.4 μmol/g PPITC + NMBA	7^{1,2} (93)	0.1 ± 0.1^{1} (99)
12	2.5 μmol/g PBITC + NMBA	100^{4} (0)	4.0 ± 0.4^{2} (40)
13	1.0 μmol/g PBITC + NMBA	93^{3,4} (7)	5.1 ± 0.7^{2,3} (24)
14	0.4 μmol/g PBITC + NMBA	93^{3,4} (7)	3.9 ± 0.7^{2} (41)
Exp. 2			
1	Vehicle Control	0^{1}	0.0^{1}
2	NMBA Control	100^{2}	7.2 ± 0.7^{2}
3	2.5 μmol/g PHITC + NMBA	100^{2} (0)	12.2 ± 0.9^{3} (-69)
4	1.0 μmol/g PHITC + NMBA	100^{2} (0)	11.6 ± 1.0^{3} (-61)
5	0.4 μmol/g PHITC + NMBA	100^{2} (0)	8.7 ± 0.8^{2} (-21)

[a] Values with different individual numerical superscripts are statistically different from each other as determined by the Chi-square test ($p < 0.05$).

[b] Values are mean ± standard error. Values within this column that have no individual numerical superscripts in common are statistically different from each other as determined by ANOVA and Newman-Keuls' ranges test ($p < 0.05$). A negative percent inhibition value indicates the percent increase in tumor multiplicity.

Table 3. Effect of dietary isothiocyanates on NMBA-induced O^6-methylguanine levels in esophageal DNA

		pmol O^6-mGua/mg DNA[a]	
Group No.	Diet	24 hours	72 hours
1	AIN-76A	19.6 ± 0.6[1]	1.0 ± 0.4[1]
2	AIN-76A + 2.5 μmol/g BITC	14.8 ± 0.5[2]	0.3 ± 0.3[1]
3	AIN-76A + 2.5 μmol/g PEITC	2.1 ± 0.2[3]	0.5 ± 0.5[1]
4	AIN-76A + 2.5 μmol/g PPITC	3.0 ± 0.3[3]	N.D.[1]
5	AIN-76A + 2.5 μmol/g PBITC	7.0 ± 0.6[4]	1.0 ± 0.4[1]
6	AIN-76A + 2.5 μmol/g PHITC	20.7 ± 1.0[1]	13.0 ± 3.6[2]

[a] Values are mean ± standard error of five samples, each of which was derived from a pool of 4 esophagi. N.D. = not detected. Values with different superscripts are statistically different from each other as determined by ANOVA and Newman-Keuls' ranges test ($p < 0.05$).

carcinogen control group. Microscopic esophageal examinations revealed marked increases in the occurrence of dysplastic leukoplakia in all three PHITC + NMBA groups as compared with the carcinogen control (data not shown).

Effects of Dietary Isothiocyanates on NMBA-Induced Methylation

Results of the DNA methylation experiment are shown in Table 3. Based on preliminary experiments, we believe that DNA methylation induced by a single dose of NMBA is at or near maximum at 24 h. At 24 h after NMBA dosing, control rats had a DNA methylation level of 19.6 mol O^6-methylguanine per mg DNA. Both PEITC and PPITC-treated rats showed significantly lower levels of O^6-methylguanine than controls at 24 h. While rats fed BITC or PBITC had levels of DNA methylation that were lower than controls, the magnitude of these differences was not as great as with PEITC or PPITC. Rats treated with PHITC + NMBA revealed no significant differences in DNA methylation when compared to controls at 24 h following carcinogen treatment. However, at 72 h following NMBA treatment, the mean O^6-methylguanine level of rats receiving PHITC + NMBA was significantly increased compared with the NMBA control group. At 72 h, O^6-methylguanine levels in control rats had declined to 1 picomole per mg of DNA. At this time interval no other differences were observed between controls and other isothiocyanate-treated groups.

DISCUSSION

The esophageal tumor response in the PEITC dose response bioassay of the rats treated only with NMBA (9.3 ± 0.9 tumors per rat) was in agreement with earlier studies conducted using this model system.[10,25] As was found previously,[17] the 3 μmol/g dietary concentration of PEITC completely inhibited NMBA-induced tumors. The 1.5 μmol/g concentration yielded a 90% reduction in tumor multiplicity and a 40% reduction of incidence while the 0.75 μmol/g diet reduced tumor multiplicity by only 39% and had no effect on esophageal tumor incidence. No significant effects were observed at the 0.325 μmol/g concentration of PEITC. Thus, the minimum inhibitory concentration of PEITC (i.e., the dose of PEITC that will yield a significant reduction in tumor multiplicity) apparently lies between 0.325 and 0.75 μmol/g.

Previous work with the A/J mouse lung tumor model demonstrated that short chain isothiocyanates (PITC and BITC) had no inhibitory effect on NNK induced lung tumors[26] and that longer chain arylalkyl isothiocyanates (PPITC, PBITC, PPeITC, and PHITC) were more effective against NNK-induced lung tumorigenesis than PEITC.[27] The trend is for greater inhibitory activity with increasing chain length. Therefore, we expected that similar structure- activity relationships for inhibitory activity might be observed in the rat esophageal model.

While the potency of inhibition of NMBA-induced esophageal tumors and preneoplastic lesions was greater for PPITC than for PEITC, PBITC had little inhibitory effect, and PHITC actually enhanced tumor formation. Thus, the structure-activity requirements for inhibitory activity of arylalkyl isothiocyanates in the rat esophagus differ from those of the strain A mouse lung. Overall, the results of the DNA methylation study were in accordance with the bioassay results. The purpose of the DNA methylation studies was to determine the possible anti-initiating activity of isothiocyanates in the rat esophageal model. Thus, it must be noted that the DNA methylation studies described in this paper only examine the effect of isothiocyanates on a single dose of NMBA. In the actual bioassay, the carcinogen was administered over a period of 15 wks and the isothiocyanates in the diet were administered from a period of 2 wks prior to NMBA administration and throughout the study; thus, it is impossible to discern at what point the initiation and promotion/progression stages of tumor formation begin and end. The reasons for the differing inhibitory activities of the isothiocyanates in the rat esophageal tumor model are not clear. Our results were not due to differences in stability of isothiocyanates in the diet, since the isothiocyanate/diet mixtures were stable up to 7 to 10 days at room temperature and up to 28 days at 4°C (data not shown).[22] Additionally, increasing alkyl chain length demonstrably increases stability of arylalkyl isothiocyanates.[27]

Isothiocyanates are known to be inhibitors of cytochrome p450 enzymes in other model systems.[28-30] It is possible that the present results are due to differences in the abilities of the various isothiocyanates to inhibit the esophageal p450 enzymes responsible for NMBA metabolism, enzymes which are not fully elucidated. Arylalkyl isothiocyanates with a chain length of 4 or more carbons may be unable to effectively inhibit these enzymes. Indeed, in a recent *in vitro* study, it was found that PEITC was more effective than PHITC in inhibiting the metabolism and DNA binding of the esophageal carcinogen, N-nitrosomethylamylamine (NMAA) in rat esophagus. However, PHITC was more effective in inhibiting the metabolism of NMAA in the liver than in the esophagus.[31]

In summary, we have found that in this model, PPITC is a more potent inhibitor of NMBA esophageal tumorigenesis than PEITC while PBITC, a highly potent inhibitor of NNK-induced lung tumorigenesis in strain A mice has little inhibitory effect and PHITC, the most active inhibitor of NNK-induced lung tumorigenesis actually increases the esophageal tumor response to NMBA. These results indicate that arylalkyl isothiocyanates with an alkyl chain length of 4 carbons or more are not useful inhibitors in the rat esophageal model. These studies provide a basis for future investigations to elucidate the mechanism(s) responsible for NMBA-induced esophageal tumorigenesis and its prevention by isothiocyanates and other chemopreventive agents.

ACKNOWLEDGMENTS

This work was supported by NIH grant CA46535.

REFERENCES

1. Parkin, D.M., Stjernsward, J., and Muir, C.S. (1984) Estimates of the worldwide frequency of twelve major cancers. *Bull. World Health Organization*, 62:163 (1984).

2. DuPlessis, L.S., Nunn, J.R., and Roach, W.A. Carcinogen in a Transkeian Bantu food additive, *Nature*, 222: 1198 (1969).
3. Warwick, G.P. and Harington, J.S. Some aspects of the epidemiology and etiology of esophageal cancer with particular emphasis on the Transkei, South Africa, *Adv. Cancer Res.*, 17: 18-229 (1973).
4. Yang, C.S. Research on esophageal cancer in China: a review. *Cancer Res.*, 40: 2633 (1980).
5. Tuyns, A.J., Pequignot, G., and Abbatuci, J.S. Oesophageal cancer and alcohol consumption: importance of type of beverage, *Int. J. Cancer*, 23:443 (1979).
6. Walker, E.A., Castegnario, M., Garren, L., Toussaint, G., and Kowalski, B. Intake of volatile nitrosamines from consumption of alcohols, *J. Natl. Cancer Inst.*, 63:947 (1979).
7. Wynder, E.L. and Bross, I.J. A study of etiological factors in cancer of the esophagus, *Cancer*, 14:389 (1961).
8. Chang, F., Syrjanen, S., Shen, Q., Ji, H.X., and Syrjanen, K. Human papillomavirus (HPV) DNA in esophageal precancer lesions and squamous cell carcinomas from China, *Int. J. Cancer,* 45:21 (1990).
9. Lijinsky, W., Saavedra, J.E., Rueber, M.D., and Singer, S.S. Esophageal carcinogenesis in F344 rats by nitroso-methylethylamines substituted in the ethyl group, J. Natl. Cancer Inst. *68*: 681 (1982).
10. Mandal, S. and Stoner, G.D. Inhibition of N-nitrosobenzylmethylamine-induced esophageal tumorigenesis in rats by ellagic acid, *Carcinogenesis* (London), 11:55 (1990).
11. Pozharriski, K.M. Tumors of the oesophagus. in: V.S. Turusov ed., "*Pathology of Tumors in Laboratory Animals*," Vol. 1. IARC Scientific Publication No. 5. Lyon, France, (1973).
12. Labuc, G.E. and Archer, M.C. Esophageal and hepatic microsomal metabolism of N-nitrosomethylbenzylamine and N-nitrosodimethylamine in the rat, *Cancer Res.*, 42:3181 (1982).
13. Peterson, L.A. Detection of benzylated DNA adducts in livers from N-nitrosomethylbenzylamine (NMBzA) treated rats, *Proc. Am. Assoc. Cancer Res.*, 36:139 (1995).
14. Wang, Y., You, M., Reynolds, S.H., Stoner, G.D., and Anderson, M.W. Mutational activation the cellular Harvey *ras* oncogene in rat esophageal papillomas induced by methylbenzylnitrosamine, *Cancer Res.* 50:1591 (1990).
15. Lozano, J.C., Nakazawa, H., Cros, M.P., Cabral, R., and Yamasaki, H. G→A mutations in p53 and Ha-*ras* genes in esophageal papillomas induced by N-nitrosomethylbenzylamine in two strains of rats, *Mol. Carcinog. 9: 33-39.* (1994)
16. Wang, Q.-S., Sabourin, C.L.K., Bijur, G.N., Robertson, F.M., and Stoner, G.D. Alterations in transforming growth factor-alpha and epidermal growth factor receptor expression duing rat esophageal tumorigenesis, *In press.*
17. Stoner, G.D., Morrissey, D.T., Heur, Y., Daniel, E.M., Galati, A.J., and Wagner, S.A. Inhibitory effects of phenethyl isothiocyanate on N-nitrosobenzylmethylamine carcinogenesis in the rat esophagus, *Cancer Res.*, 51:2063 (1991).
18. Carlson, D.G., Daxenbichler, M.E., VanEtten, C.H., Tookey, M.L., and Williams, P.H. Glucosinolates in crucifer vegetables: turnips and rutabagas, *J. Agric. Food Chem.*, 29:1235 (1981).
19. Hanley, A.B., Heaney, R.K., and Fenwick, G.R. Improved isolation of glucobrassicin and other glucosinolates, *J. Sci. Food Agric.* 34:869 (1983).
20. Morse, M.A., Wang, CX., Stoner, G.D., Mandal, S., Conran, P.B., Amin, S.G., Hecht, S.S., and Chung, FL. Inhibition of 4(methylnitrosamino)1(3pyridyl)1butanoneinduced DNA adduct formation and tumorigenicity in the lung of F344 rats by dietary phenethyl isothiocyanate, *Cancer Res.* 49:549 (1989).
21. Morse, M.A., Amin, S.G., Hecht, S.S. and Chung, FL. Effects of aromatic isothiocyanates on tumorigenicity, O^6methylguanine formation, and metabolism of the tobaccospecific nitrosamine 4(methylnitrosamino)1(3pyridyl)1butanone in A/J mouse lung, *Cancer Res.* 49:2894 (1989).
22. Wilkinson, J.T., Morse, M.A., Kresty, L.A., and Stoner, G.D. Effect of alkyl chain length on inhibition of N-nitrosomethylbenzylamine-induced esophageal tumorigenesis and DNA methylation by isothiocyanates, *Carcinogenesis* 16:1011 (1995).
23. Druckrey, H., Preussman, R., Ivankovic, S., and Schmahl, D. Organotrope carcinogene Wirkungen bei 65 verschiedenen N-nitroso-Verbindungen an BD-Ratten, *Z. Krebsforsch.*, 69:103 (1967).
24. Morse, M.A., Zu, H., Galati, A.J., Schmidt, C.J., and Stoner, G.D.) Dose-related inhibition by dietary phenethyl isothiocyanate of esophageal tumorigenesis and DNA methylation induced by N-nitrosomethylbenzylamine in rats, *Cancer Lett.,* 72:103 (1993.
25. Daniel, E.M. and Stoner, G.D. The effects of ellagic acid and 13cisretinoic acid on Nnitrosobenzylmethylamineinduced esophageal tumorigenesis in rats, *Cancer Lett.* 56:117 (1991).
26. Morse, M.A., Eklind, K.I., Amin, S.G., Hecht, S.S., and Chung, FL. Effects of alkyl chain length on the inhibition of NNKinduced lung neoplasia in A/J mice by arylalkyl isothiocyanates, *Carcinogenesis* 10:757 (1989).

27. Morse, M.A., Eklind, K.I., Hecht, S.S., Jordan, K.G., Choi, CI., Desai, D.H., Amin, S.G., and Chung, FL. Structureactivity relationships for inhibition of 4(methylnitrosamino)1(3pyridyl)1butanone (NNK) lung tumorigenesis by arylalkyl isothiocyanates in A/J mice, *Cancer Res.* 51:846 (1991).
28. Chung, F.L., Juchatz, A., Vitarius, J. and Hecht, S.S. Effects of dietary compounds on αhydroxylation of N-nitrosopyrrolidine and **N**'nitrosonornicotine in rat target tissues, *Cancer Res.* 44:2924 (1984).
29. Ishizaki, H., Brady, J.F., Ning, S.F., and Yang, C.S. Effect of phenethyl isothiocyanate on microsomal N-nitrosodimethylamine metabolism and other monooxygenase activities, *Xenobiotica* 20:255 (1990).
30. Smith, T.J., Guo, Z., Thomas, F.E., Chung, F.L., Morse, M.A., Eklind, K. and Yang, C.S. Metabolism of 4-(methylnitrosamino)1(3pyridyl)1butanone in mouse lung microsomes and its inhibition by isothiocyanates, *Cancer Res.* 50:6817 (1990).
31. Huang, Q., Lawson, T.A., Chung, F.-L., Morris, C.R., and Mirvish, S.S. Inhibition by phenylethyl and phenylhexyl isothiocyanate of metabolism of and DNA methylation by N-nitrosomethylamylamine in rats, *Carcinogenesis* 14:749 (1993).

3

PLANT PHENOLICS AS POTENTIAL CANCER PREVENTION AGENTS

Harold L. Newmark

Memorial Sloan Kettering Cancer Center
New York, New York 10021
and
Rutgers University
Piscataway, New Jersey 08855-0789

ABSTRACT

The frequent consumption of fresh fruits and vegetables is associated with a lower cancer incidence in humans, and in experimental carcinogenesis. There are several groups of substances in plant foods which may contribute to this inhibition of tumor development. Almost all fresh fruits, vegetables and cereal grains contain appreciable amounts of naturally occurring plant phenolics. A brief overview will be presented of the most common plant phenolics in human foods and their chemical and biochemical properties. Plant phenolics, originally hypothesized to inhibit mutagenesis and/or carcinogenesis by virtue of antioxidant or electrophile trapping mechanisms, can also act as potent modulators of arachidonic metabolism cascade pathways. Certain plant phenols can be effective inhibitors of chemical mutagens, *in vitro,* and/or carcinogenesis *in vivo*. The historical origins, hypotheses of actions, current status and potential adverse effects of the utility of plant phenolics to reduce risk of cancer are discussed, as well as future possibilities and needs and objectives for future research.

INTRODUCTION

It has been known for several decades that there are substances in commonly consumed foods that reduce the incidence of chemically induced carcinogenesis in laboratory rodents. In early pioneering studies, Wattenberg found that rodents on a purified or semi-synthetic diet developed more chemically-induced lesions than on a mixed "natural food" cereal-vegetable diet. Further studies elucidated many active "chemopreventive" substances in plant foods, including terpenes, aromatic isothiocyanates, organosulfur compounds, protease inhibitors, dithiolthiones and indoles.[1-3] Of particular interest as chemopreventive agents were the monophenols, polyphenols, flavones, flavonoids and tannins in foods derived from cereals, vegetables and fruits, which may be consumed in large quantities (up to 1-2 grams per day) in some human diets.

OCCURRENCE OF PLANT PHENOLIC COMPOUNDS

The phenolic compounds which occur commonly in food material may be classified into three groups, namely, simple phenols and phenolic acids, hydroxycinnamic acid derivatives and flavonoids.

The Simple Phenols and Phenolic Acids. The simple phenols include monophenols such as *p*-cresol isolated from several fruits (e.g. raspberry, blackberry),[4,5] 3-ethylphenol and 3,4-dimethylphenol found to be responsible for the smoky taste of certain cocoa beans and diphenols such as hydroquinone which is probably the most widespread simple phenol.[6]

Vanillin (4-hydroxy-3-methoxybenzaldehyde) occurs frequently and is a popular flavor.[3] Gallic acid, a triphenol, is present in an esterified form in tea catechins. Gallic acid may occur in plants in soluble form either as quinic acid esters[7] or condensed into hydrolyzable tannins (tannic acids), or ellagic acid derivatives. Some representative gallic acid derived plant phenolics in foods are shown in Figure 1.

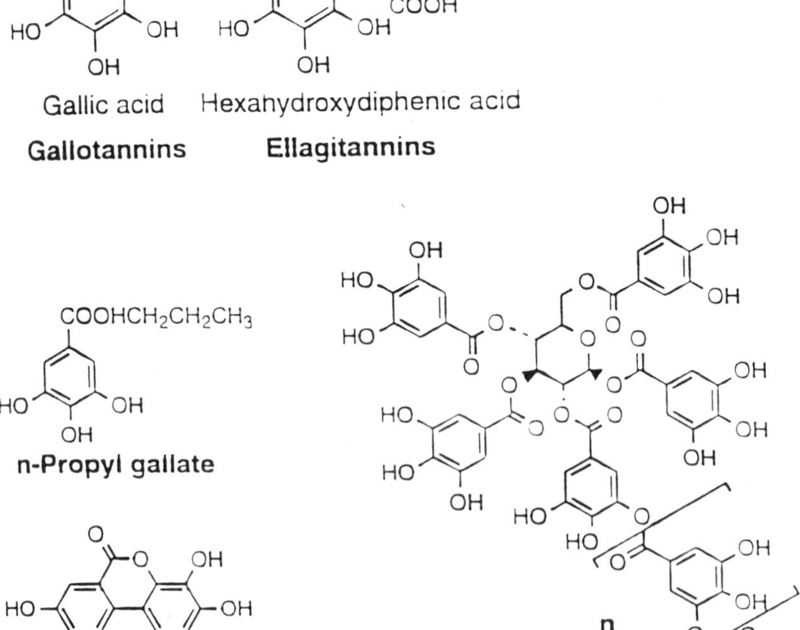

Figure 1. Chemical structures of some naturally occurring gallic acid components and derivatives present in human foods.

The Hydroxycinnamic Acid Derivatives. Hydroxycinnamic acids and their derivatives are almost exclusively derived from *p*-coumaric, caffeic, and ferulic acid. Hydroxycinnamic acids usually occur in various conjugated forms, more frequently as esters than glycosides.[8]

The most important member of this group in food material is chlorogenic acid, an ester of caffeic acid with the sugar quinic acid, which is the key substrate for enzymatic browning, particularly in apples and pears.[9] It also represents about 15% of dry instant coffee.

The Flavonoids. The most important single group of phenolics in food are flavonoids which consist mainly of catechins, proanthocyanins, anthocyanidins and flaveons, flavonols and their glycosides.

Although catechins seem to be widely distributed in plants, they are particularly rich in tea leaves where catechins may constitute up to 30% of dry leaf weight. There is much current research on antioxidative and cancer chemopreventive properties of tea and its catechin components.

Anthocyanins are almost universal plant colorants and are largely responsible for the brilliant orange, pink, scarlet, red, mauve, violet and blue colors of flower petals and fruits of higher plants.[10,11]

Flavones, flavonols and their glycosides also occur widely in the plant kingdom. Their structural variations and distribution have been the subjects of several comprehensive reviews in recent years.[12-14] It has been estimated that humans consuming high fruit and vegetable diets ingest up to 1 g of these compounds daily. The most common and biologically active dietary flavonol is quercetin and its glycoside rutin. See Figure 3 for the structures of a few representative flavonoids in human foods.

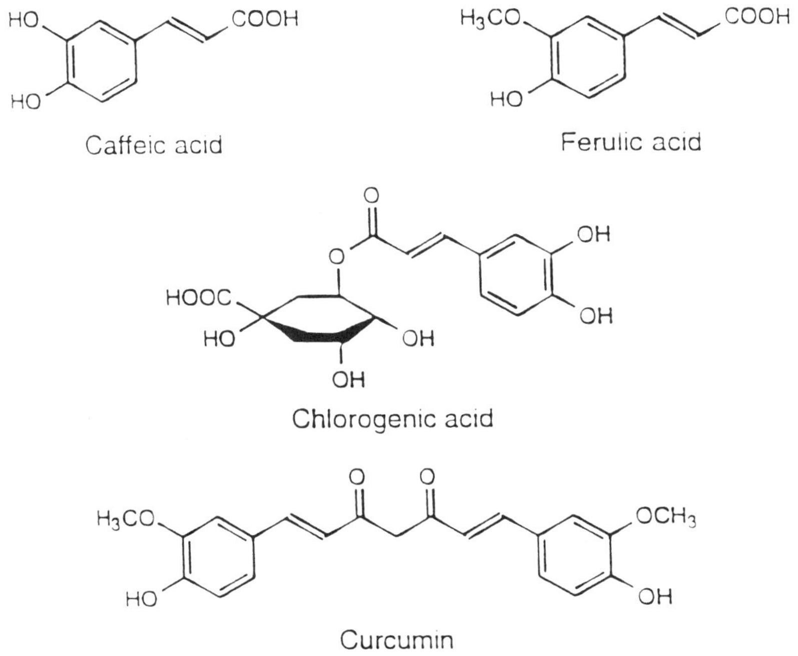

HO, HO, COOH

Caffeic acid

H_3CO, HO, COOH

Ferulic acid

O, O, OH, OH, HOOC, HO, OH, OH

Chlorogenic acid

O, O, H_3CO, HO, OCH_3, OH

Curcumin

Figure 2. Chemical structures of some naturally occurring hydroxycinnamic acid components present in human foods.

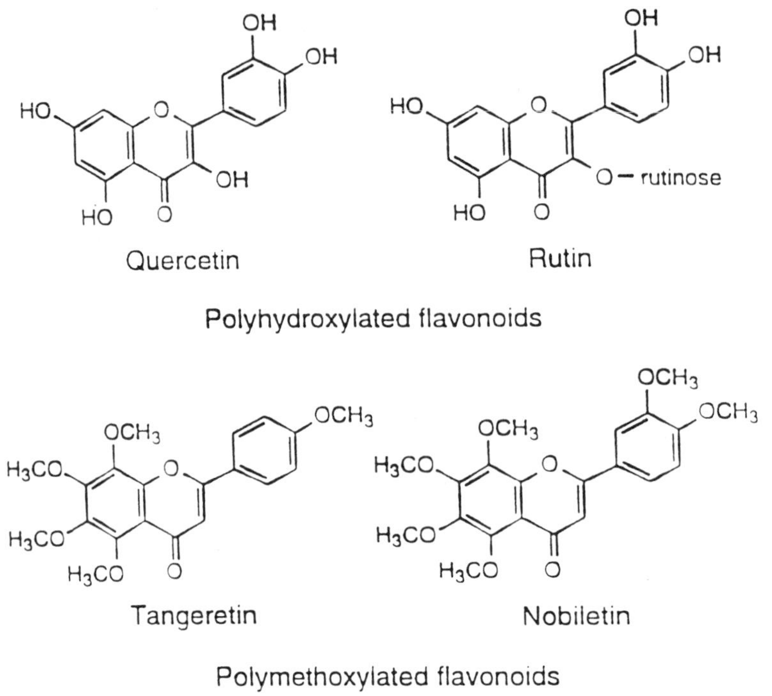

Quercetin Rutin

Polyhydroxylated flavonoids

Tangeretin Nobiletin

Polymethoxylated flavonoids

Figure 3. Structures of some citrus flavonoids.

CHEMICAL PROPERTIES OF PLANT PHENOLIC COMPOUNDS

There are many hundreds of plant phenolics in the plants used for human foods, although most are present in low concentrations. They have several key functions in the plants, as indicated in Table 1.

ANTI-CARCINOGENIC PROPERTIES

The Antioxidant Hypothesis

Wattenberg found several synthetic food antioxidants, such as butylated hydroxyanisole (BHA), that reduced the incidence of neoplasia induced by chemical carcinogens in laboratory animals, and expanded the studies to show similar effects for the plant phenolics caffeic and ferulic acids.[15]

Table 1. Some functions of phenolics in plants

A.	Antioxidant - as potent oxidant radical scavengers, and oxidation chain reaction terminators: e.g. caffeic acid, ferulic, quercetin, tocopherol
B.	Antimicrobial agents
C.	Antiviral agents
D.	Food colors
E.	Structural - hydroxycinnamic alcohols used in plants to produce cell wall lignins.

Inhibition of Nitrosation

In the mid 1970's it was found that caffeic and ferulic acids were very effective consumers of nitrite ion, particularly at acid pH.[16] This results in strong activity of these plant phenols, commonly present in many human foods, in preventing nitrosation of susceptible secondary amines and amides to form highly potent carcinogenic nitrosamines and nitrosamides in our foods, and *in vivo.*[17] Sources of nitrites in foods are almost ubiquitous, particularly in fermented or smoked foods, or added as aids to preservation, as in processed meats. In addition, nitrates naturally found in our foods are readily recycled to the saliva after ingestion and absorption, and then reduced to nitrite by buccal flora, resulting in gastric nitrosation of susceptible amines. The function of dietary plant phenolics in blocking these reactions in foods during food processing and cooking and *in vivo* has probably been underrated as a major cancer prevention process.

Electrophile Radical Trap Hypothesis

The current axiom of chemical carcinogenesis is that many, perhaps most, carcinogens are converted by either non-enzymatic means (in the case of direct acting carcinogens) or metabolic activation to highly reactive species that can attack cellular components. The best known form of the reaction species is the electrophilic reactant, possessing a positively charged group such as a carbonium ion, which reacts with electron-rich moieties chemically termed nucleophiles. Many cellular components can be targets for such electrophilic attack, but the (probably minor) attack and resultant chemical and structural alteration of DNA is believed to be a key step in carcinogenic initiation in the cell. Protection of the DNA in the cell is largely achieved by competitive efficient chemical nucleophiles such as glutathione. However, this protective effect can be overwhelmed.

In 1980, we realized that some plant phenolics, such as caffeic and ferulic acids could act as potent chemical nucleophiles, based on our previous studies of their reaction with nitrite.[16] On testing as inhibitors of mutagenesis *in vitro,* as induced by benzo[a]pyrene diol epoxide, these plant phenolics were indeed found to be potent, particularly the related ellagic acid.[18,19] The postulated mechanisms of reaction involved π bond interactions between the planar molecules involved and their capacity to act as electron-rich donors (i.e. electrophilic trap for electron-poor carcinogenic electrophiles). The originally promising studies of anti-mutagenic activity were later shown to be partly dependent on *in vitro* reactions of the phenolics with the tested mutagen (often benzo[a]pyrene) outside the cell, before cell entry (R. Chang Rutgers University, personal communication.) Furthermore, the phenolics appear highly reactive within the cells in a variety of functional systems. Thus, determination of the specificity of effective phenolics in reaching a tissue, entering the cells, and performing a useful tumor inhibitory function with adequate safety for normal cell function, will require much further study.

Arachidonic Metabolism Modulation

It has long been known that several plant phenolics such as salicylic acid, eugenol, curcumin, quercetin and others can inhibit the cyclo-oxygenase pathway of arachidonic acid metabolism to prostaglandins. Many of the same plant phenolics also inhibit lipoxygenase pathways to other eicosanoids.[20] Kato *et al.*, who demonstrated inhibition of phorbol ester promotion of mouse skin tumors by quercetin, suggested the possible involvement of lipoxygenase inhibition.[21] Indeed, several inhibitors of lipoxygenase pathways of arachidonic acid metabolism in mammalian cells, as well as cyclooxygenase inhibitors have demonstrated anti-tumor activity.[22] Modulation of arachidonic acid metabolism appears to affect promotion rather than initiation processes in carcinogenesis. Plant phenolics as such modulators can act as inhibitors of promotion processes.

There is currently great interest in inhibitors of arachidonic acid metabolism and their potential for the chemoprevention of cancer. Many laboratory animal studies have established non-steroid anti-inflammatory drugs (NSAIDS) as potent inhibitors of chemically induced carcinogenesis, e.g. of the colon or breast. These active NSAIDS, including sulindac, peroxicam, indomethacin, aspirin and others, primarily inhibit only the cyclooxygenase pathway of the arachidonic metabolic cascade, particularly cyclooxygenase 1 (called COX-1). This results in risk of gastric bleeding, as well as other side effects.

Several plant phenolics are inhibitors of both cyclooxygenase and lipoxygenase pathways, including quercetin, eugenol, curcumin and green tea.[20,23] Human consumption of these plant phenolics in foods over many centuries suggests their probable long term safety as cancer chemopreventive agents. Quercetin, rutin, curcumin and tea polyphenols have inhibited colon carcinogenesis in animal models.[24,25]

Cell Protein Kinase Inhibition

Tyrosine kinase and other protein kinases are enzymes involved in cell proliferation. Plant phenolics could be useful dietary inhibitors of these kinases by reducing hyperproliferation of epithelial cells, thus lowering cancer risk. For example, quercetin is an inhibitor of protein kinase C, tyrosine protein kinase, and a specific protein kinase in rat colonic epithelium.[26]

INHIBITION OF CARCINOGENESIS: ANIMAL EXPERIMENTS

Several plant phenolics are modest to strong inhibitors of the neoplastic effects of chemical carcinogens in rodents, as discussed below. (Note: the effects of tea phenolics will be discussed by others at this Symposium.)

Quercetin and Its Glycoside Rutin. These are among the most commonly encountered plant phenolics in human food. Kato *et al* [27] showed that topical administration of quercetin inhibited phorbol ester (TPA)-induced tumor promotion on the skins of mice. Verma *et al* [28] found dietary quercatin inhibited DMBA and NMU induced mammary carcinogenesis in rats. However, it is known that quercetin and its glycoside rutin are poorly absorbed from the gastrointestinal tract in mammals, presumably in part due to very low solubility in aqueous media. The glycoside rutin is largely hydrolyzed to quercetin by the microbial flora of the colon, suggesting enhanced local bioavailability in the colon. In short term experiments we noted that dietary quercetin and rutin clearly exhibited significant activity in reducing azoxymethane (AOM) induced focal areas of dysplasia (FADs) in the colons of mice, even with a high fat diet.[24,29] In longer studies on AOM-induced colon carcinogenesis, quercetin significantly reduced the tumor incidence.[24] AOM-induced aberrant crypts in mice were significantly inhibited by both dietary rutin and quercetin (M.J. Wargovich, private communication).

Curcumin. This natural yellow color in turmeric spice, mustard, and curry has been used as a color in other foods. It has demonstrated anti-inflammatory activity in the mouse ear test and strong antitumorigenic activity on chemically induced (DMBA) and TPA promoted mouse skin.[30] Curcumin also inhibited gastric, duodenal and colon carcinogenesis in separate studieswith animal models.[26] The anticarcinogenic and biological activities of curcumin have been reviewed.[30]

Caffeic and Ferulic Acids. These are commonly present in many fruits and vegetables, have inhibitory effects on chemically induced forestomach tumorigenesis in mice.[15] Caffeic acid also is a very potent inhibitor of the formation carcinogenic nitrosamines.[18]

Ellagic Acid. Found in walnuts, raspberries and other nuts and fruits, it has shown inhibitory effects against chemically-induced esophageal, lung, and liver carcinogenesis in animal studies.[31-33]

Genistein and Daidzein. These isoflavonoid plant phenolics in soy and soy-based products are weak phytoestrogens, appear well absorbed, and appear to bind to estrogen receptors, thus blocking binding of the more potent natural estrogens and acting as anti-estrogens. There is much active research to test the chemopreventive potential of these compounds.

PROBLEMS AND ADVERSE EFFECTS

Ito and co-workers have shown that caffeic acid (2% of diet) sesamol (2% of diet) and catechol (0.8% of diet) could induce stomach cancer in rodents. However, there is a difference in species sensitivity, rats being more sensitive than mice.[34] These effects seem to derive from a gastric mucosal hyperplasia stemming from irritation by the chronically ingested dietary phenolics.

It has long been known that phenols, especially ortho dihydroxy phenols (catechols), can readily oxidize in highly aerobic atmospheres. Trace metal ions present such as copper or iron, act as potent catalysts for oxidation of phenols *in vitro* where the rate increases as the pH rises. The oxidation reaction produces hydrogen peroxide via an intermediate superoxide. Thus, when phenolic substances are tested for "mutagenicity" *in vitro* in an Ames-type assay, under highly aerobic conditions, with appreciable levels of trace metals in the media, it is no surprise that the phenols test positive as mutagens. This is probably largely the "mutagenicity" of the hydrogen peroxide produced in the media by the conditions of testing. Consequently, addition of catalase enzyme to the *in vitro* system virtually eliminated clastogenic activity of caffeic acid, emphasizing the role of artifactual generation of hydrogen peroxide in laboratory tests for mutagenicity of plant phenolics.[35] Re-evaluation of the apparent mutagenicity of such compounds with elimination of hydrogen peroxide formation would probably result in finding most of these phenolics free of mutagenic activity. Hydrogen peroxide formation in the dietary systems used in testing by Ito may also be partly responsible for promotion of stomach cancer in rats.[34] The difference in species sensitivity to stomach irritation by plant phenols, with rats being more sensitive than mice, may be related to the higher gastric pH of rats compared to mice and humans. Metal ion catalysis of phenolic oxidation to form hydrogen peroxide is normally reduced at lower pH. Plant phenols, particularly high molecular weight polyphenols such as the gallotannins, can precipitate proteins by physico-chemical interactions. In higher concentrations in food products, such as strong black coffee without added milk as a neutralizing protein source, this can also be a source of chronic gastric irritation.

ENDOGENOUS PRODUCTION

A newly emerging area is the endogenous production of phenolic lignans (diphenolic compounds). These are produced from plant precursors (probably plant phenolics) through modification by the colon microflora, possibly the *Clostridia* group.[36] The two most common

Table 2. Multifunctional activity of phenolics: quercetin

Antioxidant:	Lipids, membranes
Antimutagen:	Electrophilic trapping (PAH, heterocyclic amines)
Anti-eicosanoid:	Inhibition of lipoxygenases and cyclooxygenases
Anti-protein kinases:	Inhibitor of tyrosine, protein and other kinases
Anti-nitrosation:	Inhibitor of nitrosation of amines, amides

mammalian lignans are enterolactone and enterodiol. In limited studies, they appear to have tumor inhibitory properties, particularly as anti-estrogens. These lignans were identified in urine from subjects with a varied range of high fiber cereal diets. Linseed (flaxseed) in the diet gave particularly high levels of urinary lignans.[37] Dietary supplementation with flaxseed has inhibited some aspects of mammary tumorigenesis in rats.[38] This approach may be of practical use in reducing mammary and possibly colon cancer risk in human studies.

MULTIFUNCTIONAL BIOCHEMICAL ACTIVITIES OF PLANT PHENOLICS

It must be emphasized that plant phenols have multifunctional biochemical activities, as illustrated by quercetin (Table 2). Many of these involve modulation of one or more processes thought to be involved in carcinogenesis. While some plant phenols may be chosen as candidates for chemoprevention of carcinogenesis based on a single hypothesized mechanism, probably several activities are involved, adding to the total anticarcinogenic potential. Quercetin inhibited both initiation with 7,12-dimethylbenz[a]anthracene (DMBA) and tumor promotion with 12-0-tetradecanoylphorbol-13-acetate (TPA) of mouse skin tumor formation.[39]

NEEDS AND OBJECTIVE FOR FUTURE RESEARCH

Plant phenolics, components of human foods, have shown interesting activities as inhibitors of mutagenic and carcinogenic processes. In order to utilize these properties for chemopreventive reduction in risk for human cancer, much further work is needed. This includes further extension of anti-cancer studies, but also fundamental studies in allied areas, including:

- Reliable food composition data of amounts of specific phenolics in fresh foods, and losses in processing, storage, etc. to obtain realistic estimates of dietary intake.
- Absorption and metabolism. Little is known about the fate of most plant phenolics after ingestion. Rutin and quercetin are poorly absorbed,[40] while caffeic acid appears well absorbed; only one-fifth of a dose has been identified as urinary metabolites in human studies.[41] However, these studies were performed with pure crystalline substances, while in foods the substances are usually present as glycosides or esters, or in solution in the terpene-lipid components of the foods, all of which affect biological absorption and distribution.
- Achievable circulating blood and tissue levels of specific plant phenolics, and their chemical nature (conjugated, free, etc.) and the relative anti-tumor activity of the circulating forms, are needed to design optimum dietary intakes for chemopreventive studies.

- Cellular reactions of the plant phenolics with mammalian tissues, including mode and chemical form of delivery to specific target tissues, effects on cell membranes, cytosolic enzymes and activation systems.
- Of theoretical interest would be information on the effects of individual plant phenolics on specific cytochrome P450 systems, and resultant effects on detoxification activities towards endogenous and xenobiotic substances.[42]
- Activity studies of important common food plant phenolics as inhibitors of specific pathways of arachidonic acid metabolism, such as cyclooxygenase 1 (COX-1), cyclooxygenase 2 (COX-2) and lipoxygenase forms.

REFERENCES

1. Wattenberg LW. Chemoprevention of cancer. Cancer Res, 1985, 45, 1-8.
2. Wattenberg LW. Inhibition of carcinogenesis by minor anutrient constituents of the diet. Proc of the Nutrition Soc, 1990, 49, 173-183.
3. Hartman PE and Shankel DM. Antimutagens and anticarcinogens: a survey of putative interceptor molecules. Env and Mol Mutagenesis, 1990, 15, 145-182.
4. Harborne JB. in Methods in Plant Biochemistry, Vol. I: Plant Phenolics; Harborne JB, Ed; Academic Press: London, UK, 1989, 1-28.
5. Van Straten S. Volatile compounds in food; Central Institute for Nutrition and Food Research: Zeist, The Netherlands, 1977.
6. Van Sumere CF. in Methods in Plant Biochemistry, Vol I: Plant Phenolics; Harborne JB, ed; Academic Press: London, UK, 1989, 29-73.
7. Nishimura H, Nonaka GI, Nishioaka I. Seven quinic acid gallates from quercus stenophylla. Phytochem. 1984, 23, 2621-2623.
8. Herrman K. Occurrence and content of hydroxycinnamic and hydroxybenzoic acid compounds in foods. CRC Crit Rev Food Sci Nutri, 1989, 28, 315-347.
9. Eskin NAM. Biochemistry of Foods; Academic Press: San Diego, CA 1990, 401-432.
10. Haslam E. Plant Polyphenols: Cambridge University Press: Cambridge, UK. 1989.
11. Harborne JB. Comparative Biochemistry of the Flavonoids; Academic Press: London, 1967.
12. Harborne JB and Mabry TJ. The Flavonoids: Advances in Research; Chapman and Hall: London, UK, 1982.
13. Harborne JB, Mabry TJ. The Flavonoids: Advances in Research Since 1980; Chapman and Hall: London, UK, 1988, Vol. 2.
14. Markham KR. In Methods in Plant Biochemistry, Vol I: Plant Phenolics, Harborne JB, Ed; Academic Press: London, UK, 1989, 197-235.
15. Wattenberg LW, Coccia JB, Lam LKT. Inhibitory effects of phenolic compounds on benzo(a)pyrene induced neoplasia. Cancer Res, 1990, 40, 2820-2823.
16. Newmark HL and Mergens WJ. α-Tocopherol (Vitamin E) and its relationship to tumor induction and development, in: Inhibition of Tumor Induction and Development; Zedek MS and Lipkin M. eds. Plenum Press, New York, NY 1981, 127-168.
17. Kuenzig W, Chang J, Norkus E, Holowaschenko H, Newmark HL, Mergens W and Conney AH. Caffeic and ferulic acid as blockers of nitrosamine formation. Carcinogenesis, 1984, 5: 309-313.
18. Newmark HL. A hypothesis for dietary components as blocking agents of chemical carcinogenesis: plant phenolics and pyrrole pigments. Nutrition and Cancer, 1984, 6: 58-70.
19. Newmark HL. Plant phenolics as inhibitors of mutational and precarcinogenic events. Can J Physiol Pharmacol, 1987, 65: 461-466.
20. Dewhirst FE. Structure-activity relationships for inhibition of prostaglandin cyclooxygenase by phenolic compounds. Prostaglandins, 1980, 20: 209-222.
21. Nakadate T, Yamamoto S, Ishu M, Kato R. Inhibition of 0-tetradecanoylphorbol-13-acetate induced epidermal ornithine decarboxylase by lipoxygenase inhibitors: possible role of product(s) of lipoxygenase pathway. Carcinogenesis, 1982, 3: 1411-1414.
22. Karmali RA. Lipid nutrition, prostaglandins and cancer, in: Biochemistry of Arachidonic Acid Metabolism; Lands WEM, Ed: Martinas Nijhoff Publishing: Boston, MA, 203-212.

23. Katiyar S, Agarwal R, Wood GS et al. Inhibition of 12-0-tetradecanoylphorbol-13-acetate-caused tumor promotion in 7,12-dimethylbenz(a)anthracene-initiated SENCAR mouse skin by a polyphenolic fraction isolated from green tea. Cancer Res, 1992, 52:6890-6897.
24. Deschner EE, Ruperto J, Wong G, Newmark HL. Quercetin and rutin as inhibitors of azoxymethanol-induced colonic neoplasia. Carcinogenesis, 1991, 12: 1193-1196.
25. Huang MT, Lou YR, Ma W, Newmark HL, Reuhl KR, Conney AH. Inhibitory effects of dietary curcumin on forestomach, duodenal and colon carcinogensis in mice. Cancer Research, 1994, 54: 5841-5847.
26. Schwartz B, Fraser GM, Levy J, Sharoni Y, Guberman R, Krawiec J, Lamprecht SA. Differential distribution of protein kinases along the crypt-to-lumen regions of rat colonic epithelium. Gut, 1988, 29, 1213-1221.
27. Kato R, Nakadate T, Yamamoto S, Sugimura T. Inhibition of 12-0-tetradecanoylphorbol-13-acetate induced tumor promotion and ornithine decarboxylase activty by quercetin: possible involvement by lipoxygenase inhibition. Carcinogenesis, 1983, 4: 1301-1305.
28. Verma AK, Johnson JA, Gould MN, Tanner MA. Inhibition of 7,12-dimethylbenz(a)anthracene and N-nitrosomethylurea-induced rat mammary cancer by dietary flavonol quercetin. Cancer Res, 1988, 48, 5754-5758.
29. Deschner EE, Ruperto JF, Wong GY, Newmark HL. The effect of dietary quercetin and rutin on AOM-induced acute colonic epithelial abnormalities in mice fed a high-fat diet. Nutrition and Cancer, 1993, 20: 199-204.
30. Huang MT, Robertson FM, Lysz T, Ferraro T, Wang ZY, Georgiadis CA, Laskin JD, Conney AH. Inhibitory effects of curcumin on carcinogenesis in mouse epidermis. in: MT Huang, CT Ho, CY Lee. (eds) Phenolic Compounds in Food and Their Effects on Health II: 1992. Antioxidants and Cancer Prevention Society, 339-349, Washington DC: American Chemical Society.
31. Tanaka T, Yoshima N, Sugie S, Mori H. Protective effects against liver, colon and tongue carcinogenesis by plant phenols. In: MT Huang, CT Ho, and CY Lee (Eds). Phenolic Compounds in Food and Their Effects on Health II: Antioxidants and Cancer Prevention. pp. 326-337. Washington DC: American Chemical Society, 1192.
32. Lesca P. Protective effects of ellagic acid and other plant phenols on benzo(a)pyrene-induced neoplasia in mice. Carcinogenesis, 1983, 6: 1651-1653.
33. Mandal S, Stoner GD. Inhibition of methylbenzylnitrosamine-induced esophageal tumors in rats by ellagic acid. Carcinogenesis, 1990, 11: 55-61.
34. Hirose M, Fukushima S, Shirai T, Hasegawa R, Kato T, Tanaka H, Asakawa E, and Ito N. Stomach carcinogenicity of caffeic acid, sesamol and catechol in rats and mice. 1990. Jpn J Cancer Res, 81: 207-212, 1990.
35. Hanham AF, Dunn BP, Stich HF. Clastogenic activity of caffeic acid and its relationship to hydrogen peroxide generated during antooxidation. Mutation Res, 1982, 116, 333-339.
36. Aldercreutz M. Western diet and western diseases: some hormonal and biochemical mechanisms and associations. Scand J Lab Invest, 1990, 50, Suppl 201, 3-23.
37. Horwitz C, Walker APR. Lignans - additional benefits from fiber? Nutr Cancer, 1984, 6: 73-76.
38. Serraino M, Thompson LU. The effect of flaxseed supplementation on the initiation and promotional stages of mammary carcinogenesis. Nutr Cancer, 1992, 17: 153-159.
39. Verma AK. Modulation of mouse skin carcinogenesis and epidermal phospholipid biosynthesis by the flavonol quercetin. In: MT Huang, CT Ho and CY Lee (Eds) Phenolic Compounds in Food and Their Effects on Health II: Antioxidants and Cancer Prevention. pp. 350-364. Washington DC. American Chemical Society, 1992.
40. Gugler R, Leschick M, Dengler HJ. Disposition of quercetin in man after single oral and intravenous doses. Europ J Clin Pharmacol, 1975, 9: 229-234.
41. Jacobson EA, Newmark HL, Baptista J, Bruce WR. A preliminary investigation of the metabolism of dietary phenolics in humans. Nutr Reports Intern, 1983, 28:1409-1417.
42. Smith TJ and Yang CS. Effects of food phytochemicals on xenobiotic metabolism and tumorigenesis. *In:* Food Phytochemicals for Cancer Prevention I. (Huang M-T, Osawa T, Ho C-T, Rosen RT, eds.) pp. 17-48, ACS Symposium Series 546, Washington, DC, Chapter 2, 1994.

4

CANCER CHEMOPREVENTION BY POLYPHENOLS IN GREEN TEA AND ARTICHOKE

Rajesh Agarwal[1] and Hasan Mukhtar[2]

[1] Department of Dermatology
University Hospitals of Cleveland
[2] Skin Diseases Research Center
Case Western Reserve University
Cleveland, Ohio 44106

INTRODUCTION

Cancer is a major disease accounting for over 7 million deaths per year worldwide.[1] Though recent advances in cancer diagnosis, its early detection, and therapy have improved the quality of life for cancer patients, there is little, if any, effect on the mortality rates for most cancers. To reduce cancer related deaths, four strategies are possible. These include: early diagnosis and intervention, improved management of non-localized cancers, successful treatment of localized cancers, and prevention of cancer occurrence in the first place. Development of successful approaches to cancer prevention appears the most practical strategy for reduction in cancer related deaths. One approach is "chemoprevention" in which the progress of this disease can be slowed, completely blocked or reversed by the administration of one or more naturally occurring or synthetic chemical agents. It is increasingly appreciated that an ideal cancer chemopreventive agent for human use must fulfill the following criteria: little or no toxic effects, high efficacy against multiple sites, capability of oral consumption, a known mechanism of action, low cost, and human acceptability. In recent years, considerable efforts have been directed to identify agents that may have the ability to inhibit, retard or reverse one or more stages of multistage carcinogenesis, which is comprised of initiation, promotion and progression.[2-4] Since cancer usually evolves in a prolonged manner, agents that inhibit or retard one or more of these stages could affect the overall cancer induction. In this context, a few naturally occurring micronutrients present in human diets have been found to possess potent cancer chemopreventive effects.[5-11] Fruits, vegetables, common beverages as well as several herbs and plants with diversified pharmacological properties are all rich sources of cancer chemopreventive agents. At present over thirty classes of chemicals with verified cancer chemopreventive effects, at least in the experimental model systems, have been described.[5-11] Among these classes, polyphenolic compounds present in a variety of foods including fruits, vegetables, and beverages consumed by the human population are receiving increasing attention.[10] In this article, we have summarized the studies showing that polyphenolic compounds present in green tea possess significant cancer chemopreventive effects in animal tumor bioassay systems.[5, 12-14] Epidemiological studies are beginning to show that these agents may also prove useful for the human population. Emphasis has also been placed on another agent, silymarin, recently identified from our laboratory as a very strong cancer chemopreventive agent in the mouse skin tumorigenesis model.

CANCER CHEMOPREVENTION BY GREEN TEA

General Background

Tea (*Camellia sinensis*) is one of the most popular beverages consumed worldwide. The term "green tea" relates to its manufacturing from fresh tea leaves and the bud of the plant by steaming or drying at elevated temperatures with the precaution to avoid oxidation of the polyphenolic compounds which mainly include epicatechins, the flavonol skeleton. Of all the commercially produced tea annually (~2.5 million metric tons), about 20% is green tea which is mainly consumed in Asian countries like Japan, China, Korea and India; about

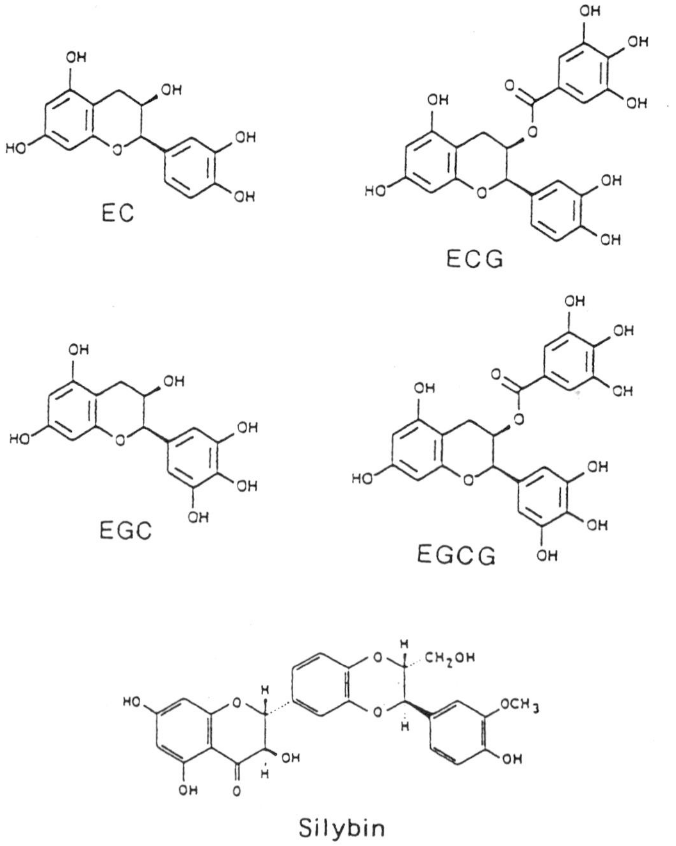

EC

ECG

EGC

EGCG

Silybin

Figure 1. Chemical structures of epicatechin derivatives present in green tea, and silybin - the major constituent in silymarin.

78% is black tea, mainly consumed in the Western countries and some Asian countries; and about 2% is oolong tea, mainly produced and consumed in southeastern China.[15] With regard to the chemical composition of green tea, its major constituents are flavonols, flavonoids, and phenolic acids. These compounds may account for up to 30% of the dry weight of green tea leaves. Most of the polyphenolic compounds present in green tea are flavonols, commonly known as epicatechins. Some major epicatechins identified qualitatively and quantitatively in green tea include (-)-epicatechin (EC), (-)-epicatechin-3-gallate (ECG), (-)-epigallocatechin (EGC), and (-)-epigallocatechin-3-gallate (EGCG). The chemical structures of these compounds are given in Figure 1. Unlike green tea, a typical black tea beverage contains 3-10% epicatechins, 3-6% theaflavins, 12-18% thearubigins and other components.

Initial studies showing that water extract of green tea (WEGT) is antimutagenic in bacterial test systems were done in Japan.[16,17] Subsequently, we evaluated the anti-carcinogenic effects of WEGT and a polyphenolic fraction isolated from green tea (GTP) employing mouse skin multi-stage carcinogenesis protocols.[13] We also conducted studies to delineate the mechanism of anti-carcinogenic effects of GTP and individual epicatechin derivatives present therein.[13] To date many laboratories have reported significant cancer chemopreventive effects of GTP, WEGT and/or individual epicatechin derivatives (ECDs) employing many organ-site specific animal tumor protocols where various test carcinogens were employed.[12-14] More recently, studies from our laboratory in cell culture systems have also suggested that GTP and ECDs may have the potential to inhibit prostate cancer development.[18] Taken together, it can be concluded that "green tea does prevent cancer in many tumor bioassay systems". With regard to the prevention of cancers in human populations, recent epidemiological studies show that animal data may be valid for at least some forms of human cancers.

Anti-Carcinogenic Effect of Green Tea in Mouse Skin Model of Carcinogenesis

Over the last fifty years or so it has been recognized that carcinogenesis in mouse skin and possibly in other tissues is a stepwise process of initiation, promotion and progression stages.[2-4] Laboratory studies have shown that green tea inhibits each of these three stages.[13] In the mouse skin tumorigenesis model, the carcinogenic process can be initiated by both chemical agents such as polycyclic aromatic hydrocarbons (PAHs) and some nitrosamines, as well as physical agents such as ultraviolet B (UVB) radiation.[2,3,19,20] With regard to PAH-type carcinogens, it has become clear that they are first metabolized by a cytochrome P450-dependent monooxygenase enzyme system which leads to the formation of a highly reactive electrophilic metabolite, termed the ultimate carcinogen. This binds to target DNA resulting in a mutation in a cellular gene; *ras* proto-oncogene in most cases.[2-4,21] Benzo(a) pyrene (BP), DMBA, and 3-methylcholanthrene (3MC) are some of the PAH-type carcinogens used in mouse skin carcinogenesis protocols.[2-4,21] Among these, BP is ubiquitously present in the environment due to smoking and automobile exhaust. In the case of nitrosamines, they act directly at target DNA leading to a genetic change in cellular genes, e.g. *ras* proto-oncogenes and/or p53 tumor suppressor gene.[2-4,20] UV radiation has been identified as the major cause of non-melanoma skin cancer, both basal cell and squamous cell carcinomas, in humans.[19,22,23] Studies employing mouse skin models have shown that UV radiation results in a major oxidative burst in the skin which ultimately leads to mutation in target genes.[22-24] While earlier studies showed that mutation in both *ras* proto-oncogenes and p53 tumor suppressor gene occurs following UV-induced tumorigenesis in mouse skin, recent studies from our laboratory, as well as from others, have shown that mutations in *ras* proto-oncogenes are rare events in UV-induced tumorigenesis in mouse skin, suggesting a major role of p53 in the causation of such tumors.[22,23,25,26] Based on these studies, it can be stated that agents which a) inhibit cytochrome P450-dependent metabolic activation of carcinogens, the DNA-binding of ultimate carcinogenic metabolites or direct acting carcinogens, and b) increase the anti-oxidant state of target tissue, skin in the present context, could be strong protective agents against skin tumor initiation.

Consistent with this mechanistic approach, we have shown that the addition of individual ECDs or GTP to microsomes prepared from rat liver resulted in a dose-dependent inhibition of P-450-dependent aryl hydrocarbon hydroxylase, 7-ethoxycoumarin-O-deethylase, and 7-ethoxyresorufin-O-deethylase activities.[27] We also demonstrated that the administration of GTP, either topically or orally, to SENCAR mice inhibits carcinogen-DNA adduct formation in epidermis after topical application of [^{3}H]BP or [^{3}H]DMBA.[28] In another study, epidermal aryl hydrocarbon hydroxylase activity and epidermal enzyme-mediated binding of BP and DMBA to DNA were inhibited by ECDs and GTP.[29] Further, chronic oral administration of GTP to mice for four weeks resulted in moderate to significant enhancement in glutathione peroxidase, catalase, NADPH-quinone oxidoreductase, and glutathione S-transferase activities in small bowel, lung and liver.[30] Enhancement by green tea of these enzymatic pathways, which play a role in detoxification of carcinogenic metabolites after formation by P-450 and other enzymes, and/or its ability to inhibit enzymatic pathways that are key determinants for tumor initiation may be expected to have protective functions against carcinogenesis. In addition, other studies have shown that pre-application of GTP or oral feeding of WEGT to mice prior to UV irradiation leads to a different spectrum of p53 tumor suppressor gene mutations.[31]

With regard to long-term protocols, tumors are generally induced in mouse skin by a two-stage initiation-promotion carcinogenesis protocol in which a single topical application of the initiating agent, in most cases DMBA, is followed a week later by repeated topical applications of a promoting agent, usually 12-*O*-tetradecanoylphorbol-13-acetate (TPA).[2,3] A single topical application of a large dose of an initiator on murine skin also results in cutaneous tumor induction, a complete carcinogenesis protocol. In another modification of the two-stage protocol, the repeated applications of subcarcinogenic doses (the initiating dose) of a carcinogen also induces cutaneous tumors and results in a higher number of neoplasms.[2,3] As with chemical carcinogens, UV radiation also acts as an initiator and a promoter as well as a complete carcinogen.[19] Employing these protocols, several studies have shown the preventive potential of GTP, WEGT and EGCG. We found that skin application of GTP (1.2 mg in 0.2 ml acetone) for seven days prior to the application of 3MC in a complete carcinogenesis protocol, afforded significant protection against skin tumorigenesis in terms of both tumor incidence as well as tumor multiplicity.[28] We observed that chronic oral feeding of GTP (0.1%, w/v) in drinking water to SKH-1 hairless mice during the entire period of a UVB-carcinogenesis protocol, resulted in significantly lower tumor incidence and multiplicity as compared to the animals which did not receive GTP. Topical application of GTP before UVB irradiation also afforded some protection against photocarcinogenesis; but the effect was lower than after oral administration of GTP in drinking water.[32]

In other studies using the two-stage carcinogenesis protocol, topical application of GTP (10 mg in 0.2 ml acetone) for 7 days prior to a single application of DMBA followed by twice weekly applications of TPA as the promoting agent also afforded significant protection in terms of both skin tumor incidence and multiplicity,[28] as collaborated later by Huang *et al.* Similarly, oral feeding (equivalent of four cups of tea) of GTP in drinking water for 50 days prior to the start of DMBA-TPA treatment or its continuous feeding during the entire experimental period also resulted in significantly fewer skin tumors per mouse, compared with non-GTP fed animals.[28] Similarly, the protective effects of GTP were also observed in an identical tumor protocol when (±)-7β,8α-dihydroxy-9α,10α-epoxy-7,8,9,10-

tetrahydrobenzo(a)pyrene (BPDE-2) was the tumor initiating agent.[33] In the DMBA-TPA protocol, topical application of EGCG (5 μ mol in 0.2 ml acetone per animal per day up to seven days) prior to challenge with DMBA on the skin of SENCAR mice yielded significant protection against DMBA-caused tumor initiation in terms of decrease in percentage of mice with tumors, cumulative number of tumors, and tumors per mouse.[35] In another study it was shown that feeding WEGT (1.25%, w/v) as the sole source of drinking water to SKH-1 hairless mice affords protection, in a dose-dependent manner, against UVB radiation-induced intensity of red color and area of skin lesions, as well as UVB radiation-induced skin tumor initiation (TPA was used as tumor promoter).[36]

Though chemopreventive agents can be targeted for intervention at either of the initiation, promotion or progression stages of multi-stage carcinogenesis, intervention at the promotion stage appears to be most appropriate. The major reason is that tumor promotion is a reversible event, at least in the early stages, and requires repeated and prolonged exposure of a promoting agent.[2-4] In contrast, initiation can be accomplished by a single exposure to a carcinogen and is a rapid and irreversible process.[2-4] Since humans are constantly exposed to environmental carcinogens, it can be implied that the initiation step of carcinogenesis is inevitable. Furthermore, progression, a critical and late step of tumorigenesis, which involves the conversion of a preneoplastic cell to a cancer cell, has not been extensively investigated.[2-4] Thus a practical approach for cancer chemoprevention strategies should be based on identification of anti-tumor promoting agents and their mechanism of action.

The process of tumor promotion in murine skin involves a combination of different mechanisms. The topical application of a tumor promoter, specifically of the phorbol ester TPA-type, to mouse skin results in a number of cellular, histological, biochemical and molecular changes. These include induction of epidermal hyperplasia and inflammation, increase in the number of dark basal keratinocytes, induction of epidermal cyclooxygenase, lipoxygenase and ornithine decarboxylase (ODC) activities, depletion of the anti-oxidant defense system in epidermis, activation of PKC, and modulation of inflammatory cytokines such as interleukin-1α (IL-1α) expression.[2,3] The cancer chemopreventive agents which exert their effects against tumor promotion usually must inhibit one, several or even all these events. Whether inhibition of any particular event involved in promotion is obligatory, sufficient and/or mandatory to exert maximum to complete anti-tumor promoting effects is not clear. However, most (>95%), if not all, of the anti-skin tumor promoting agents significantly inhibit tumor promoter-caused induction of epidermal ODC activity.[37,38] With regard to the inhibitory effects of green tea we have shown that pre-application of GTP on the dorsal skin of mice results in significant inhibition of TPA-caused epidermal edema and hyperplasia.[39] Similarly, single or multiple applications of GTP to SENCAR mouse ear skin prior to or after the application of TPA, also affords significant protection against TPA-caused ear edema and infiltration of polymorphonuclear leukocytes in the ear skin,[40] as confirmed by Huang *et al.* using both GTP as well as EGCG.[33] Single or multiple applications of GTP to mouse skin prior to that of TPA, result in a significant inhibition of TPA-induced epidermal cyclooxygenase and lipoxygenase activities.[39] We found that GTP and individual ECDs present therein inhibit differently skin tumor promoter-caused induction of epidermal ODC activity and mRNA expression.[41] Also, GTP, EGCG, ECG, EGC and EC inhibit differently the skin tumor promoter-caused induction of epidermal interleukin-1α mRNA expression in SENCAR mice.[42] Moreover, GTP possesses protective effects against UVB radiation-caused changes in murine skin.[43] Chronic oral feeding of 0.2% GTP (w/v) as the sole source of drinking water for 30 days to SKH-1 hairless mice followed by irradiation with UVB (900 mJ/cm^2) resulted in significant protection against UVB radiation-caused cutaneous edema, depletion of antioxidant-defense systems in the epidermis, and of induction of epidermal ODC and cyclooxygenase activities, in a time-dependent manner.[43]

In long-term studies, we demonstrated the preventive potential of GTP against TPA-caused tumor promotion in DMBA-initiated SENCAR mouse skin.[39] Topical application of GTP at different doses 30 min prior to that of TPA to the skins of DMBA-initiated SENCAR mice, resulted in highly significant protection against TPA-induced skin tumor promotion in a dose-dependent manner. The animals pretreated with GTP showed comparatively fewer tumors per animal, decrease in tumor volume per mouse and average tumor size per tumor as compared to non-GTP-treated animals,[39] as also demonstrated in CD-1 mice, by Huang *et al.*[33] In SKH-1 hairless mice, oral feeding of WEGT in drinking water had a protective effect on TPA-caused tumor promotion in DMBA-initiated skin.[36] The antiskin-tumor promoting action of EGCG has also been shown against teleocidin or okadaic acid-caused tumor promotion in DMBA-initiated mouse skin.[44] We also demonstrated the preventive effect of GTP on both stage I and stage II tumor promotion in SENCAR mouse skin. Application of GTP concurrently with each application of either TPA or mezerein in a stage I or stage II murine skin tumor promotion protocol, respectively, resulted in significant protection in terms of both tumor multiplicity and growth. However, more profound and sustained protective effects of GTP were evident when it was applied continuously during both stage I and stage II tumor promotion concurrent to TPA and mezerein, respectively.[45] The water extract of black tea (WEBT) also possesses significant cancer chemopreventive effects in the mouse skin tumor model.[46]

With regard to tumor progression, we have shown the protective effect of topical application of GTP on the conversion of benign skin papillomas to malignant carcinomas in mouse skin.[47] Pre-application of GTP 30 min prior to that of 4-nitroquinoline N-oxide or benzoyl peroxide resulted in significant protection against malignant conversion of DMBA-TPA-induced benign skin papillomas to squamous cell carcinomas as enhanced by these agents.[47] Topical application of the same dose of GTP also showed some protection against spontaneous malignant conversion.[47] The protective effects of GTP were evident by reduction in number of carcinomas per mouse, percentage of mice with carcinomas and percentage malignant conversion rate. Though not related to the anti-tumor progression protocol, others have shown that oral feeding or intraperitoneal administration of varying doses of WEGT, GTP or EGCG to mice bearing experimentally-induced skin papillomas results in partial regression of such tumors.[48]

Taken together, these results can be summarized as "in mouse skin chemical carcinogenesis and photocarcinogenesis protocols, green tea possesses considerable effects against tumor initiation, tumor progression and regression of established papillomas, and remarkably significant effects against tumor promotion." In Table 1 we have outlined the studies showing the cancer chemopreventive potential of green tea preparations in long-term skin tumorigenesis protocols.

Anti-Carcinogenic Effect of Green Tea in Internal Body Organs

Several studies have also been conducted to assess the anti-carcinogenic effects of systemic administration of WEGT, GTP and EGCG against induced tumors of internal organs. In a complete carcinogenesis protocol, oral feeding of WEGT as the sole source of drinking water to A/J mice during the entire experiment resulted in 55-70% protection against BP and N-nitrosodiethylamine (DEN)-induced lung and forestomach neoplasia formation.[49] Likewise, oral feeding of WEGT to A/J mice resulted in significant protection in a dose-dependent manner, against DEN-induced lung and forestomach tumorigenesis in A/J mice.[50] Giving WEGT or GTP as the sole source of drinking water during the initiation, post-initiation and entire period of tumorigenesis protocols resulted in significantly fewer tumors per mouse in both lung and forestomach.[51] Similar protective effects were also evident when GTP was intubated orally before the challenge with carcinogens.[52] Oral feeding of WEGT

Table 1. Anti-carcinogenic effect of green tea in mouse skin model of carcinogenesis[1]

Carcinogenesis Protocol/Mouse strain	Carcinogen Employed	Tea preparation and mode of administration	Reference
Complete			
Balb/C	3-MC	GTP, topical	28
SKH-1	UVB radiation	GTP, topical/drinking water	32
Multistage - During Initiation			
SENCAR	DMBA/TPA	GTP, topical/drinking water	28
SENCAR	BPDE-II/TPA	GTP, topical	33
SENCAR	DMBA/TPA	EGCG, topical	34
CD-1	DMBA/TPA	GTP, topical	35
SKH-1	UVB/TPA	WEGT, drinking water	36
SKH-1	DMBA/UVB	WEGT/WEBT, drinking water	46
Multistage - During Promotion			
SENCAR	DMBA/TPA	GTP, topical	39
CD-1	DMBA/TPA	GTP, topical	35
CD-1	DMBA/Teleocidin	EGCG, topical	44
CD-1	DMBA/Okadaic acid	EGCG, topical	28
SENCAR	DMBA/TPA (Stage I)	GTP, topical	45
SENCAR	DMBA/mezerein	GTP, topical (Stage II)	45
Multistage - During Progression			
SENCAR	4-NQO or BPO[2]	GTP, topical	47

[1] All the abbreviations are defined in the running text. The details of carcinogenesis protocol, and the dose and mode of treatment with GTP/WEGT/EGCG are elaborated in the running text.
[2] Used to enhance the rate of malignant conversion.

or WEBT prior to or after the challenge with 4-(methylnitrosamino)-1-(3-pyridyl)-1-butanone (NNK) up to the end of the experiment resulted in significant protection in terms of lung tumor incidence as well as tumor multiplicity;[50] similar results with WEGT and EGCG against NNK-induced lung tumorigenicity in A/J mice were noted.[53]

The protective effects of GTP, WEGT and/or EGCG are evident against experimental tumors of the duodenum, esophagus, colon, liver, pancreas and mammary gland. EGCG had preventive effects against N-ethyl-N'-nitro-N-nitrosoguanidine (ENNG)-induced duodenal tumors in C57BL/6 mice.[54] Oral GTP in drinking water after a challenge with azoxymethane (AOM), up to the termination of the experiment, resulted in significant protection against AOM-induced colon cancer in Fischer rats.[55] Likewise, green tea at a low dose in drinking water protected against N-nitrosomethylurea (NMU)-induced colon tumorigenesis in F344 rats.[56] Green tea also afforded significant preventive effects against N-nitrosomethylbenzylamine (NMBA)- and nitrososarcosine-induced esophageal tumorigenesis in rats[57] and mice, respectively.[58] Green tea has also protected against aflatoxin B_1 (AFB_1)- and DEN-induced hepatocarcinogenesis.[59,60] Green tea extracts also showed an anti-promotion action on pancreatic cancer induced by N-nitroso-bis(2-oxopropyl)amine in golden hamsters,[61] and against mammary cancer in female rats pretreated with DMBA.[62] Furthermore, oral feeding of EGCG (0.05% in drinking water) for 15 weeks after the administration of N-ethyl-N'-nitro-N-nitrosoguanidine (MNNG) resulted in 50% reduction in the percentage of rats with glandular stomach tumors.[63] Taken together, these studies clearly show that green tea possesses cancer preventive effects in several animal tumor bioassay systems and that the polyphenolic epicatechin compounds present in green tea are responsible. In Table 2 we have outlined briefly all the studies showing the cancer chemopreventive potential of green tea preparations in long-term tumorigenesis protocols of internal organs.

Table 2. Experimental studies showing anti-carcinogenic effect of green tea in internal body organs[1]

Animal species/ Carcinogen used	Organ site	Tea preparation and mode of administration	Reference
A/J Mice: BP/DEN	Forestomach and Lung	WEGT/GTP - drinking water/oral intubation	49,50,51,52
A/J Mice: NNK	Lung	WEGT/WEBT/EGCG-drinking water	50,53
C57BL/6 Mice: ENNG	Duodenum	EGCG-oral feeding	54
Rats: MNNG	Glandular stomach	EGCG - drinking water	63
Fischer/F344 rats: AOM/NMU	Colon	GTP - drinking water	55,56
Rats/mice: NMBA/nitrososarcosine	Esophagus	GTP - drinking water	57,58
Rats: AFB_1-/DEN-induced	Liver	GTP - drinking water	59,60
Golden hamster: N-nitroso-bis (2-oxopropyl)amine	Pancreas	GTP - drinking water	61
Rats: DMBA	Breast	GTP - drinking water	62

[1] All the abbreviations are defined in the running text. The details of carcinogenesis protocol, and the dose and mode of treatment with GTP/WEGT/EGCG are elaborated in the running text.

Potential Anti-Carcinogenic Effect of Green Tea in Prostate Cancer

After lung, prostate cancer is the major cause of cancer-related deaths in United States.[64,65] In recent years, much emphasis has been placed on identification of agents which could inhibit the occurrence or retard the development of prostate cancer.[66-71] It has been demonstrated that elevated levels of circulating testosterone play a major role in prostate cancer induction.[72-80] Based on these reports, we formulated a working hypothesis that elevated levels of testosterone, in addition to other biological effects, would result in an increase in polyamine biosynthesis and induction of ODC which would then lead to uncontrolled proliferation and differentiation of the prostate, ultimately causing prostate cancer. The major support for this hypothesis was based on the fact that compared to all the body sites, prostate has the highest levels of ODC and polyamines.[81]

We performed studies where the modulation by testosterone of ODC activity was assessed in androgen-dependent (LNCaP cells and CWR22 tissue) and -independent (DU-145 and PC-3 cells, and CWR22i tissue) human prostate carcinoma cell lines and tissue xenografts.[82,83] The treatment of androgen-dependent LNCaP cells and CWR22 tissue in culture with testosterone resulted in a significant induction of ODC activity in a dose- and time-dependent manner,[18] with maximum induction at the dose of 50 nM testosterone in culture medium in both cases.[18] The pre-treatment of cultures with GTP for 1 hr prior to the addition of testosterone caused a significant inhibition of testosterone-caused induction of ODC activity in both LNCaP cells and CWR22 tissue. The inhibitory response of GTP was dose dependent, and was maximum at the dose of 40 μg/ml in case of LNCaP cells and 60 μg/ml in case of CWR22 tissue.[18] Interestingly, the androgen-independent cell lines -DU-145 and PC-3, and CWR22i did not show any testosterone-mediated increase in ODC activity. Further studies were also performed where the inhibitory effect of GTP was assessed against testosterone-caused increase in the CWR22 cell colony formation, employing the soft agar assay. In these studies, compared to vehicle-treated controls showing 100 colonies/well, treatment of CWR22 with testosterone led to 400 colonies/ well. The pre-treatment with

Table 3. Potential anti-carcinogenic effect of green tea in prostate cancer

Test System and Parameters	Control (vehicle treated)	Testosterone (50 nM in alcohol)	GTP + testosterone	% Inhibition
LNCaP cells: ODC activity[1]	296 ± 23	1202 ± 89	305 ± 29	99[3]
CWR22 tissue: ODC activity	452 ± 39	1324 ± 109	498 ± 46	95[4]
CWR22 tissue: Soft agar assay[2]	100 ± 9	400 ± 35	250 ± 29	50

[1] ODC activity is expressed as p mol CO_2 released/hr/mg protein.
[2] The data are expressed as number of colonies/well.
[3] The dose of GTP employed in the culture was 40 μg/ml medium.
[4] The dose of GTP employed in the culture was 60 μg/ml medium.

GTP, however, resulted in more than 50% reduction in the testosterone-caused increase in colony formation. These data are summarized in Table 3.

Epidemiological Studies on Tea Consumption and Human Cancer

With regard to epidemiological studies related to the effects of tea consumption on human cancer, a case-control study in Japan has indicated that individuals consuming green tea tend to have a lower risk for gastric cancer.[84] Another study in Shizuoka Prefecture in Japan showed that the stomach cancer death rate in this tea producing and consuming area was lower than the national average.[85] According to the International Agency for Research on Cancer (IARC) 1991 position, which included publications up to 1989, it was concluded that "there is no evidence to implicate tea or tea component(s) as human carcinogen". However, based largely on recently published studies, a suggestion has been made that "green tea contains an interesting and potentially important group of chemicals with anticarcinogenic effect".[86] Many studies since 1990, however, have shown lower cancer risk in populations consuming tea. Gao *et al.* did a case-control study and concluded that "the population based case control study of esophageal cancer in urban Shanghai, Peoples's Republic of China, suggests a protective effect of green tea consumption. Although these findings are consistent with studies in laboratory animals, indicating that green tea can inhibit esophageal carcinogenesis, further investigations are definitely needed".[87] Analyses of the levels of tea polyphenols in plasma and urinary samples of human subjects may be important for determining the effect of tea consumption on human cancers.[88] In blood samples from nonsmokers, smokers and smokers consuming green tea, the frequencies of sister-chromatid exchange (SCE) in mitogen-stimulated peripheral lymphocytes from each group were determined and statistically analyzed.[89] While the SCE rates were elevated significantly in smokers compared to nonsmokers, the frequency of SCE in smokers who consumed green tea was comparable to that of nonsmokers, indicating that consumption of green tea can block the cigarette-induced increase in SCE frequency.[89]

Since green tea has been shown to afford cancer chemopreventive effects in a broad spectrum of animal tumor model systems,[12-14] identification of its anticarcinogenic constituent is urgently needed. The preventive agent, once identified, could be supplemented in various food items we consume routinely. Thus, one day it is possible that humans may be able to obtain "designer products" such as gums, candies, snacks, ice-cream, puddings, other food items and cosmetic products, which can be supplemented with polyphenolic compounds from green tea.

CANCER CHEMOPREVENTION BY SILYMARIN

General Background

Silymarin, a flavonoid isolated from the artichoke (*Silybum marianum* (L.) Gaertn),[90] is composed of three isomers: silybin, silydianin and silichristin; the chemical structure of the major constituent (>98%) silybin is shown in Figure 1. Chemical analysis has shown that silymarin is present to an extent of 0.7% (w/w) in artichoke.[91] For the past two decades silymarin has been used clinically in Europe for the treatment of alcoholic liver diseases.[92] In laboratory studies, *in vivo* pretreatment of experimental animals with silymarin protected against hepatotoxicity induced by a wide range of toxicants including allyl alcohol, carbon tetrachloride, galactosamine, phalloidin, thioacetamide, and microcystin-LR.[90,92,93] The *in vitro* studies with hepatocytes and liver microsomes showed that silymarin affords protection against lipid peroxidation induced by several xenobiotics.[94,95] More defined mechanistic studies in rodents and in cell cultures indicated that silymarin is a strong antioxidant capable of scavenging free-radicals.[96-100] In addition, limited *in vitro* studies demonstrated that silymarin inhibits a) the formation of transformed rat tracheal epithelial cell colonies induced by exposure to BP,[101] b) TPA-induced anchorage-independent growth of the JB6 mouse epidermal cells,[102] and c) DMBA-initiated and TPA-promoted mammary lesion formation in organ culture.[103]

Most antioxidants afford protection against tumor promotion by inhibiting the oxidative stress induced by tumor promoters.[24,37,38] An increase in the expression of ODC has been suggested as a prerequisite, though not obligatory, for tumor promotion.[2,3] Since silymarin is an antioxidant,[96-100] we assessed the effect of preapplication of silymarin on TPA-induced epidermal ODC activity and ODC mRNA levels in SENCAR mice. Topical application of silymarin prior toTPA resulted in a highly significant inhibition of TPA-induced epidermal ODC activity in a dose-and time-dependent manner.[104] Northern blot analysis revealed that pre-application of silymarin led to almost complete inhibition of TPA-induced epidermal ODC mRNA.[104] Silymarin also inhibited significantly epidermal ODC activity induced by several other tumor promoters, including free-radical-generating compounds.[104] In total, these results suggested that silymarin could be a useful anti-tumor promoting agent capable of ameliorating the effects of a wide range of tumor promoters. Long-term tumor studies, therefore, were conducted to evaluate the cancer chemopreventive effects of silymarin.

Anti-Carcinogenic Effect of Silymarin in Mouse Skin Model of Carcinogenesis

First, studies were carried out to assess the anti- tumor initiating and anti-tumor promoting potential of silymarin in a mouse skin two-stage initiation-promotion protocol. Topical application of silymarin at the dose of 6 mg per mouse per application for seven days followed by tumor initiation with DMBA and one week later tumor promotion with TPA, resulted in considerable protection in terms of both skin tumor incidence and multiplicity. For anti-tumor promotion studies, tumor initiation was achieved in SENCAR mice by a single topical application of DMBA, a week later animals were either promoted with TPA twice a week or first treated with silymarin at the dose of 6 mg per mouse per application and then 1 hr later with TPA. Under this treatment protocol, silymarin showed highly significant protection against tumor promotion which was evident in terms of percentage of mice with tumors, number of tumors per mouse and the total tumor volume per mouse as well as per tumor.[105,106] Recently we also assessed the preventive effect of silymarin against both stage I and stage II tumor promotion in SENCAR mouse skin. Application of silymarin prior to each application of either TPA or mezerein in a stage I or stage II murine skin tumor promotion protocol, respectively, resulted in highly significant protection against stage I tumor promotion only. The protective effect of silymarin observed against stage I tumor promotion was almost comparable to that observed when it was applied continuously during both stage I and stage II tumor promotion concurrent to TPA and mezerein, respectively (Agarwal et al., unpublished observation). The results of this study clearly demonstrate that silymarin is a potent inhibitor of stage I tumor promotion; this inhibition is sufficient to account for the very strong anti-tumor promoting effect of silymarin in complete as well as both stage I and stage II tumor promotion studies.

Therefore, further studies were carried out to assess the preventive effect of silymarin against UVB radiation-caused tumor initiation, tumor promotion and complete carcinogenesis.[105-107] Similar to chemical carcinogenesis protocols, the dose of silymarin employed in all cases was 6 mg per mouse per application. In an anti-tumor initiation protocol employing UVB as initiating agent, topical application of silymarin for seven days prior to UVB exposure afforded considerable protection against UVB-caused tumor initiation. In the anti-tumor promotion protocol with UVB as the agent, application of silymarin prior to each UVB irradiation resulted in highly significant protection against UVB-caused tumor promotion in terms of tumor incidence, tumor multiplicity as well as tumor volume. In a complete carcinogenesis protocol using UVB as the carcinogen, application of silymarin prior to each UVB irradiation resulted in highly significant protection, as evident by reduced tumor incidence, tumor multiplicity and as tumor volume.[105-107]

Thus, these studies clearly demonstrated that silymarin possesses strong potential as an anti-tumor promoting agent, as summarized in Table 4.

Based on the results of the long-term tumor experiments with the photocarcinogenesis model of mouse skin, short-term studies were also performed to define the mechanism associated with the preventive effect of silymarin against UVB-caused tumorigenesis.[107] Application of silymarin 30 min prior to that of UVB irradiation resulted in significant protection against UVB-caused skin edema as measured by bi-fold skin thickness and ear punch weight. Pre-application of silymarin before UVB irradiation also resulted in highly

Table 4. Anti-carcinogenic effect of silymarin in mouse skin model of carcinogenesis[1]

Carcinogenesis Protocol/ Mouse strain	Carcinogen Employed	Inhibitory response: Delay in latency period[2]	Inhibitory response: % Reduction[3]	Reference
Complete				
SKH-1 hairless	UVB	Yes	92	105-107
Multistage - During Initiation				
SENCAR	DMBA/TPA	No	26	105,106
SKH-1 hairless	UVB/TPA	No	66	105-107
Multistage - During Promotion				
SENCAR	DMBA/TPA	Yes	85	105,106
SKH-1 hairless	DMBA/UVB	Yes	78	105-107
SENCAR	DMBA/TPA (Stage I)	Yes	95[4]	–
SENCAR	DMBA/mezerein (Stage II)	No	71[4]	–

[1] Silymarin was applied topically at the dose of 6 mg per application in each experiment. Other details of experimental protocols are described in reference cited.

[2] The inhibitory response in terms of appearance of first tumor following tumor initiation.

[3] Percent reduction in number of tumors per animal at the termination of the experiment.

[4] Agarwal et al., unpublished data.

significant protection against UVB-caused induction of epidermal ODC and cyclooxygenase activities, and UVB-caused induction of epidermal ODC mRNA expression.[107] We have also observed that prior application of silymarin to that of UVB radiation results in almost 80% reduction in UVB-caused sun burn cell formation in mouse epidermis (Agarwal et al., unpublished observation).

Based on all these studies, it can be concluded that silymarin may possibly possess significant protective effects against chemical carcinogen- as well as solar radiation-induced non-melanoma skin cancers in humans.

ACKNOWLEDGMENTS

The original studies with green tea are supported by American Institute for Cancer Research grants 92B35 and 93A30, and United States Public Health Service Grant ES-1900 and P-30-AR-39750. The studies with silymarin are supported by United States Public Health Service Grant CA 64514.

REFERENCES

1. B.N. Ames, L.S. Gold, W.C. Willett, The causes and prevention of cancer, *Proc Natl Acad Sci USA* 92: 5258-5265 (1995).
2. J. DiGiovanni, Multistage carcinogenesis in mouse skin, *Pharmac Ther* 54: 63-128 (1992).
3. R. Agarwal, H. Mukhtar, Cutaneous chemical carcinogenesis. In Pharmacology of the skin. H. Mukhtar (Ed.), CRC Press, Boca Raton, pp. 371-387, 1991.
4. H.C. Pitot, Y.P. Dragan, Facts and theories concerning the mechanisms of carcinogenesis, *FASEB J* 5: 2280-2286 (1991).
5. G.D. Stoner, H. Mukhtar, Polyphenols as cancer chemopreventive agents, *J Cellular Biochem (Suppl)* 22: 169-180 (1995)
6. C.W. Boone, G.J. Kelloff, W.E. Malone, Identification of candidate cancer chemopreventive agents and their evaluation in animal models and human clinical trials, *Cancer Res* 50: 2-9 (1990).
7. M.A. Morse, G.D. Stoner, Cancer chemoprevention: Principles and prospects, *Carcinogenesis* 14: 1737-1746 (1993).
8. H. Sumiyoshi, M.J. Wargovich, Chemoprevention of 1,2-dimethylhydrazine-induced colon cancer in mice by naturally occurring organosulfur compounds, *Cancer Res* 50: 5084-5087 (1990).
9. H. Wei, L. Tye, E. Bresnick, D.F. Birt, Inhibitory effect of apigenin, a plant flavonoid, on epidermal ornithine decarboxylase and skin tumor promotion in mice, *Cancer Res* 50: 499-502 (1990).
10. H.F. Stich, The beneficial and hazardous effects of simple phenolic compounds, *Mutation Res* 259: 307-324 (1991).
11. L.W. Wattenberg, Inhibition of carcinogenesis by naturally occurring and synthetic compounds. In Antimutagenesis and Anticarcinogenesis, Mechanisms II. Y. Kuroda, D.M. Shankel and M.D. Waters (Eds.), Plenum Publishing Corp, New York, pp. 155-166, 1990.
12. H. Mukhtar, Z.Y. Wang, S.K. Katiyar, R. Agarwal, Tea components: Antimutagenic and anticarcinogenic effects, *Preventive Medicine* 21: 351-360 (1992).
13. H. Mukhtar, S.K. Katiyar, R. Agarwal, Green tea and skin - anticarcinogenic effects, *J Invest Dermatol* 102: 3-7 (1994).
14. C.S. Yang, Z.Y. Wang, Tea and cancer, *J Natl Cancer Insti* 85: 1038-1049 (1993).
15. T. Yaminishi, I. Tomita, Proceedings of the International Symposium on Tea Sciences, 1992.
16. T. Kada, K. Kaneko, S. Matsuzaki, T. Matsuzaki, Y. Hara, Detection and chemical identification of natural bio-antimutagens. A case of green tea factor, *Mutation Res* 150: 127-132 (1985).
17. S.J. Cheng, C.T. Ho, H.Z. Lou, Y.D. Bao, Y.Z. Jian, M.H. Li, Y.N. Gao, G.F. Zhu, J.F. Bai, S.P. Guo, X.Q. Li, A preliminary study on the antimutagenicity of green tea antioxidants, *Acta Biol Exp Sinica* 19: 427-431 (1986).
18. R.R. Mohan, S.G. Khan, R. Agarwal, H. Mukhtar, Testosterone induces ornithine decarboxylase (ODC) activity and mRNA expression in human prostate carcinoma cell line LNCaP: inhibition by green tea. *Proc Amer Assoc Cancer Res* 36:274 (1995).

19. P.T. Strickland, Tumor induction in Sencar mice in response to ultraviolet radiation, *Carcinogenesis* 3: 1487-1489 (1982).
20. H. Mukhtar,(Ed.) Skin Cancer: Mechanisms and Human Relevance, CRC Press, Boca Raton, FL, 1995.
21. A.H. Conney, Induction of microsomal enzymes by foreign chemicals and carcinogenesis by polycyclic aromatic hydrocarbons: GHA Clowes Memorial Lecture, *Cancer Res* 42: 4875-4917 (1982).
22. H.N. Ananthaswamy, W.E. Pierceall, Molecular mechanisms of ultraviolet radiation carcinogenesis. *Photochem Photobiol* 52: 1119-1136 (1990).
23. A.J. Nataraj, J.C. Trent II, H.N. Ananthaswamy, p53 gene mutations and photocarcinogenesis, *Photochem Photobiol* 62: 218-230 (1995).
24. R. Agarwal, H. Mukhtar, Oxidative stress in skin chemical carcinogenesis. In Oxidative stress in dermatology (Fuchs, J. and Packer, L. eds), Marcel Dekker, Inc, New York, NY, pp 207-241, 1993.
25. S.G. Khan, R.R. Mohan, S.K. Katiyar, G.S. Wood, D.R. Bickers, H. Mukhtar, R. Agarwal, Mutations in ras oncogenes are rare events in ultraviolet B radiation induced mouse skin tumorigenesis, Mol carcinogenesis, In Press, 1995.
26. J.P.G. Volpe, L. Wang, J.H. Epstein, T.J. Slaga, J.E. Cleaver, Absence of ras mutations in UV-induced skin tumors in mice. *Proc Am Assoc Cancer Res* 36:187 (1995).
27. Z.Y. Wang, M. Das, D.R. Bickers, H. Mukhtar, Interaction of epicatechins derived from green tea with rat hepatic cytochrome P-450, *Drug Metab Dispos* 16: 98-103 (1988).
28. Z.Y. Wang, W.A. Khan, D.R. Bickers, H. Mukhtar, Protection against polycyclic aromatic hydrocarbon-induced skin tumor initiation in mice by green tea polyphenols. *Carcinogenesis 10*: 411-415 (1989).
29. Z.Y. Wang, S.J. Cheng, Z.C. Zhou, M. Athar, W.A. Khan, D.R. Bickers, H. Mukhtar, Antimutagenic activity of green tea polyphenols. *Mutation Res* 223: 273-289 (1989).
30. S.G. Khan, S.K. Katiyar, R. Agarwal, H. Mukhtar, Enhancement of antioxidant and phase II enzymes by oral feeding of green tea polyphenols in drinking water to SKH-1 hairless mice: possible role in cancer chemoprevention, *Cancer Res* 52: 4050-4052 (1992).
31. Q. Liu, Y. Wang, K.A. Crist, M.T. Huang, A. Conney, M. You, Analysis of p53 and H-ras mutations in UV- and UV-green tea-induced tumorigenesis in the skin of SKH-1 mice. *Proc Am Assoc Cancer Res* 36:591 (1995).
32. Z.Y. Wang, R. Agarwal, D.R. Bickers, H. Mukhtar, Protection against ultraviolet B radiation-induced photocarcinogenesis in hairless mice by green tea polyphenols, *Carcinogenesis* 12: 1527-1530 (1991).
33. M.-T. Huang, C.-T. Ho, Z.Y. Wang, T. Ferraro, T. Finnegan-Olive, Y.-R. Lou, J.M. Mitchell, J.D. Laskin, H. Newmark, C.S. Yang, A.H. Conney, Inhibitory effect of topical application of a green tea polyphenol fraction on tumor initiation and promotion in mouse skin, *Carcinogenesis* 13: 947-954 (1992).
34. W.A. Khan, Z.Y. Wang, M. Athar, D.R. Bickers, H. Mukhtar, Inhibition of the skin tumorigenicity of (+)-7ß,8α-dihydroxy-9α,10α-epoxy-7,8,9,10-tetrahydrobenzo(a)pyrene by tannic acid, green tea polyphenols and quercetin in SENCAR mice, *Cancer Lett* 42: 7-12 (1988).
35. S.K. Katiyar, R. Agarwal, Z.Y. Wang, A.K. Bhatia, H. Mukhtar,(-)-Epigallocatechin-3-gallate in *Camellia sinensis* leaves from Himalayan region of Sikkim: inhibitory effects against biochemical events and tumor initiation in SENCAR mouse skin, *Nutr Cancer* 18: 73-83 (1992).
36. Z.Y. Wang, M.-T. Huang, T. Ferraro, C.-Q. Wong, Y.-R. Lou, M. Iatropoulos, C.S. Yang, A.H. Conney, Inhibitory effect of green tea in the drinking water on tumorigenesis by ultraviolet light and 12-O-tetradecanoylphorbol-13-acetate in the skin of SKH-1 mice, *Cancer Res* 52: 1162-1170 (1992).
37. J.-P. Perchellet, E.M. Perchellet, Antioxidants and multistage carcinogenesis in mouse skin, *Free Radical Biol Med* 7: 377-408 (1989).
38. H.U. Gali, E.M. Perchellet, J.-P. Perchellet, Inhibition of tumor promoter-induced ornithine decarboxylase activity by tannic acid and other polyphenols in mouse epidermis in vivo, *Cancer Res* 51: 2820-2825 (1991).
39. S.K. Katiyar, R. Agarwal, G.S. Wood, H. Mukhtar, Inhibition of 12-*O*-tetradecanoylphorbol-13-acetate-caused tumor promotion in 7,12-dimethylbenz[a]anthracene-initiated SENCAR mouse skin by a polyphenolic fraction isolated from green tea, *Cancer Res* 52: 6890-6897 (1992).
40. S.K. Katiyar, R. Agarwal, S. Ekker, G.S. Wood, H. Mukhtar, Protection against 12-O-tetradecanoylphorbol-13-acetate-caused inflammation in SENCAR mouse ear skin by polyphenolic fraction isolated from green tea, *Carcinogenesis* 14: 361-365 (1993).
41. R. Agarwal, S.K. Katiyar, S.I.A. Zaidi, H. Mukhtar, Inhibition of tumor promoter-caused induction of ornithine decarboxylase activity in SENCAR mice by polyphenolic fraction isolated from green tea and its individual epicatechin derivatives, *Cancer Res* 52: 3582-3588 (1992).
42. S.K. Katiyar, C.O. Rupp, N.J. Korman, R. Agarwal, H. Mukhtar, Inhibition of 12-O-tetradecanoylphorbol-13-acetate and other skin tumor promoter-caused induction of epidermal interleukin-1α mRNA and protein expression in SENCAR mice by green tea polyphenols, *J Invest Dermatol* In Press, 1995.

43. R. Agarwal, S.K. Katiyar, S.G. Khan, H. Mukhtar, Protection against ultraviolet B radiation-induced effects in the skin of SKH-1 hairless mice by a polyphenolic fraction isolated from green tea, *Photochem Photobiol* 58: 695-700 (1993).
44. S. Yoshizawa, T. Horiuchi, H. Fujiki, T. Yoshida, T. Okuda,T. Sugimura, Antitumor Promoting Activity of (-)-Epigallocatechin Gallate, the Main Constituent of "Tannin" in Green Tea, *Phytotherapy Res* 1: 44-47 (1987).
45. S.K. Katiyar, R. Agarwal, H. Mukhtar, Inhibition of both stage I and stage II tumor promotion in SENCAR mice by a polyphenolic fraction isolated from green tea: Inhibition depends on the duration of polyphenols treatment, *Carcinogenesis* 14: 2641-2643 (1993).
46. Z.Y. Wang, M.-T. Huang, Y.-R. Lou, J.-G. Xie, K.R. Reuhl, H.L. Newmark, C.-T. Ho, C.S. Yang, A.H. Conney, Inhibitory effects of black tea, green tea, decaffeinated black tea, and decaffeinated green tea on ultraviolet B light-induced skin carcinogenesis in 7,12-dimethylbenz(a)anthracene-initiated SKH-1 mice, *Cancer Res* 54: 3428-3435 (1994).
47. S.K. Katiyar, R. Agarwal, H. Mukhtar, Protection against malignant conversion of chemically-induced benign skin papillomas to squamous cell carcinomas in SENCAR mice by a polyphenolic fraction isolated from green tea, *Cancer Res* 53: 5409-5412 (1993).
48. Z.Y. Wang, M.-T. Huang, C.-T. Ho, R. Chang, W. Ma, T. Ferraro, K.R. Reuhl, C.S. Yang, A.H. Conney, Inhibitory effect of green tea on the growth of established skin papillomas in mice, *Cancer Res* 52: 6657-6665 (1992).
49. Z.Y. Wang, R. Agarwal, W.A. Khan, H. Mukhtar, Protection against benzo(a)pyrene and N-nitrosodiethylamine-induced lung and forestomach tumorigenesis in A/J mice by water extracts of green tea and licorice, *Carcinogenesis* 13: 1491-1494 (1992).
50. Z.Y. Wang, J.Y. Hong, M.-T. Huang, K.R. Reuhl, A.H. Conney, C.S. Yang, Inhibition of N-nitrosodiethylamine- and 4-(methylnitrosamino)-1-(3-pyridyl)-1-butanone-induced tumorigenesis in A/J mice by green tea and black tea, *Cancer Res* 52: 1943-1947 (1992).
51. S.K. Katiyar, R. Agarwal, M.T. Zaim, H. Mukhtar, Protection against N-nitrosodiethylamine and benzo(a)pyrene-induced forestomach and lung tumorigenesis in A/J mice by green tea, *Carcinogenesis* 14: 849-855 (1993).
52. S.K. Katiyar, R. Agarwal, H. Mukhtar, Protective effects of green tea polyphenols administered by oral intubation against chemical carcinogen-induced forestomach and pulmonary neoplasia in A/J mice, *Cancer Lett* 73: 167-172 (1993).
53. Y. Xu, C.-T. Ho, S.G. Amin, C. Han, F.L. Chung, Inhibition of tobacco-specific nitrosamine-induced lung tumorigenesis in A/J mice by green tea and its major polyphenol as antioxidants, *Cancer Res* 52: 3875-3879 (1992).
54. Y. Fujita, T. Yamane, M. Tanaka, K. Kuwata, J. Okuzumi, T. Takahashi, H. Fujiki, T. Okuda, Inhibitory Effect of (-)-Epigallocatechin Gallate on Carcinogenesis with N-Ethyl-N'-Nitro-N-Nitrosoguanidine in Mouse Duodenum, *Jpn J Cancer Res (Gann)* 80: 503-505 (1989).
55. T. Yamane, N. Hagiwara, M. Tateishi, S. Akachi, M. Kim, J. Okuzumi, Y. Kitao, M. Inagake, K. Kuwata, T. Takahashi, Inhibition of azoxymethane-induced colon carcinogenesis in rat by green tea polyphenol fraction, *Jpn J Cancer Res (Gann)* 82: 1336-1340 (1991).
56. T. Narasiwa, Y. Fukaura, A very low dose of green tea polyphenols in drinking water prevents N-methyl-N-nitrosourea-induced colon carcinogenesis in F344 rats, *Jpn J Cancer Res)* 84: 1007-1009 (1993).
57. Y. Xu, C. Han, The effect of Chinese tea on the occurrence of esophageal tumors induced by N-nitrosomethylbenzylamine formed in vivo, *Biomed Environ Sci* 3: 406-412 (1990).
58. G.D. Gao, L.F. Zhou, G. Qi, Initial study of antitumorigenesis of green tea: animal test and flow cytometry, *Tumor* 10: 42-44 (1990).
59. Z.Y. Chen, R.Q. Yan, G.Z. Qin, Effect of six edible plants on the development of aflatoxin B_1-induced γ-glutamyltranspepidase positive hepatocyte foci in rats, *Chung Hua Chung Liu Tsa Chih* 9: 109-111 (1987).
60. Y. Li, Comparative study on the inhibitory effect of green tea, coffee and levamisole on the hepatocarcinogenic action of diethylnitrosamine, *Chung Hua Chung Liu Tsa Chih* 13: 193-195 (1991).
61. N. Harada, F. Takabayashi, I. Oguini, Anti-promotion effect of green tea extracts on pancreatic cancer in golden hamster induced by N-nitroso-bis(2-oxopropyl)amine, Int Sympo on Tea Sci (Japan), 1991.
62. M. Hirose, T. Hoshiya, K. Akagi, M. Futakuchi, N. Ito, Inhibition of mammary gland carcinogenesis by green tea catechins and other naturally occurring antioxidants in female Sprague-Dawley rats pretreated with 7,12, dimethylbenz(a)anthracene, *Cancer Lett* 83: 149-156 (1994).
63. T. Yamane, T. Takahashi, K. Kuwata, K. Oya, M. Inagake, Y. Kitao, M. Suganuma, H. Fujiki, Inhibition of N-methyl-N'-nitro-N-nitrosoguanidine-induced carcinogenesis by (-)-epigallocatechin gallate in the rat glandular stomach, *Cancar Res* 55: 2081-2084 (1995).

64. C.C. Boring, T.S. Squires, T. Tong, Cancer Statistics, 1993. *CA Cancer J Clin* 43:7-26 (1993).
65. C.C. Boring, T.S. Squires, T. Tong, S. Montgomery, Cancer Statistics, 1994. *CA Cancer J Clin* 44:22-23 (1994).
66. D.G. Bostwick, G.J. Kelloff, P.T. Scardino, C.W. Boone, Chemoprevention of premalignant and early malignant lesions of the prostate, *J Cellular Biochem Supplement* 16H (1992).
67. C.W. Boone, G.J. Kelloff, Qualitative pathology in chemoprevention trails: standardization and quality control of surrogate endpoint biomarker assays for colon, brest, and prostate, *J Cellular Biochem Suppl* 19 (1994).
68. D.G. Bostwick, H.B. Burke, T.M. Wheeler, L.W.K. Chung, R. Bookstein, T.G. Pretlow, R.B. Nagle, R. Montirino, M.M. Lieber, R.W. Veltri, W.E. Grizzel, D.J. Grignon, The most promising surrogate endpoint biomarkers for screening candidate chemopreventive compounds for prostatic adenocarcinoma in short-term phase II clinical trials, *J Cellular Biochem Suppl* 19:283-289 (1994).
69. G.K. Kelloff, C.W. Boone, J.A. Crowell, V.E. Steele, R. Lubet, L.A. Doody, Surrogate endpoint biomarkers from phase II cancer chemoprevention trials, *J Cellular Biochem Suppl* 19:1-9 (1994).
70. M. Pollard, P.H. Luckert, M.B. Sporn, Prevention of primary prostate cancer in Lobound-Wistar rats by N-(4-hydroxyphenyl) retinamide, *Cancer Res* 51:3610-3611 (1991).
71. K.Slavin, D. Kadmon, S.H. Park, P.T. Scardino, M. Anzano, M.B. Sporn, Thompson TC: Dietary fenritinide, a synthetic retinoid, decreases the tumor incidence And the tumor mass of ras+myc-induced carcinomas in the mouse prostate reconstitution model system, *Cancer Res* 53:4461-4465 (1993).
72. M.B. Garnick, The dilemmas of prostate cancer, *Scientific American* 72-81 (1994).
73. M.C. Bosland, H.C. Dreef-Van Der Meulen, S. Sukumar, P. Ofner, I. Leav, Multistage prostate carcinogenesis: The Role of Hormones, In Multistage Carcinogenesis, CC Harris, S. Hirohashi, N Ito, HC Pitot, T Sugimura (eds). Boca Raton, FL, CRC Press, pp 109-123, 1992.
74. R.K. Ross, B.E. Henderson, Do diet and andorgens alter prostate cancer risk via a common etiologic pathway? *J Natl Cancer Inst* 86:252-254 (1994).
75. C. Muir, J. Waterhouse, T. Mack, (eds): Cancer Incidence in Five Continents. V. Lyon IARC, 1987.
76. H. Yu, R.E. Harris, Y.T. Gao, T. Gao, E.L. Wynder, Comparative epidemiology of cancers of the colon, rectum, prostate and breast in Shanghai, China versus the United States, *Int J Epidermiol* 30:76-81(1991).
77. E.L. Wynder,D.P. Rose , L.A. Cohen, Nutrition and prostate cancer: a proposal for dietary intervention, *Nutrition and Cancer* 22:1-10(1994).
78. D.P. Rose, A.P. Boyar, E.L. Wynder, International comparisons of mortality rates for cancer of the breast, ovary, prostate and colon and per capita food consumption, *Cancer* 58:2363-2371 (1986).
79. R.K. Rose, H. Shimizu, A. Paganini-Hill, G. Honda, Case-control studies of prostate cancer in blacks and whites in southern California, *J Natl Cancer Inst* 78:869-874 (1987).
80. K.J. Pienta, P.S. Esper, Risk factors for prostate cancer, *Ann Intern Med* 118:793-803 (1993).
81. L.J. Marton, A.E. Pegg, Polyamines as targets for therapeutic intervention, *Annu Rev Pharmacol Toxicol* 35:55-91 (1995).
82. M.A. Wainstein, F. He, D. Robinson, H.-J. Kung, S. Schwartz, J.M. Giaconia, N.L. Edgehouse, T.P. Pretlow, D.R. Bodner, E.D. Kursh, M.I. Resnick, A. Seftel, T.G. Pretlow, CWR22: Androgen-dependent xenograft model derived from a primary human prostatic carcinoma, *Cancer Res* 59:6049-6052 (1994).
83. J.S. Horoszewicz, S.S. Leong, E. Kawinski, J.P. Karr, H. Rosenthal, T.M. Chu, E.A. Mirand, G.P. Murphy, LNCaP model of human prostatic carcinoma. Cancer Res 43:1809-1818 (1983).
84. S. Kono, M. Ikeda, S. Tokudome, M. Kuratsune, A case-control study of gastric cancer and diet in Northern Kyushu, *Jpn J Cancer Res* 79: 1067-1074 (1988).
85. I. Oguni, K. Nasu, S. Yammamoto, T. Nomura, On the antitumor activity of fresh green tea leaf, *Agric Biol Chem* 52: 1879-1880 (1988).
86. IARC Monographs on the evaluation of the carcinogenic risk to humans: coffee, tea, mate, methylxanthines and methylglyoxal. International Agency for Research on Cancer Working Group, Vol. 51, 1991.
87. Y.T. Gao, J.K. McLaughlin, W.J. Blot, B.T. Ji, Q. Dai, J.F. Fraumeni, Jr., Reduced risk of esophageal cancer associated with green tea consumption, *J Natl Cancer Inst* 86: 855-858 (1994).
88. M.J. Lee, Z.Y. Wang, H. Li, L. Chen, Y. Sun, S. Gobbo, D.A. Balentine, C.S. Yang, Analysis of plasma and urinary tea polyphenols in human subjects, *Cancer Epidemiol Biomarkers Prevention* 4: 393-399 (1995).
89. J.S. Shim, M.H. Kang, Y.H. Kim, J.K. Roh, C. Roberts, I.P. Lee, Chemopreventive effect of green tea (*Camellia sinensis*) among cigarette smokers, *Cancer Epidemiol Biomarkers Prevention* 4: 387-391 (1995).
90. K.A. Mereish, D.L. Bunner, D.R. Ragland, D.A. Creasia, Protection against microcystin-LR-induced hepatotoxicity by silymarin: biochemistry, histopathology, and lethality, *Pharmaceutical Res* 8:273-277 (1991).

91. H. Wagner, O. Seligmann, L. Horhammer, R. Munster, The chemistry of silymarin (silybin), the active principle of the fruits of *Silybum marianum* (L) Gaertn. (*Carduus marianus*) (L), *Arzneimittelforsch* 18:688-696 (1968).
92. P. Letteron, G. Labbe, C. Degott, A. Berson, B. Fromenty, M. Delaforge, D. Larrey, D. Pessayre Mechanism for the protective effects of silymarin against carbon tetrachloride-induced lipid peroxidation and hepatotoxicity in mice, *Biochem Pharmacol* 39:2027-2034 (1990).
93. M. Mourelle, P. Muriel, L. Favari, T. Franco, Prevention of CCl_4-induced liver cirrhosis by silymarin, *Fundam Clin Pharmacol* 3:183-191 (1989).
94. E. Bosisio, C. Benelli, O. Pirola, Effect of the flavanolignans of *Silybum marianum* L. on lipid peroxidation in rat liver microsomes and freshly isolated hepatocytes, *Pharmacol Res* 25:147-154 (1992).
95. R. Carini, A. Comoglio, E. Albano, G. Poli, Lipid peroxidation and irreversible damage in the rat hepatocyte model. Protection by the silybin-phospholipid complex IdB 1016, *Biochem Pharmacol* 43:2111-2115 (1992).
96. K. Racz, J. Feher, G. Csomos, I. Varga, R. Kiss, E. Glaz, An antioxidant drug, silibinin, modulates steroid secretion in human pathological adrenocortical cells, *J Endocrinol* 124:341-345 (1990).
97. A. Valenzuela, R. Guerra, L.A. Videla, Antioxidant properties of the flavonoids silybin and (+)-cyanidanol-3: Comparison with butylated hydroxyanisole and butylated hydroxytoluene, *Planta Medica* 5:438-440 (1986).
98. A. Garrido, A. Arancibia, R. Campos, A. Valenzuela, Acetaminophen does not induce oxidative stress in isolated rat hepatocytes: Its probable antioxidant effect is potentiated by the flavonoid silybin, *Pharmacol Toxicol* 69:9-12 (1991).
99. A. Comoglio, G. Leonarduzzi, R. Carini, D. Busolin, H. Basaga, E. Albano, A. Tomasi, G. Poli, P. Morazzoni, M.J. Magistretti, Studies on the antioxidant and free radical scavenging properties of IdB1016 a new flavanolignan complex, *Free Rad Res Comms* 11:109-115 (1990).
100. G. Muzes, G. Deak, I. Lang, K. Nekam, P. Gergely, J. Feher, Effect of the bioflavonoid silymarin on the in vitro activity and expression of superoxide dismutase (SOD) enzyme, *Acta Physiologica Hungarica* 78:3-9 (1991).
101. V.E. Steele, G.J. Kelloff, B.P. Wilkinson, J.T. Arnold, Inhibition of transformation in cultured rat tracheal epithelial cells by potential chemopreventive agents, *Cancer Res* 50:2068-2074 (1990).
102. C.J. Rudd, K.D. Suing, K. Pardo, G. Kelloff, Evaluation of potential chemopreventive agents using a mouse epidermal cell line, JB6, *Proc Am Assoc Cancer Res* 31:127 (1990).
103. R.G. Mehta, R.C. Moon, Characterization of effective chemopreventive agents in mammary gland *in vitro* using an initiation-promotion protocol, *Anticancer Res* 11:593-596 (1991).
104. R. Agarwal, S.K. Katiyar, D.W. Lundgren, H. Mukhtar, Inhibitory effect of silymarin, an anti-hepatotoxic flavonoid, on 12-O-tetradecanoylphorbol-13-acetate-induced epidermal ornithine decarboxylase activity and mRNA in SENCAR mice, *Carcinogenesis* 15:1099-1103 (1994).
105. R. Agarwal, S.K. Katiyar, H. Mukhtar, Protection against tumor promotion in mouse skin by silymarin, *Proc Am Assoc Cancer Res* 36:593 (1995).
106. R. Agarwal, S.K. Katiyar, H. Mukhtar, Protective effects of silymarin against ultraviolet B radiation-induced tumorigenesis in SKH-1 hairless mouse skin. *J Invest Dermatol* 104:635 (1995).
107. R. Agarwal, S.K. Katiyar, H. Mukhtar, Protection against photocarcinogenesis in SKH-1 hairless mice by silymarin, *Photochem Photobiol* 61s:14s (1995).

5

EFFECTS OF TEA ON CARCINOGENESIS IN ANIMAL MODELS AND HUMANS*

Chung S. Yang,† Laishun Chen, Mao-Jung Lee, and Janelle M. Landau

Laboratory for Cancer Research
College of Pharmacy, Rutgers University
Piscataway, New Jersey 08855

INTRODUCTION

The effects of tea consumption on cancer is an area of great scientific and public interest, and this topic was reviewed in 1993 in an article entitled "Tea and Cancer."[1] Since then more than one hundred new research articles related to this topic have been published. In this chapter, we review the available evidence from laboratory and epidemiological studies concerning the effects of tea on cancer formation. Possible mechanisms by which tea components inhibit carcinogenesis are discussed based on the results of some new studies.

The chemistry of tea has been reviewed previously.[1,2] Although it remains to be substantiated further, the major anti-cancer components in tea are believed to be the polyphenols, also known as catechins. In green tea, (-)-epigallocatechin gallate (EGCG), (-)-epigallocatechin (EGC), (-)-epicatechin-3-gallate (ECG), and (-)-epicatechin (EC) are the major polyphenols. In the manufacturing of black tea, most of these compounds are oxidized and polymerized to form theaflavins which give the reddish color of black tea and thearubingins which are not well characterized chemically. The structures of some of the major tea polyphenols are shown in Figure 1.

INHIBITION OF CARCINOGENESIS IN ANIMAL MODELS BY TEA

Tea extracts and tea polyphenols have been demonstrated to inhibit tumorigenesis in a variety of animal models and in different organ sites. Some recent studies from our and other laboratories are reviewed here.

---

* *Abbreviations used:* EGCG, (-)-epigallocatechin gallate; EGC, (-)-epigallocatechin; ECG, (-)-epicatechin-3-gallate; EC, (-)-epicatechin; NDEA, *N*-nitrosodiethylamine; DGT, decaffeinated green tea; DBT, decaffeinated black tea; NNK, 4-(methylnitrosamino)-1- (3-pyridyl)-1-butanone; NMBzA, *N*-nitrosomethylbenzylamine; UVB, ultraviolet B light; TPA, 12-*O*-tetradecanoylphorbol-13-acetate.

† Fax: 908-445-0687; Tel: 908-445-5361.

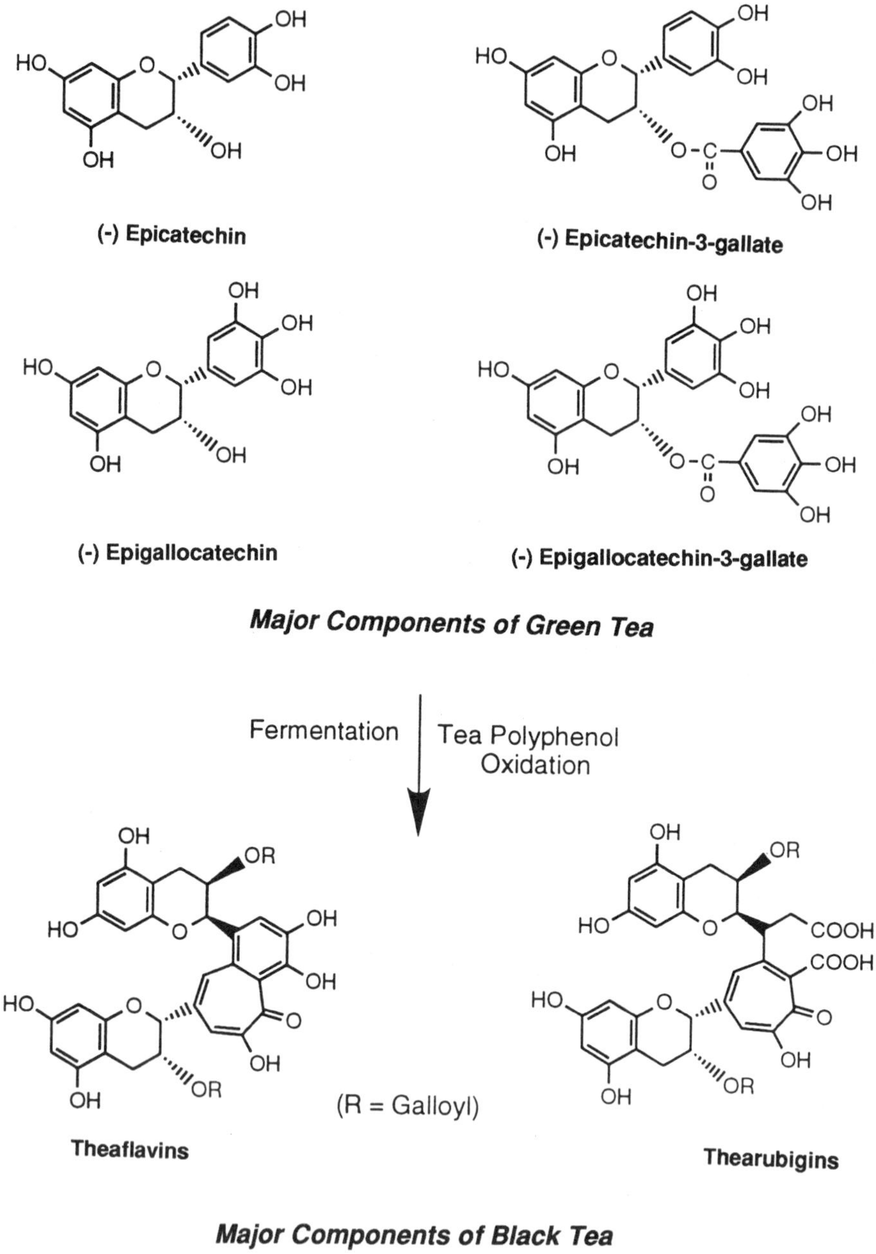

(-) Epicatechin

(-) Epicatechin-3-gallate

(-) Epigallocatechin

(-) Epigallocatechin-3-gallate

Major Components of Green Tea

Fermentation

Tea Polyphenol Oxidation

(R = Galloyl)

Theaflavins

Thearubigins

Major Components of Black Tea

Figure 1. Structures of major tea polyphenols in green tea and black tea.

Lung Tumorigenesis

Administration of green tea infusion (e.g. 1.25 g of tea leaves brewed in 100 ml of boiling water), as the sole source of drinking fluid to A/J mice, significantly decreased *N*-nitrosodiethylamine (NDEA) (10 mg/kg, i.g., once weekly for 8 weeks) induced lung tumor incidence (by 36% to 44%) and tumor multiplicity (by 44% to 60%).[3] When administered as the drinking fluid to A/J mice, decaffeinated green tea (DGT) or decaffeinated black tea (DBT) extracts (e.g. 0.6 g of dehydrated tea extract reconstituted in 100 ml of distilled water) also inhibited lung tumorigenesis caused by 4-(methylnitrosamino)-1-(3-pyridyl)-1-butanone (NNK) (103 mg/kg, i.p., one dose).[3] In this experiment, tumor multiplicity was reduced by 65% to 85% and tumor incidence was inhibited to a lesser extent, showing 14% to 30% inhibition in some of the experiments. In both models, tea preparations were effective when administered to mice either during the carcinogen treatment period (starting 2 weeks prior to the treatment until 1 week after the treatment) or after the treatment (starting 1 week after the treatment until the termination of the experiment). In order to gain mechanistic information, mice were given DGT in different time periods before or after the NNK treatment.[4] Treatment with DGT starting 2 weeks before and until 1 week after the NNK injection was more effective in the reduction of tumor multiplicity (56%) than treatment for 2 days before NNK injection (31%). When tea was given after NNK treatment for a period of 1 week, tumor multiplicity was also reduced (by 20%). Tumor multiplicity was inhibited (by 54%) even when DGT administration was started at 5 weeks after NNK treatment.

Similar inhibitory activity was also observed by other investigators in lung tumorigenesis induced by NDEA, benzo(a)pyrene,[5,6] and NNK.[7]

Esophageal Carcinogenesis

When 0.6% DGT or DBT extracts were given to male Sprague-Dawley rats as the drinking fluid during a *N*-nitrosomethylbenzylamine (NMBzA) treatment period (2.5 mg/kg, s.c., twice weekly for 5 weeks), esophageal tumor incidence and multiplicity were reduced by approximately 70% [8]. When the tea preparations were given after the NMBzA treatment period, the esophageal papilloma incidence and multiplicity were reduced by approximately 50%. The volume per tumor was also reduced in rats which received DBT after the carcinogen treatment period. In a second experiment, the rats received 0.9% regular green tea (0.9 g of dehydrated tea extract reconstituted in 100 ml of water) or DGT after the NMBzA treatment period (3.5 mg/kg, s.c., twice weekly for 5 weeks); the tumor multiplicity was decreased by more than 55% at 16 weeks after the first dose of NMBzA. The volume per tumor was reduced by approximately 60% in the rats receiving 0.9% regular green tea. Histological analysis indicated that both the incidence and multiplicity of esophageal carcinoma were decreased by treatment with either regular green tea or DGT.[8]

These results confirmed the observations by Han and Xu[9] who demonstrated the inhibitory activity of three brands of green tea, one brand of oolong tea, and one brand of black tea against esophageal tumorigenesis in Wistar rats as induced by NMBzA which was administered during the entire experimental period. Similar inhibitory activity was also observed when precursors of NMBzA (*N*-methylbenzylamine and sodium nitrite) were used as the carcinogen.[10] Green tea extracts have also been shown to inhibit esophageal tumorigenesis in mice induced by precursors of *N*-nitrososarcosine ethyl ester.[11] Our recent results[8] demonstrated that tea administration was effective when given either during the initiation or post-initiation stage, and that decaffeinated green tea is also effective.

Gastric Carcinogenesis

In the previously described NDEA-induced carcinogenesis model with A/J mice, forestomach tumorigenesis was also significantly inhibited by green tea infusion.[3] In this model, hyperplasia, papillomas, carcinoma *in situ*, and squamous cell carcinomas in the forestomach were observed. Tea treatment inhibited the tumor multiplicity (up to 63% inhibition) more effectively than tumor incidence (up to 26% inhibition). The tumor volume was lower in mice in the tea treatment group than in the control group.[3] Similar inhibitory activity by green tea preparations has also been observed in mouse forestomach tumors induced by NDEA or benzo(a)pyrene[5,6] and by precursors of nitrosamines.[11,12] Inhibition of *N*-methyl-*N*′-nitro-*N*-nitrosoguanidine-induced gastric cancer in rats by green tea extract[13] and EGCG[14] has also been reported. In the latter study, EGCG (0.05% in water) administration in the drinking fluid for 15 weeks to rats after a 28-week treatment with the carcinogen caused a 50% reduction in glandular tumor incidence.[14]

Formation and Growth of Skin Tumors

Skin tumorigenesis initiated by ultraviolet B light (UVB) and promoted by 12-*O*-tetradecanoylphorbol-13-acetate (TPA) was inhibited by the administration of 1.25% green tea as the drinking fluid prior to and during the 10 days of the UVB treatment period. In other experiments, using 7,12-dimethylbenz(a)anthracene as an initiator and UVB as the promotor, administration of 1.25% green tea as the drinking fluid prior to and during the 30 weeks of UVB treatment decreased the number and size of the skin tumors.[15] Using a similar protocol, oral administration of green tea, black tea, DGT, or DBT also had a marked inhibitory effect on UVB-induced skin carcinogenesis in 7,12-dimethylbenz(a)anthracene -initiated SKH-1 mice.[16] When mice bearing chemically-induced or UVB light-induced skin papillomas were given green tea, green tea polyphenols, or EGCG, partial tumor regression or marked inhibition of tumor growth was observed in 10 experiments.[17] Other studies on the inhibition of skin tumorigenesis are reviewed by Mukhtar *et al.* in this volume.

Other Laboratory Studies

More than 20 laboratories in several countries have reported the inhibitory effect of tea against tumorigenesis in animal models at different organ sites, such as skin, esophagus, forestomach, stomach, duodenum and small intestine, colon, lung, liver, pancreas, and mammary gland. Some of this work was reviewed previously;[1,18] some newer studies are listed in the reference section.[19-24]

EPIDEMIOLOGICAL STUDIES ON TEA AND CANCER

In a 1991 monograph of the International Agency for Research on Cancer,[25] it was concluded that "There is inadequate evidence for the carcinogenicity in humans and experimental animals of tea drinking." In our review on "Tea and Cancer" in 1993, we concluded that no clear-cut conclusion could be drawn concerning the protective effects of tea against human cancers.[1] The conclusion still held true after inclusion of the 17 new case-control studies[18,26-42] published after our review. A summary of case-control studies on the relationship between tea consumption and human cancers is shown in Table 1.

Tea Consumption and Esophageal Cancer in Humans

It has been suggested previously that excessive consumption of tea may be a causative factor for esophageal cancer.[43-45] Several case-control studies, however, showed that there was no association between drinking of tea at normal temperature and esophageal cancer, but ingestion of very hot tea was associated with a two- to three-fold increase in risk for esophageal cancer.[46-50] The high temperature of hot tea, rather than the chemicals in tea, was suggested to be an important etiological factor in human esophageal cancer.[25] A recent small case-control study conducted in northern China,[30] reported that high temperature of meals

Table 1. Case-control studies on the association between tea consumption and human cancers

	Number of reports		
Organs	Negative	Positive	No relationship
Bladder and urinary tract	–	–	17
Breast	–	–	6
Colon and rectum	4	1	6
Esophagus	1	7*	6
Kidney	–	1	5
Liver	–	–	1
Lung	–	1	1
Nasopharynx	–	–	3
Oral and tongue	1	–	1
Pancreas	2	1	8
Prostate	–	–	1
Stomach	4	1	8

*6 studies were with hot tea.

and drinks had a strong association with esophageal cancer risk, but after adjustment of the temperature, the tea consumption was still a risk factor. On the other hand, a recent report by Gao *et al.* [28] indicated that, in a case-control study on 902 patients and 1552 controls in Shanghai, frequent consumption of green tea was associated with a lower incidence of esophageal cancer. After adjustment for age and other possible confounding factors, a protective effect of green tea drinking on esophageal cancer was observed among women: odds ratio, 0.50; 95% confidence interval, 0.30-0.83, and this risk decreased (p for trend $\leq$ 0.01) as the quantity of tea consumed increased. Among men, the protective effect was not statistically significant. Among nonsmokers and non-alcohol-drinkers, statistically significant decreases in risk were observed among tea drinkers for both men (69 cases and 192 controls) and women (194 cases and 564 controls).

Tea Consumption and Stomach Cancer in Humans

Early correlative and cohort studies indicated that tea consumption was either negatively or positively associated with stomach cancer.[51-53] Many case-control studies on stomach cancer and tea consumption conducted in Buffalo,[54] Kansas City,[55] Nagoya (Japan),[56] Piraeus (Greece),[57] Milan (Italy),[47] Spain,[58] and Turkey[59] all indicated that there was no significant association between tea consumption and cancer of the stomach. A very small case-control study in Taiwan (10 cases and 14 controls) suggested that green tea consumption is a risk factor for gastric cancer.[60] On the other hand, some recent case-control studies suggested that tea consumption reduced the risk of stomach cancer in Kyushu (Japan),[61] Shanghai (China),[62] northwestern Turkey[39] and central Sweden.[34] For example, in Kynshu (139 cases and 2574 controls), individuals consuming green tea more frequently or in larger quantities (10 or more cups a day) tended to have a lower risk for gastric cancer.[61] In a population-based case-control study (669 cases and 880 controls) in central Sweden, tea (mostly black tea) consumption was associated with lower incidence of gastric cancer.[34]

Different etiological factors and molecular mechanisms may be involved even in cancers of the same organ site. Therefore, it is not surprising that a clear-cut conclusion on the relationship between tea consumption and human cancer has not been found. Many confounding factors, such as cigarette smoking, alcohol drinking, and the temperature of tea, have been adjusted in many newer studies, and these results tend to be more informative.

Table 2. Plasma levels of tea polyphenols in humans, rats, and mice ingesting green tea

Polyphenols	Humans[a]	Rats[b] *ng/ml*	Mice[c]
EGCG	120±55	37±6	124±25
EGC	148±26	55±16	62±19
EC	55±8	20±3	10±3

[a]Peak concentration in humans who ingested 1.5 g of decaffeinated green tea solids in 200 ml of water (equivalent to 2-3 cups of tea); mean ± S.E. of 4 individuals.
[b]Sprague-Dawley male rats received decaffeinated green tea (9 mg solids/ml) for 3 weeks; mean ± S.E. of 5 rats.
[c]SKH-1 female mice received regular green tea (9 mg solids/ml) for 6 weeks, mean ± S.E. of 5 mice.

Another key factor which needs to be incorporated in future studies is the quantity and type of tea consumed.

ACTIVE COMPONENTS AND POSSIBLE MECHANISMS FOR THE INHIBITION OF TUMORIGENESIS BY TEA

The inconsistencies of the protective effect of tea consumption on human cancers in epidemiological studies contrast with the repeated laboratory observations on such an effect in animal models. One factor to be considered is the quantities of tea used in animal studies in comparison to those consumed by humans. In order to conduct quantitative analyses, we developed a method for the quantification of plasma and tissue levels of tea polyphenols.[63] Plasma EGCG, EGC, and EC exist in free and conjugated (glucuronide and sulfate) forms. Table 2 shows that the plasma tea polyphenol levels in rats and mice in our anti-carcinogenesis experiments were comparable to the peak levels in humans after consuming 2 or 3 cups of tea. The peak plasma polyphenol levels were observed 2 h after ingestion, and EGCG, EGC, and EC had half-lives of 3 to 5 h.

In a preliminary experiment, after administration of regular green tea in drinking fluid to rats, substantial amounts of EGC and EC were detected in the lung (220 and 46 ng/g wet weight, respectively) and esophagus (850 and 280 ng/g, respectively). EGCG was detected in the esophagus (410 ng/g), but was not in the lung. The EGCG, EGC, and EC levels in the small intestine and intestinal contents were rather high (1.5 to 5.5 μg/g) due to the unabsorbed and biliary excreted glucuronides of polyphenols in the intestine. High EGC and EC levels were also observed in the colonic tissues (1.8 and 0.3 μg/g, respectively). Due to possible glucuronidase and esterase activities, most of the EGC and EC are in the free form, and EGCG was not detectable. EGCG has been usually considered to be the active anti-carcinogenic component in tea since it is the polyphenol with the highest concentration in tea. Indeed, inhibition of lung and colon carcinogenesis by EGCG has been reported.[24,64,65] Our results on tissue levels suggest that EGCG is converted to EGC, and thus EGC rather than EGCG may be the main active compound involved in both models. EGC and EGCG had similar potency (IC_{50} of 30-50 μM) in inhibiting the growth of colon cancer cell line HT-29 (unpublished results). These results demonstrate the importance of studies on the tissue levels of tea polyphenols, especially when a mixture of compounds is used.

In black tea, the levels of catechins are about 30% that of green tea, but the inhibitory activity against tumorigenesis of black tea was comparable to green tea in several animal models.[3,8,16] The effective components in black tea have not been delineated. Theaflavins and thearubigins contain multiple hydroxyl groups and possess antioxidant activities. The antioxidative and antimutagenic effects of theaflavins from black tea have been reported.[66] Other non-polyphenolic constituents may also play a role in the anti-tumorigenesis activities of tea. Tumorigenesis studies indicated that decaffeinated tea displayed similar inhibitory activity to regular green tea in some experiments,[3,8] but was less effective in other experiments,[16] suggesting that caffeine also possesses inhibitory activities in some animal models. Xu *et al.* reported that oral feeding of caffeine to A/J mice also inhibited NNK-induced lung tumorigenesis.[7]

The most noteworthy properties of tea polyphenols that may affect carcinogenesis are their antioxidative activities, inhibitory action against nitrosation, modulation of carcinogen-metabolizing enzymes, trapping of ultimate carcinogens, and inhibition of cell proliferation-related activities.[1] Although inhibition of carcinogen activation by tea or green tea polyphenol fractions could be demonstrated *in vitro*, and in certain cases *in vivo*,[67,68] this mechanism was not demonstrated for NNK *in vivo*.[4,7] Oral administration of tea preparations to animals has been reported to enhance moderately the activities of glutathione peroxidase, catalase, glutathione *S*-transferase, NADPH-quinone oxidoreductase, UDP-glucuronosyl transferase, and methoxyresorufin *O*-dealkylase.[69-71] The effects of these enzyme inductions on carcinogenesis are not clear.

Reactive oxygen species may play important roles in carcinogenesis through damaging DNA, altering gene expression, or affecting cell growth and differentiation.[72,73] The anticarcinogenic activities of tea polyphenols are believed to be closely related to their antioxidative properties. The findings that green tea preparations inhibited TPA-induced hydrogen peroxide formation in mouse epidermis[74] and NNK-induced 8-hydroxydeoxyguanosine formation in mouse lung[7] are consistent with this concept. Inhibition of tumor promotion-related enzymes, such as TPA-induced epidermal ornithine decarboxylase,[74,75] protein kinase C,[76] lipoxygenase and cyclooxygenase[77] by tea preparations has also been demonstrated. Mechanisms relating to the quenching of activated carcinogens,[78] antiviral activity,[79] and enhancing immune functions[80] have also been suggested, but their relevance to carcinogenesis is not clear. Inhibition of nitrosation by tea preparations has been demonstrated *in vitro* and in humans,[81,82] and this may be an important factor in preventing certain cancers, e.g., gastric cancer, if the endogenously formed *N*-nitroso compounds are causative factors. Our recent unpublished results suggest that the anti-proliferative effect of tea is important for the anti-carcinogenic activity. One may speculate that tea polyphenols may inhibit growth-related signal transduction pathways.

CONCLUDING REMARKS

Although the anti-carcinogenic activity of tea has been demonstrated repeatedly in studies with animals, such activity has not been clearly demonstrated in humans. More epidemiological investigations, especially prospective studies concerning the effect of tea consumption on human cancer risk, are needed. Because the causative factors may be different for different cancers and for the same cancer in different populations, tea consumption may inhibit carcinogenesis only in selected situations. Therefore, in future epidemiology studies, it is important to consider the etiological factors of the specific cancers and to collect specific information on the quantitative aspects of tea consumption. Urinary excretion of tea polyphenols may be used as an exposure biomarker. Definitive information on the protective effects of tea consumption on cancer will come from intervention studies. Based on the suggestive epidemiological data from Shizuoka, Kyushu, Shanghai, northwestern Turkey, and central Sweden, intervention studies on gastric and esophageal cancers among higher risk populations should be very worthwhile.

ACKNOWLEDGMENTS

The work was supported by NIH grant CA56673 and Center Grant ES05022. The authors thank Ms. Dorothy Wong for excellent secretarial service.

REFERENCES

1. Yang, C. S., Wang, Z.-Y., Tea and Cancer: A review, *J Natl. Cancer Inst.(Invited Review)* 58:1038-1049 (1993).
2. Graham, H. N., Green tea composition, consumption, and polyphenol chemistry, *Prev. Med.* 21:334-350 (1992).
3. Wang, Z. Y., Hong, J.-Y., Huang, M.-T., Reuhl, K., Conney, A. H., Yang, C. S., Inhibition of *N*-nitrosodiethylamine and 4-(methylnitrosamino)-1-(3-pridyl)-1-butanone-induced tumorigenesis in A/J mice by green tea and black tea, *Cancer Res.* 52:1943-1947 (1992).
4. Shi, S. T., Wang, Z.-Y., Smith, T. J., Hong, J.-Y., Chen, W.-F., Ho, C.-T., Yang, C. S., Effects of green tea and black tea on 4-(methylnitrosamino)-1-(3-pyridyl)-1-butanone bioactivation, DNA methylation, and lung tumorigenesis by green tea and black tea in A/J mice, *Cancer Res.* 54:4641-4647 (1994).
5. Wang, Z.-Y., Kan, W. A., Mukhtar, H., Protection against benzo(a)pyrene and *N*-nitrosodiethylamine-induced lung and forestomach tumorigenesis in A/J mice by water extracts of green tea and licorice, *Carcinogenesis* 13:1491-1493 (1992).
6. Katiyar, S. K., Agarwal, R., Zaim, M. T., Mukhtar, H., Protection against *N*-nitrosodiethylamine and benzo(a)pyrene-induced forestomach and lung tumorigenesis in A/J mice by green tea, *Carcinogenesis* 14:849-855 (1993).
7. Xu, Y., Ho, C.-T., Amin, S. G., Han, C., Chung, F.-L., Inhibition of tobacco-specific nitrosamine-induced lung tumorigenesis in A/J mice by green tea and its major polyphenol as antioxidants, *Cancer Res.* 52:3875-3879 (1992).
8. Wang, Z.-Y., Wang, L.-D., Lee, M.-J., Ho, C.-T., Huang, M.-T., Conney, A. H., Yang, C. S., Inhibition of *N*-nitrosomethylbenzylamine-induced esophageal tumorigenesis in rats by green and black tea, *Carcinogenesis* 16:2143-2148 (1995).
9. Han, C., Xu, Y., The effect of Chinese tea on the occurrence of esophageal tumors induced by *N*-nitrosomethylbenzylamine in rats, *Biomed. Environ. Sci.* 3:35-42 (1990).
10. Xu, Y., Han, C., The effect of Chinese tea on the occurrence of esophageal tumors induced by *N*-nitrosomethylbenzylamine formed *in vivo*, *Biomed. Environ. Sci.* 3(4):406-412 (1990).
11. Cheng, S.-J., Ding, L., Zhen, Y.-S., Lin, P.-Z., Zhu, Y.-J., Chen, Y.-Q., Hu, X.-Z., Progress in studies on the antimutagenicity and anticarcinogenecity on green tea epicatechins, *Chinese Med. Sci. J.* 6:233-238 (1991).
12. Gao, G. D., Zhou, L. F., Qi, G., Wang, H. S., Li, T. S., Studies of antitumorigenesis by green tea (animal test and flow cytometry), *Tumor (Chinese)* 10:42-44 (1990).
13. Yan, Y. S., The experiment of tumor-inhibiting effect of green tea extract in animals and humans, *Chung Hua Yu Fang I Hsueh Tsa Chih* 27(3):129-131 (1993).
14. Yamane, T., Takahashi, T., Kuwata, K., Oya, K., Inagake, M., Kitao, Y., Suganuma, M., Fujiki, H., Inhibition of *N*-methyl-*N'*-nitro-*N*-nitrosoguanidine-induced carcinogenesis by (-)-epigallocatechin gallate in the rat glandular stomach, *Cancer Res.* 55:2081-2084 (1995).
15. Wang, Z. Y., Huang, M. T., Ferraro, T., Wong, C. Q., Lou, Y. R., Reuhl, K., Iatropoulos, M., Yang, C. S., Conney, A. H., Inhibitory effect of green tea in the drinking water on tumorigenesis by ultraviolet light and 12-*O*-tetradecanoylphorbol-13-acetate in the skin of SKH-1 mice, *Cancer Res* 52:1162-1170 (1992).
16. Wang, Z. Y., Huang, M.-T., Lou, Y.-R., Xie, J.-G., Reuhl, K., Newmark, H. L., Ho, C.-T., Yang, C. S., Conney, A. H., Inhibitory effects of black tea, green tea, decaffeinated black tea and decaffeinated green tea on ultraviolet B light-induced skin carcinogenesis in 7,12-dimethylbenz(a)anthracene-initiated SKH-1 mice, *Cancer Res.* 54:3428-3435 (1994).
17. Wang, Z. Y., Huang, M.-T., Ho, C.-T., Chang, R., Ma, W., Ferraro, T., Reuhl, K. R., Yang, C. S., Conney, A. H., Inhibitory effect of green tea on the growth of established skin papillomas in mice, *Cancer Res.* 52:6657-6665 (1992).
18. Wang, Z.-Y., Chen, L., Lee, M.-J., Yang, C. S., Tea and cancer prevention, in: *Hypernutritions Food* (Finley, J. W., Armstrong, D., Robinson, S. F., Nagy, S., eds.), American Chemical Society Symposium, (1995) (in press).

19. Hirose, M., Hoshiya, T., Akagi, K., Takahashi, S., Hara, Y., Ito, N., Effects of green tea catechins in a rat multi-organ carcinogenesis model, *Carcinogenesis* 14:1549-1553 (1993).
20. Hirose, M., Hoshiya, T., Akagi, K., Futakuchi, M., Ito, N., Inhibition of mammary gland carcinogenesis by green tea catechins and other naturally occurring antioxidants in female Sprague-Dawley rats pretreated with 7,12,dimethylbenz(a)anthracene, *Cancer Lett.* 83:149-156 (1994).
21. Narasiwa, T., Fukaura, Y., A very low dose of green tea polyphenols in drinking water prevents *N*-methyl-*N*-nitrosourea-induced colon carcinogenesis in F344 rats, *Jpn. J. Cancer Res.* 84:1007-1009 (1993).
22. Mao, R., The inhibitory effects of epicatechin complex on diethylnitrosamine induced initiation of hepatocarcinogenesis in rats, *Chung-Hua-Yu-Fang-I-Hsueh-Tsa-Chih* 27:201-204 (1993).
23. Nishida, H., Omori, M., Fukutomi, Y., Ninomiya, M., Nishiwaki, S., Suganuma, M., Moriwaki, H., Mutao, Y., Inhibitory effects of (-)-epigallocatechin gallate on spontaneous hepatoma in C3H/NeNCrj mice and human hepatoma-derived PLC/PRF/5 cells, *Jpn. J. Cancer Res.* 85(3):221-225 (1994).
24. Yin, P., Zhao, J., Cheng, S., Hara, Y., Zhu, Q., Liu, Z., Experimental studies of the inhibitory effects of green tea catechin on mice large intestinal cancers induced by 1,2-dimethylhydrazine, *Cancer Lett.* 79:33-38 (1994).
25. WHO International Agency for Research on Cancer, *IARC Monogor. Eval. Carcinog. Risk Hum.* 51:207-271 (1991).
26. Wakai, K., Ohno, Y., Obata, K., Aoki, K., Prognostic significance of selected lifestyle factors in urinary bladder cancer, *Jpn. J. Cancer Res.* 84(12):1223-1229 (1993).
27. Shibata, A., Mack, T. M., Paganini, H. A., Ross, R. K., Henderson, B. E., A prospective study of pancreatic cancer in the elderly, *Int J Cancer* 58:46-49 (1994).
28. Gao, Y. T., McLaughlin, J. K., Blot, W. J., Ji, B. T., Dai, Q., Fraumeni, J. J., Reduced risk of esophageal cancer associated with green tea consumption, *J Natl. Cancer Inst.* 86:855-858 (1994).
29. Mellemgaard, A., Engholm, G., McLaughlin, J. K., Olsen, J. H., Risk factors for renal cell carcinoma in Denmark I. Role of socioeconomic status, tobacco use, beverages and family history, *Cancer Causes Control* 5(2):105-113 (1994).
30. Hu, J., Nyren, O., Wolk, A., Bergstrom, R., Yuen, J., Adami, H. O., Guo, L., Li, H., Huang, G., Xu, X., Risk factors for esopahgeal cancer in northeast China, *Int. J. Cancer* 57(1):38-46 (1994).
31. Slattery, M. L., West, D. W., Smoking, alcohol, coffee, tea, caffeine and theobromine: risk of prostate cancer in Utah (United States), *Cancer Causes Control* 4(6):559-563 (1993).
32. Ewertz, M., Breast cancer in Denmark. Incidence risk factors, and characteristics of survival, *Acta. Oncol.* 32(6):595-615 (1993).
33. Dhar, G. M., Shan, G. N., Naheed, B., Hafiza, Epidemiological trend in the distribution of cancer in Kashmir Valley, *J. Epidemiol. Community Health* 47:290-292 (1993).
34. Hansson, L. E., Nyren, O., Bergstrom, R., Wolk, A., Lindgren, A., Baron, J., Adami, H. O., Diet and risk of gastric cancer. A population-based case-control study in Sweden, *Int J Cancer* 55:181-189 (1993).
35. Olsen, J., Kronborg, O., Coffee, tobacco and alcohol as risk factors for cancer and adenoma of the large intestine, *Int. J. Epidemiol.* 22(3):398-402 (1993).
36. Mashberg, A., Boffetta, P., Winkelman, R., Garfinkel, L., Tobacco smoking, alcohol drinking, and cancer of the oral cavity and oropharynx among U.S. veterans, *Cancer* 72(4):1369-1375 (1993).
37. Zatonski, W. A., Boyle, P., Przewozniak, K., Maisonneuve, P., Drosik, K., Walker, A. M., Cigarette smoking, alcohol, tea and coffee consumption and pancreas cancer risk: a case-control study from Opole, Poland, *Int. J. Cancer* 53(4):601-607 (1993).
38. Huang, C., Zhang, X., Qiao, Z., Guan, L., Peng, S., Liou, J., Xie, R., Zheng, L., A case-control study of dietary factors in patients with lung cancer, *Biomed. Environ. Sci.* 5(3):257-265 (1992).
39. Memik, F., Nak, S. G., Gulten, M., Ozturk, M., Gastric carcinoma in northwestern Turkey: epidemiologic characteristics, *J Environ. Pathol. Toxicol. Oncol.* 11:335-338 (1992).
40. Mizuno, S., Watanabe, S., Nakamura, K., Omata, M., Oguchi, H., Ohashi, K., Ohyanagi, H., Fujiki, T., Motojima, K., A multi institute case-control study on the risk factors of developing pancreatic cancer, *Jpn. J. Clin. Oncol.* 22(4):286-291 (1992).
41. Franceschi, S., Serraino, D., Risk factors for adult soft tissue sarcoma in northern Italy, *Ann Oncol.* 3(2):85-88 (1992).
42. Baron, J. A., Gerhardsson de Verdier, M., Ekbom, A., Coffee, tea, tobacco, and cancer of the large bowel, *Cancer Epidemiol. Biomarkers Prev.* 3:565-570 (1994).
43. Segi, M., Tea-gruel as a possible factor for cancer of the esophagus, *Gann* 66:199-202 (1975).
44. Joint Iran-International Agency for Research on Cancer Study Group. Esophageal cancer studies in the Caspian Littoral of Iran: results of population studies-a prodrome, *J Natl Cancer Inst.* 59:1127-1138 (1977).

45. Kapadia, G. J., Rao, S., Morton, J. F., Herbal tea consumption and esophageal cancer, In: *Carcinogens and Mutagens in the Environment,* (Stich, H. F., ed.), 3, pp. 3-12, CRC Press. Inc., Boca Raton, Florida, (1983).
46. Victora, C. G., Munoz, N., Day, N. E., Barcelos, I. B., Pecein, D. A., Braga, N. M., Hot beverages and oesophageal cancer in southern Brazil: a case-control study, *Int. J. Cancer* 39:710-716 (1987).
47. La Vecchia, C., Negri, E., Franceschi, S., D'Avanzo, B., Boyle, P., Tea consumption and cancer risk, *Ntr. Cancer* 17:27-31 (1992).
48. Kaufman, B. D., Liberman, I. S., Tyshetsky, V. I., Some data concerning the incidence of oesophageal cancer in the Gurjev region of the Kazakh SSR (Russ.), *Vopr. Onkol.* 11:78-85 (1965).
49. Bashirov, M. S., Nugmanov, S. N., Kolycheva, N. I., Epidemiological study of oesophageal cancer in the Akhtubinsk region of the Kazakh Socialist Republic (Russ.), *Vopr. Onkol.* 14:3-7 (1968).
50. Cook-Mozaffari, P. J., Azordegan, F., Day, N. E., Ressicaud, A., Sabai, C., Aramesh, B., Oesophageal cancer studies in the Caspian Littoral of Iran: results of a case-control study, *Br. J. Cancer* 39:293-309 (1979).
51. Stocks, P., Cancer mortality in relation to national consumption of cigarettes, solid fuel, tea and coffee, *Br. J. Cancer* 24:215-225 (1970).
52. Oguni, I., Chen, S. J., Lin, P. Z., Hara, Y., Protection against cancer risk by Japanese green tea, *Prevent. Med.* 21:332 (1992).
53. Kinlen, L. J., McPherson, K., Pancreas cancer and coffee and tea consumption: a case-control study, *Br. J. Cancer* 49:93-96 (1984).
54. Graham, S., Lilienfeld, A. M., Tidings, J. E., Dietary and purgation: factors in the epidemiology of gastric cancer, *Cancer* 20:2224-2234 (1967).
55. Higginson, J., Etiological factors in gastrointestinal cancer in man, *J. Natl. Cancer Inst.* 37:527-545 (1966).
56. Tajima, K., Tominaga, S., Dietary habits and gastro-intestinal cancers: a comparative case-control study of stomach and large intestinal cancers in Nagoya, *Jpn. J. Cancer Res. (Gann)* 76:705-716 (1985).
57. Trichopoulos, D., Ouranos, G., Day, N. E., Tzonou, A., Manousos, O., Papadimitriou, C., Trichopoulos, A., Diet and cancer of the stomach: a case-control study in Greece, *Int. J. Cancer* 36:291-297 (1985).
58. Agudo, A., Gonzalez, C. A., Marcos, G., Sanz, M., Saigi, E., Verge, J., Boleda, M., Ortego, J., Consumption of alcohol, coffee, and tobacco, and gastric cancer in Spain, *Cancer Causes Control* 3(2):137-143 (1992).
59. Demirer, T., Icli, F., Uzunalimoglu, O., Kucuk, O., Diet and stomach cancer incidence. A case-control study in Turkey, *Cancer* 65(10):2344-2348 (1990).
60. Lee, H. H., Wu, H. Y., Chuang, Y. C., Chang, A. S., Chao, H. H., Chen, K. Y., Chen, H. K., Lai, G. M., Huang, H. H., Chen, C. J., Epidemiologic characteristics and multiple risk factors of stomach in Taiwan, *Anticancer Res.* 10(4):875-881 (1990).
61. Kono, S., Green tea and colon cancer, *Jpn. J. Cancer Res.* 83(6):669 (1992).
62. Yu, G. P., Hsie, C. C., Risk factors for stomach cancer: a population-based case-control study in Shanghai, *Cancer Causes Control* 2:169-174 (1991).
63. Lee, M.-J., Wang, Z.-Y., Li, H., Chen, L., Sun, Y., Gobbo, S., Balentine, D. A., Yang, C. S., Analysis of plasma and urinary tea polyphenols in human subjects, *Cancer Epidemiol. Biomark. Prev.* 4:393-399 (1995).
64. Yamane, T., Hagiwara, N., Tatcishi, M., Akachi, S., Kim, M., Okuzuni, J., Kitao, Y., Inagake, M., Kuwata, K., Takahashi, T., Inhibition of azoxymethane-induced colon carcinogenesis in rat by green tea polyphenol fraction, *Jpn. J. Cancer Res.* 82:1336-1339 (1991).
65. Narisawa, T., Fukaura, Y., A very low dose of green tea polyphenols in drinking water prevents *N*-methyl-*N*-nitrosourea-induced colon carcinogenesis in F344 rats, *Jpn. J. Cancer Res.* 84:1007-1009 (1993).
66. Shiraki, M., Hara, Y., Osawa, T., Kumon, H., Nakayama, T., Kawakishi, S., Antioxidative and antimutagenic effects of theaflavins from black tea, *Mutat. Res.* 323:29-34 (1994).
67. Wang, Z. Y., Khan, W. A., Bickers, D. R., Protection against polycyclic aromatic hydrocarbon-induced skin tumor initiation in mice by green tea polyphenols, *Carcinogenesis* 10:411-415 (1989).
68. Chen, J.-S., The effects of Chinese tea on the occurrence of esophageal tumors induced by *N*-nitrosomethylbenzylamine in rats, *Prev. Med.* 21:385-391 (1992).
69. Khan, S. G., Katiyar, S. K., Agarwal, R., Enhancement of antioxidant and phase II enzymes by oral feeding of green tea polyphenols in drinking water to SKH-1 hairless mice: possible role in cancer chemoprevention, *Cancer Res.* 52:4050-4052 (1992).

70. Sohn, O. S., Surace, A., Fiala, E. S., Richie, J. P. J., Colosimo, S., Zang, E., Weisburger, J. H., Effects of green and black tea on hepatic xenobiotic metabolizing systems in the male F344 rat, *Xenobiotica* 24:119-127 (1994).
71. Bu-Abbas, A., Clifford, M. N., Walker, R., Ioannides, C., Selective induction of rat hepatic CYP1 and CYP4 proteins and of peroxisomal proliferation by green tea, *Carcinogenesis* 15:2575-2579 (1994).
72. Cerutti, P. A., Mechanisms of action of oxidant carcinogens, *Cancer Detection and Prevention* 14:281-284 (1989).
73. Feig, D. I., Reid, T. M., Loeb, L. A., Reactive oxygen species in tumorigenesis, *Cancer Res.* 54 (Suppl.):1890s-1894s (1994).
74. Huang, M.-T., Ho, C.-T., Wang, Z.-Y., Ferraro, T., Finnegan-Oliver, T., Lou, Y.-R., Mitchell, J. M., Laskin, J. D., Newmark, H., Yang, C. S., Conney, A. H., Inhibitory effect of topical application of a green tea polyphenol fraction on tumor initiation and promotion in mouse skin, *Carcinogenesis)* 13:947-954 (1992).
75. Wang, Z.-Y., Agarwal, R., Bickers, D. R., Mukhtar, H., Protection against ultraviolet B radiation-induced photocarcinogenesis in hairless mice by green tea polyphenols, *Carcinogenesis* 12:1527-1530 (1991).
76. Yoshizawa, S., Horiuchi, T., Sugimura, M., Penta-o-galloyl-β-D-glucose and (-)-epigallocatechin gallate: cancer prevention agent, in: *Phenolic Compounds in Foods and Health II: Antioxidant & Cancer Prevention* (Huang, M.-T., Ho, C.-T., Lee, C. Y., eds.), pp. 316-325, American Chemical Society, Washington, D.C., (1992).
77. Katiyar, S., Agarwal, R., Wood, G. S., Inhibition of 12-*O*-tetradecanoylphorbol-13-acetate-caused tumor promotion in 7,12-dimethylbenz(a)anthracene-initiated SENCAR mouse skin by a polyphenolic fraction isolated from green tea, *Cancer Res.* 52:6890-6897 (1992).
78. Khan, W. A., Wang, Z. Y., Athar, M., Bickers, D. R., Mukhtar, H., Inhibition of the skin tumorigenicity of (±)7β, 8α-epoxy-7, 8, 9, 10-tetrahydrobenzo(a)pyrene by tannic acid, green tea polyphenols and quercetin in Sencar mice, *Cancer Letters* 42:7-12 (1988).
79. Shimamura, T., Inhibition of influenza virus infection by tea polyphenols, in: *Food Phytochemicals for Cancer Prevention II* (Huang, M.-T., Osawa, T., Ho, C.-T., Rosen, R. T., eds.), pp. 97-101, ACS Symposium Series 546, Washington, D.C., (1992).
80. Yan, Y. S., Effect of Chinese tea extract on the immune function of mice bearing tumor and their antitumor activity, *Chung Hua Yu Fang I Hsueh Tsa Chih* 26:5-7 (1992).
81. Nakamamura, M., Kawabata, T., Effect of Japanese green tea on nitrosamine formation *in vitro*, *J. Food Sci.* 46:306-307 (1981).
82. Stich, H. F., Teas and tea components as inhibitors of carcinogen formation in model system and man, *Prev. Med.* 21:377-384 (1992).

6

ESTROGENS, PHYTOESTROGENS, AND BREAST CANCER

Robert Clarke,[1,2] Leena Hilakivi-Clarke,[1,3] Elizabeth Cho,[1] Mattie R. James,[1] and Fabio Leonessa[1,2]

[1] Vincent T Lombardi Cancer Center
[2] Department of Physiology and Biophysics
[3] Department of Psychiatry
Georgetown University Medical School
3970 Reservoir Rd. NW, Washington, DC 20007

INTRODUCTION

Estrogens have been widely implicated in the genesis and progression of breast cancer. Over 200 years ago, the Italian physician Ramazzini observed an increased incidence of breast cancer among nuns. Almost 100 years ago the Scottish physician Beatson described the beneficial effects of ovariectomy on the progress of breast cancer in premenopausal women.[1] Nevertheless, the precise role(s) of estrogens and estrogenic stimuli in breast cancer remains unknown. Indeed, it is difficult to provide a universal definition of either an estrogen or an estrogenic response.

Epidemiological evidence in women indicates that the length of exposure of the mammary glands to ovarian-derived estrogenic stimuli and to other dietary, genetic and environmental factors is directly proportional to breast cancer risk. Early menarche, late menopause and late age at first full-term pregnancy can all increase the total exposure of the breast to estrogenic stimuli from natural estrogens. Risk is decreased with early menopause and childbearing but is increased by first-trimester abortion.[2,3] Estrogens are widely reported to increase the proliferation of some human breast cancer cell lines *in vivo* and *in vitro*.[4-6] The menstrual dependent proliferation of normal breast epithelium tends to occur during the luteal phase of the cycle,[7] when progesterone levels are highest and the second (smaller) estrogen peak occurs (see Figure 1). These mitogenic effects may partly account for the association between estrogens and breast cancer risk.

In addition to the promotional effects of estrogens on cellular proliferation, estrogens may directly damage DNA. The synthetic estrogen diethylstilbestrol (DES) is oxidized by a peroxidase-mediated reaction, and can produce short lived epoxide and semiquinone intermediates that appear to be carcinogenic. However, it is unclear whether these reactive species are sufficiently stable to account for the carcinogenic potential of DES.[8] DES can induce other effects, including perturbations in mitotic spindle formation or function.[9]

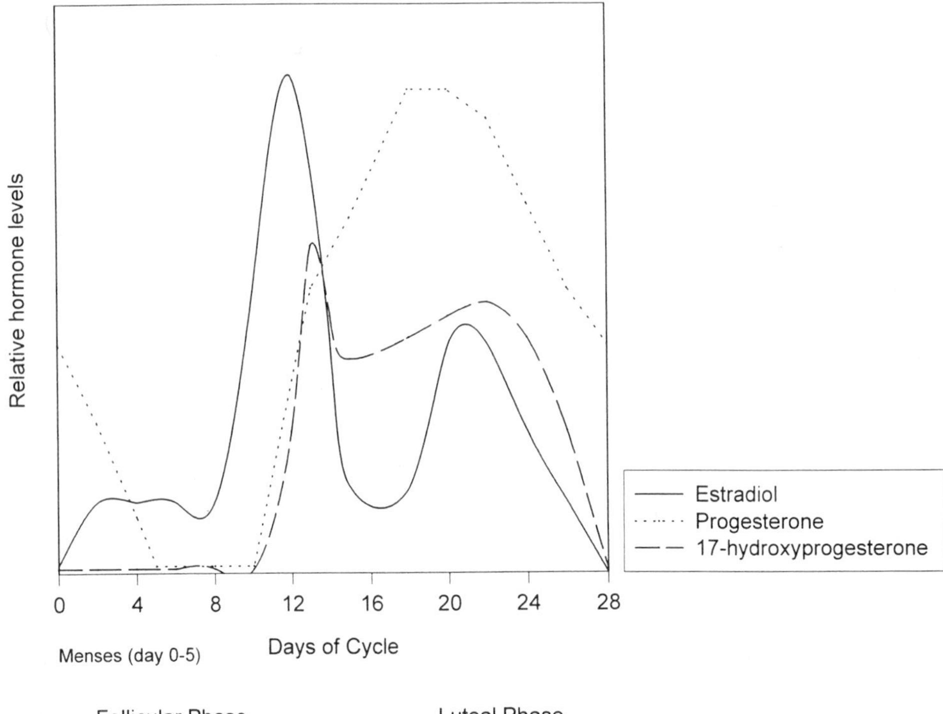

Figure 1. Schematic representation of the endocrine changes that occur during a normal menstrual cycle. Note that the relative concentrations for 17β-estradiol range from approximately 20-65 pg/ml; the range for progesterone and 17-hydroxyprogesterone is approximately 0-20 ng/ml. Adapted from: Rhoades, R.A. & Tanner, G.A. "Medical Physiology" Little, Brown and Company, Boston, pp769, 1995.

Carcinogenic reactive species may be generated by the direct metabolism of the natural estrogens. Hydroxylation of 17β-estradiol (E2) at positions C2 and C16 produces reactive metabolites that can attack both DNA and proteins.[9-11] Since estrogens can act by both inducing genomic damage (initiation) and increasing cellular proliferation (promotion), it is likely that estrogens function as both carcinogens and as tumor promoters. Many of the effects induced by the natural estrogens and their metabolites also could be induced by compounds with similar structure/function attributes. Plant-derived estrogenic compounds, generally called phytoestrogens, could influence breast cancer risk by mimicking or inhibiting these effects. We will discuss the possible structure-function attributes of the natural estrogens and phytoestrogens, and attempt to bring the diverse activities of these compounds together from the perspective of breast cancer.

ESTROGENS AND PHYTOESTROGENS

What Is an Estrogen?

In the chemical sense, an estrogen must have a steroidal nucleus, essentially limiting estrogens to the three main "natural" estrogens, estrone (E1), 17β-estradiol (E2) and estriol (E3), and their metabolites, *e.g.,* the estrogen sulfates. However, this is too restrictive, since many chemically unrelated compounds appear capable of eliciting an estrogenic response. In the broadest sense, an estrogen is any compound capable of eliciting an estrogenic response. However, this definition includes compounds that are indirectly estrogenic, perhaps by regulating gene pathways that also are regulated by estrogens.

Perhaps a more rational approach would be to define an estrogen in terms of the biological activity of the most potent of the natural estrogens. Thus, an estrogen could be considered to be any compound that can regulate gene expression that is mediated by an estrogen response element (ERE), in a manner comparable to 17β-estradiol, where this regulation is the result of direct binding of the compound to the estrogen receptor (ER). We have chosen to use this as a working definition, and have applied it throughout this article. This definition would include many compounds that do not have a classical steroid nucleus. For example, the heavy metal cadmium appears to fulfil all the requirements of this definition of an estrogen, since it binds to ER, and can activate expression of a reporter gene linked to an ERE.[12] Nevertheless, this is a potentially inadequate definition for all circumstances and all activities, since it excludes the non-genomic activities of steroids, which also can be important in generating an estrogenic response.[13,14]

By applying this definition of an estrogen it is possible to derive a working definition for an antiestrogen. An antiestrogen is essentially a compound that can regulate gene expression that is mediated by an ERE, in a direction opposite to that induced by 17β-estradiol, as a result of either direct binding to, and/or competition for, ER. This working definition would include both partial agonists like Tamoxifen (TAM)[15] and "pure" antagonists like ICI 182,780.[16] However, it excludes other interactions, including the ability of TAM to inhibit directly both calmodulin[17] and protein kinase C,[18] and to bind to the so-called antiestrogen binding sites.[19]

What Is a Phytoestrogen?

If we choose to define an estrogen in terms of its ability to bind ER and, as a direct consequence, regulate gene expression, then we also can loosely define a phytoestrogen. Thus, a phytoestrogen could be any plant derived compound that can regulate gene expression that is mediated by an ERE, in a manner either comparable or apparently antagonistic to 17β-estradiol, as a result of direct binding to ER. However, many compounds that can fulfil this definition also have other properties, *e.g.*, genistein appears to be a phytoestrogen (Table 3) and both a tyrosine kinase[20] and topoisomerase II inhibitor.[21] In many cases it may be difficult to distinguish between these mechanisms, particularly when an estrogenic pathway that includes the regulation of tyrosine kinase activities is involved. Indeed, it may ultimately prove more valuable to classify compounds as having "phytoestrogenic properties", rather than assigning them as "phytoestrogens."

We have chosen our working definition of a phytoestrogen, which essentially combines both estrogenic and antiestrogen activities, for purely practical reasons. For the purposes of this article we will consider phytoestrogen and phytoestrogenic as essentially interchangeable concepts. Until we have a better understanding of the origin and function of all plant derived compounds that may fulfil this working definition, it is unclear whether attempting to delineate between "phytoestrogenic" and "phytoantiestrogenic" is a worthwhile exercise. However, it does seem likely that, from the perspective of breast cancer research, there are "good" phytoestrogens and "bad" phytoestrogens. This concept is further developed in the following sections.

Where Are Compounds with Phytoestrogenic Activity?

There are several groups of plant estrogens, including the lignans, isoflavones (*e.g.*, genistein) and mycotoxins derived from fungal molds (*e.g.*, zearalenone). Sources for lignans and isoflavone phytoestrogens and their metabolites include whole grain products, fruits,

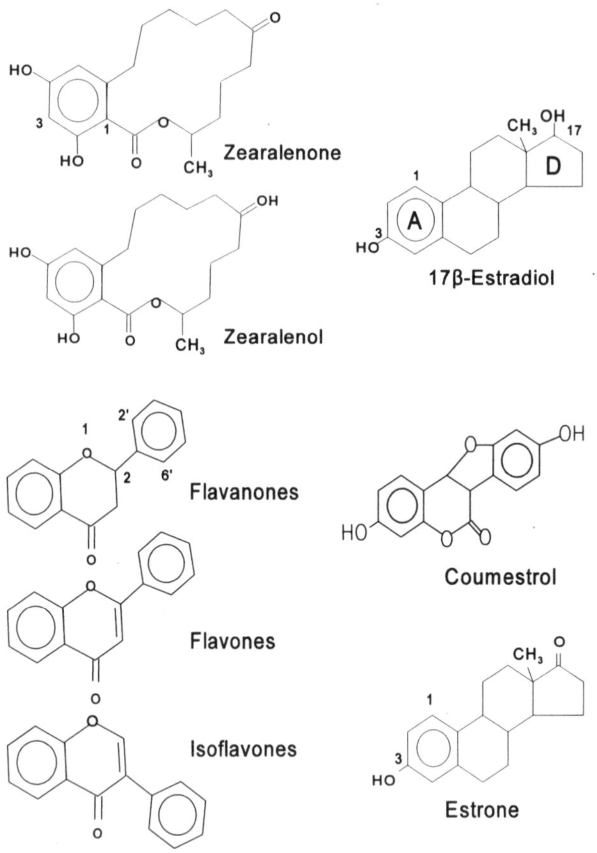

Figure 2. Structure of some phytoestrogenic compounds. Representations are not to scale.

berries and soy products.[22] Isoflavonoids are generally considered "good" phytoestrogens. They may reduce the risk for developing breast cancer.[22,23] In support for this suggestion, plant flavonoids are antitumorigenic against some experimental tumors.[22] Many synthetic and naturally occurring flavonoids inhibit the *in vitro* growth of MCF-7 and ZR-75-1 human breast carcinoma cells.[23,24] However, some investigators have found that lignans and isoflavones stimulate the growth of different breast cancer cell lines.[24,25] The structures of several phytoestrogenic compounds are provided in Figure 2.

The mycotoxin zearalenone is mainly produced by *Fusarium graminearum*[26] that is found in a variety of host plants and debris from soil in the United States.[27] Zearalenone may be present in barley, wheat, corn, corn flakes, rice and maize at concentrations varying from 35-115 mg/kg.[26, 28-30] A sample of US corn contained 18 mg/kg zearalenone.[30] The overall estimated tolerable daily intake for zearalenone in humans is 100-500 ng per kg body weight.[31] Zearalenone has been used as a contraceptive,[32] as estrogen therapy in postmenopausal women,[33,34] and as an anabolic agent to enhance growth in cattle and lambs.[35,36] Despite these uses of zearalenone, it may be a "bad" phytoestrogen. Mycotoxins are linked to the occurrence of human esophageal cancer in China.[28] Zearalenone stimulates the growth of the MCF-7 human breast cancer cells,[25] and has been reported to enlarge the mammary gland and induce spontaneous mammary tumors in mice.[37] These effects likely reflect ER agonist properties of zearalenone.

THE INTERACTIONS OF ESTROGENS AND PHYTOESTROGENS WITH ESTROGEN RECEPTORS

Structure-Activity Relationships of Estrogen Receptor Ligands

The ability of the normal ER to function as a nuclear transcription factor is dependent upon its association with an appropriate ligand, to bind to an ERE and facilitate assembly of a functional transcription complex. The first step in this process is the binding of an appropriate ligand. The ER is capable of binding several structurally diverse compounds (Figure 2), including the natural estrogens (Table 2), triphenylethylenes, *e.g.*,TAM,[15] C7-substituted estrogens, *e.g.*, ICI 182,780, ICI 164,384,[16,38] resorcyclic lactone acids, *e.g.*, zearalenone (Table 2), isoflavones, *e.g.*, genistein (Table 2), and coumestrol (Table 2). For these compounds to be able to function through ER binding, they should share some structural similarities. Thus, phytoestrogenic compounds must have some apparent structural similarities to the natural estrogens.

The structural requirements for ER binding have been widely investigated.[39-43] Almost all ER ligands have at least two important components, in the form of rings, one of which probably must be aromatic.[44] The spacing between these rings,[39] and both whether and how they are functionalized, appear important in conferring the nature of the ligand-receptor interaction. All of the compounds in Figure 2 possess these two structural components in one form or another. For the natural estrogens (estrone and 17β-estradiol are shown), the A-ring (1st component) is aromatic and the D-ring (2nd component) is a cyclopentano ring, whereas the flavanones and coumestrol have two aromatic rings (Figure 2). The resorcyclic acid lactones have an aromatic ring for the first component. While the second component is not a simple ring structure, the structural similarity of the region bearing the ketone (zearalenone) or hydroxyl (zearalenol) to the other ligands is immediately apparent (Figures 2 and 3).

The role of a functional substitution to either of the rings is clear upon comparison of the natural estrogens and the resorcyclic lactone acids. In almost all cases, the critical substitution that confers the greatest affinity for ER is hydroxylation. For example, in E2 both the A and D rings are hydroxylated at positions C3 and C17 respectively. Thus, both components are probably capable of hydrogen bonding to residues on the ER. At position C3, the OH likely functions as a H-bond donor.[40] For E1, position C17 has a ketone substitution, significantly reducing the ability of the D-ring to form an analogous association with the ER. The consequence is that E2 has a higher affinity, and is a more potent estrogen, when compared with E1. A similar relationship occurs with the resorcyclic lactone acids. Conversion of the ketone to a hydroxyl confers a significant increase in affinity for ER upon zearalenol over zearalenone (Figure 2; Table 2).

It has been suggested that binding to the ER occurs in two independent but interrelated steps, receptor recognition and stabilization of the receptor-ligand complex.[40] The two structural components (substituted rings) described above appear to confer these functions upon ligands. The likely interactions between receptor and ligand are schematically represented in Figure 3. Receptor recognition probably occurs first, representing the initial interaction between receptor and ligand. For the estrogens, this function appears to be conferred by the hydroxyl at position C3, which results from the aromatization of the A-ring in testosterone and/or androstenedione. It seems most likely that the hydroxyl is capable of hydrogen bonding with the side chain of an amino acid in the binding domain of the ER.[40,45] Sulfation at this position, produced by the steroid sulfotransferase enzyme, effectively inactivates all three of the natural estrogens. The presence of a ketone at position C3, *e.g.*, in the androgens and progestins, is not sufficient to facilitate significant binding to ER.

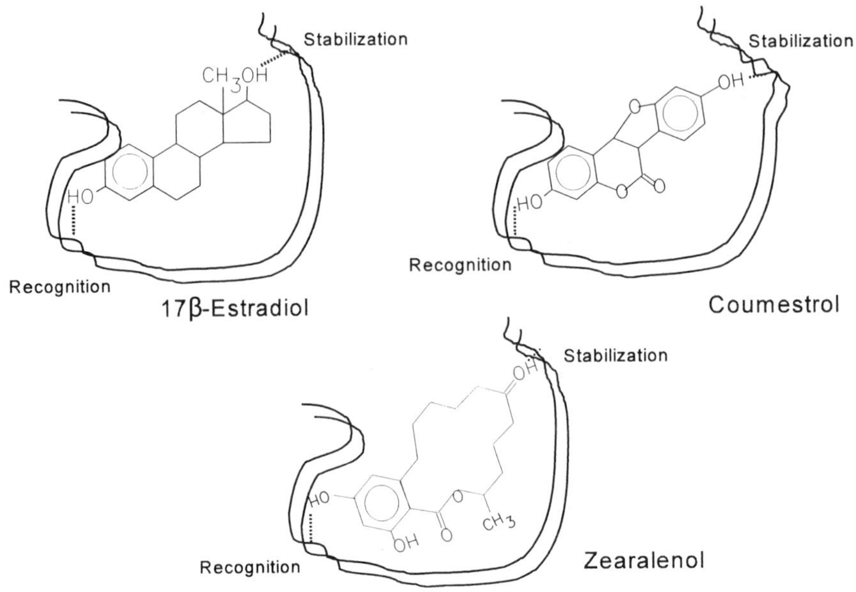

Figure 3. Schematic representation of the putative interactions between 17β-estradiol, coumestrol and zearalenol and the estrogen receptor binding site.

The stability of the ligand-receptor interaction is conferred by the second structural component.[40] Again, hydroxylation appears to be the more potent substitution. For both estrone and zearalenol, a ketone is sufficient to produce a stable and active receptor complex (Figure 3), but both the affinity of the ER interaction and the biological potency of the complex are significantly reduced (Table 2; Figure 3).

Relative Potencies

In addition to the different substitutions at the second component, the molecules in Figure 2 will likely exhibit other properties that can alter the nature of their interactions with ER. In general, all of the ligands in Figure 2 will probably be relatively flexible structures.

Table 1. The major sources of estrogen during a normal woman's life

Life cycle	Origin of estrogens	Major circulating estrogen	#Potency
Fetus	Feto-placental unit	All estrogens†	Low-High
Puberty⇨Menopause	Ovary and Placenta**	17β-Estradiol	High
Postmenopause	Adipose Tissues	Estrone	Low
Throughout Life	Other extragonadal synthesis*	17β-Estradiol	High
Throughout Life	Environment	Various, mostly non-steroidal	Mostly Low
Throughout Life	Diet	Various, mostly non-steroidal	Mostly Low

#Relative to 17β-estradiol.

*This source is likely to be most important in terms of local production; it is unclear that significant quantities of extragonadal estrogens are secreted in an endocrine manner, e.g., from brain.

**The contribution from the placenta occurs during any normal pregnancies.

†The major surge of estriol levels occurs during pregnancy {{2446,2388}}. 17β-estradiol is approximately 10-fold more potent than estrone and 80-fold more potent than estriol.[115]

Table 2. Estimates of dietary intake of some phytoestrogenic compounds

Compound	Urinary excretion nmol/day
Natural Estrogens[#]	~200
Lignans[#]	1,000-3,000
Isoflavones[#]	0-300
Compound	Dietary intake mg/day
All flavonoids**	1,004
Zearalenone*	1-5

[#]Adapted from reference [138]
*adapted from reference [31]
**adapted from reference [139]

However, it seems more likely that the second component, *e.g.,* in the flavanones, will be less rigidly constrained than in the other ligands due to the single C-C bond at positions 2⇨1' (Figure 2). The relative distances between positions C3 and C17 in the natural estrogens may be slightly different from the comparable ER binding components in the other ligands. These potential differences in ligand structure also may contribute to the apparent affinities of each ligand for ER, and generate subtle differences in the conformation of the occupied receptor. The consequence of this could also explain the relative biological potencies of each ligand, which can appear independent of receptor affinity (Table 2).

The biological importance of the relative affinities and relative potencies differs. A compound with a relatively high affinity and a relatively high agonist potency, *e.g.,* zearalenol (Table 2) could effectively compete with the endogenous natural estrogens for binding and activation of ER when present in sufficient concentrations. In postmenopausal women, where the predominant estrogen is the lower potency estrone, exposure to a high concentration of zearalenol could result in a significant proportion of ER being occupied and activated by a potent agonist. A hormone-dependent breast tumor would be more effectively driven in this environment by the estrogenic stimuli from zearalenol than by estrone. This would result in more rapid proliferation of the tumor, and the patient likely having a worse prognosis.

We can see the effect of biological potency by comparing zearalenone with genistein. From Table 2, genistein's apparent ability to regulate gene expression through an ERE is ~50-fold less than zearalenone but its affinity is only 2-3 fold less. Thus, the extent of receptor occupancy by genistein would be broadly comparable to zearalenone, but the ability to induce an agonist response would be significantly less. A hormone-dependent breast tumor would likely grow more rapidly in the presence of a comparable exposure to zearalenone than with genistein. If we were to compare the growth of breast cancer cells exposed to a combination of $1x$ zearalenone $+1x$ genistein with $2x$ zearalenone alone, the genistein would appear antiestrogenic. This is because the receptors now occupied by genistein are less potent than when occupied by zearalenone, and the combination would exhibit a lower potency than zearalenone alone.

Bioavailability

In addition to the effects of the structure/activity relationships, the biologic activity of the ER ligands is significantly influenced by bioavailability. For the natural estrogens, bioavailability is directly regulated by metabolism and binding to serum proteins. The effects of metabolism are primarily mediated through the relative activities of the sulfotransferase (inactivates) and sulfatase (activates) enzymes. These enzymes respectively sulfate and desulfate estrogens at position C3. The major serum protein to which estrogens bind is sex hormone binding globulin (SHBG). The data in Table 2 clearly indicate that the bioavailability of the phytoestrogenic compounds, at least in terms of their sequestration by SHBG, is significantly less than for the natural estrogens. Some phytoestrogenic compounds can be substrates for steroid metabolizing enzymes. For example, zearalenone is a substrate for several steroid metabolizing enzymes,[46] including estrogen hydroxylase.[47]

Bioavailability also is affected by other pharmacokinetic parameters. The longer the serum/tissue half-life, the greater the availability for binding ER. Detailed pharmacokinetics are not available for many phytoestrogenic compounds. However, it is apparent that some compounds may have complex pharmacokinetics, and exhibit longer half-lives, than previously anticipated (see Barnes *et al.*, this volume). It is clear that, until we fully understand the bioavailability of phytoestrogenic compounds, assessing the true estrogenic exposure arising from dietary intake may be difficult.

Agonists, Partial Agonists, and Antagonists

In the most simple terms, antagonists negatively regulate the activity of agonists, whereas partial agonists exhibit properties of both agonists and antagonists. The classical agonist for ER is 17β estradiol. The triphenylethylene "antiestrogens", *e.g.,* tamoxifen (TAM) are actually partial agonists. At low concentrations and/or in the absence of other estrogens, TAM functions as an agonist. The nature of the effect induced by TAM is both tissue and species specific (reviewed in ref. 15). In endocrine responsive breast tumors, TAM generally appears to function primarily as an antagonist.[15]

Some weak agonists can appear to function as antagonists by occupying receptor in the presence of more potent ligands. While these ligands alone can produce an estrogenic response, they do so with a lower potency. When significant numbers of receptors are occupied with low potency ligands, in the presence of lower concentrations of higher potency ligands, they will appear to be antagonizing the effects of the higher potency compounds. However, they will still be functioning as agonists in that they are still inducing estrogenic effects. This may be one way in which an ER ligand can produce the effects of a partial agonist. Several factors will affect the ability of a lower affinity ligand to occupy receptor in the presence of higher affinity ligands, including the relative concentrations of each ligand and the magnitude of the difference in the respective affinities for the receptor. The kinetics for ligand/receptor association-disassociation also can influence occupancy, since ligands with a slower "off-rate" will remain on the receptor for a longer time than ligands with a rapid "off-rate".[48,49]

The nature of the response induced by an ER ligand is likely dependent upon the structural conformation of the protein induced by the receptor-ligand interaction. For example, it has been shown that agonists, partial agonists and antagonists each induce different conformations in the ER.[50] Since the ER protein is a nuclear transcription factor, it seems likely that these different conformations are partly responsible for altering the function of the transcription complex assembled around the estrogen response element/promoter of the target genes. We have previously reviewed, in detail, the mechanisms by which antiestrogens can regulate gene expression.[15,51,52]

THE NATURE AND DIVERSITY OF ESTROGENIC RESPONSES

What Is an Estrogenic Response?

In principle, an estrogenic response is any specific biological response that occurs following administration of an estrogen. However, we have chosen to limit our working definition of an estrogen to those transcriptional events mediated by the transcription-activating functions of an ER acting at an ERE, and this also restricts our definition of an estrogenic response. Responses to estrogenic stimuli are remarkably diverse, perhaps more so than for many other hormones. Many of the effects induced by estrogens require the participation of other steroids or hormones, with the magnitude or nature of the ultimate response reflecting the relative ratio of a complex endocrine system. For example, progesterone is required to complete or appropriately modify an estrogenic response in several tissues. Androgens also can function as a modifying component in regulating responses to estrogens. Thus, defining any effect as being "estrogenic" may imply/include the modifying effects of other known or unknown factors.

While often considered "female hormones" estrogens regulate many functions that are not specifically associated with sexual differentiation and that occur in both males and females such as closure of the epiphyses of bones during puberty.[53] The effects upon tissues and organs that are specific for either sex, *e.g.*, ovaries and endometrium in females, are not considered sexual differentiation effects.

The Diversity of Estrogenic Responses

The brain is a major target organ for estrogenic stimuli, and it is the effects of estrogens in the brain that probably represent their major effect upon sexual differentiation in humans. ER and progesterone receptors (PGR) are present in many of the structures of the brain, including the hypothalamic and limbic nuclei and the pituitary.[54-57] Estrogens regulate diverse and complex behaviors, including sexual,[58] aggressive,[59] depressive behaviors,[60] and can influence food consumption.[61] The synthesis, secretion, degradation and uptake of several neurotransmitters can be regulated by estrogens, *e.g.*, serotonin,[62] and aromatization to estrogen is detected in the medial preoptic nucleus, the bed nucleus of the stria terminalis, and the tuberal hypothalamus.[63] The regulation of estrogen and progesterone secretion from the ovaries through the release of the gonadotropins and gonadotropin releasing hormone are regulated by estrogen. Both meningiomas[64,65] and gliomas[66] express ER, with some gliomas exhibiting a clinical response to TAM.[67]

The ovaries and endometrium are major targets for estrogenic stimuli. Normal and malignant cells of both ovarian[68-70] and endometrial tissues[71-73] express ER and PGR. Estrogenic stimuli are required for the development and proliferation of normal endometrial epithelium. Estrogen alone is generally sufficient to initiate puberty, although cyclic changes in both estrogens and progesterone are required to successfully complete puberty.

Estrogens regulate laryngeal development during puberty and several components of immunity. For example, estrogens inhibit T-suppressor cell function and T-helper maturation.[74,75] Pokeweed mitogen-induced Ig synthesis of B-lymphocytes is stimulated by estrogen.[74,76] Estrogen produces a biphasic effect upon natural killer cell activity,[77-80] with long term/high dose treatments eventually inhibiting NK cells. In contrast, TAM stimulates natural killer cell function.[81] Normal and neoplastic hepatocellular tissues express ER,[82-84] as does colon epithelium.[85-87] There is a high expression of ER in the pancreas[88,89] but pancreatic tumors do not appear to consistently respond to TAM.[89,90]

What Is a Phytoestrogenic Response?

Our working definition of a phytoestrogen leads us to treat "estrogenic" and "phytoestrogenic" as being essentially similar but not identical. There are three specific caveats that bear reemphasis in this regard. Firstly, our definition of estrogenic excludes antiestrogenic effects. Thus, it will ultimately be necessary to determine which of these activities (agonist *vs* antagonist) is associated with the exposure to each phytoestrogen. From the perspective of breast cancer, compounds that are potent agonists will likely be "bad" phytoestrogens, while partial agonists/antagonists may prove to be "good" phytoestrogens. Secondly, compounds described as phytoestrogens also have effects that are clearly not mediated through an ER pathway, *e.g.*, the ability of genistein to inhibit some tyrosine kinase[20] and topoisomerase II activities.[21] These effects are not considered phytoestrogenic responses, *i.e.*, every biological pathway/system perturbed by a phytoestrogen does not constitute a phytoestrogenic response. Finally, we have included only genomic effects and cannot exclude the possibility that estrogens and phytoestrogens will produce some similar non-genomic effects.[13,14]

We have described the diversity of tissues and responses to estrogenic stimuli. Some occur throughout life, others at specific times in the human life cycle. It seems likely that, if estrogenic exposure from phytoestrogenic compounds is biologically relevant, the effects of a consumption of a diet high in phytoestrogens also will influence these other endpoints. Many of these effects will have an important impact upon breast cancer. For example, the effects on depressive behavior,[91] immunity,[77-79] and breast cancer cell growth[4,92] could each impact disease progression.

ESTROGENIC STIMULI IN THE NORMAL BREAST

Estrogens and the Normal Breast

In mice and rats, where pregnancy lasts for 21 days, mammary buds can first be seen on either day 10 or 11 of fetal life.[93] The mammary epithelial cells begin to proliferate around fetal day 16 to form the primitive mammary epithelial tree. However, the growth is relatively slow. Only by the 3rd postnatal week do mammary epithelial cells begin to proliferate from the nipple region, forming the epithelial ducts that will eventually fill the mammary fat pad (Figure 4). At 5 weeks of age, the tips of branching epithelial ducts contain many terminal end buds (TEBs) and they approach the lymph node located in the middle of the mammary fat pad (see Fig 4). TEBs are the primary site of ductal and myoepithelial proliferation.[94] At the age of 8 weeks, epithelial structures have filled the fat pad. At 16 weeks, most TEBs have differentiated into end buds (EBs) or LAUs.[95]

Estrogen and Mammary Gland Development

Multiple factors regulate the growth of mammary epithelial cells.[96-99] Evidence that estrogens play an important role during early life comes from studies showing that environmentally-derived hormones affect breast development and mammary tumor growth. Exogenous estrogens accelerate differentiation of the nipple and cause extensive proliferation of the surrounding mesenchyma in the 16-day-old female mouse fetus.[100]

Postnatal development of the mammary gland is regulated by hormones originating from pituitary, ovaries, and adrenals,[101] and estrogen is clearly important. Mice treated with 2-75 μg E2 during the first five postnatal days of life have increased branching of the mammary epithelium at 33-42 days of age,[102-104] and at the age of 12-24 months exhibit dilated ducts and abnormal lobuloalveolar development.[105] We found that female mice treated with doses as low as 2-4 μg E2 during postnatal days one to three, have increased numbers of TEBs but decreased numbers of LAUs at 4-16 weeks of age.[104] Thus, early treatment with E2 first stimulates epithelial mammary cell growth, but at the age when these cells begin to differentiate, E2-treated animals show reduced epithelial differentiation. E2 also increases the size of the total mammary fat pad area, suggesting that the stroma and/or basement membrane also are affected by the early hormonal treatment. Since the TEB is the site for transformation in the rodent mammary gland,[95] these effects would increase the susceptibility of the mammary gland to subsequent carcinogenesis.

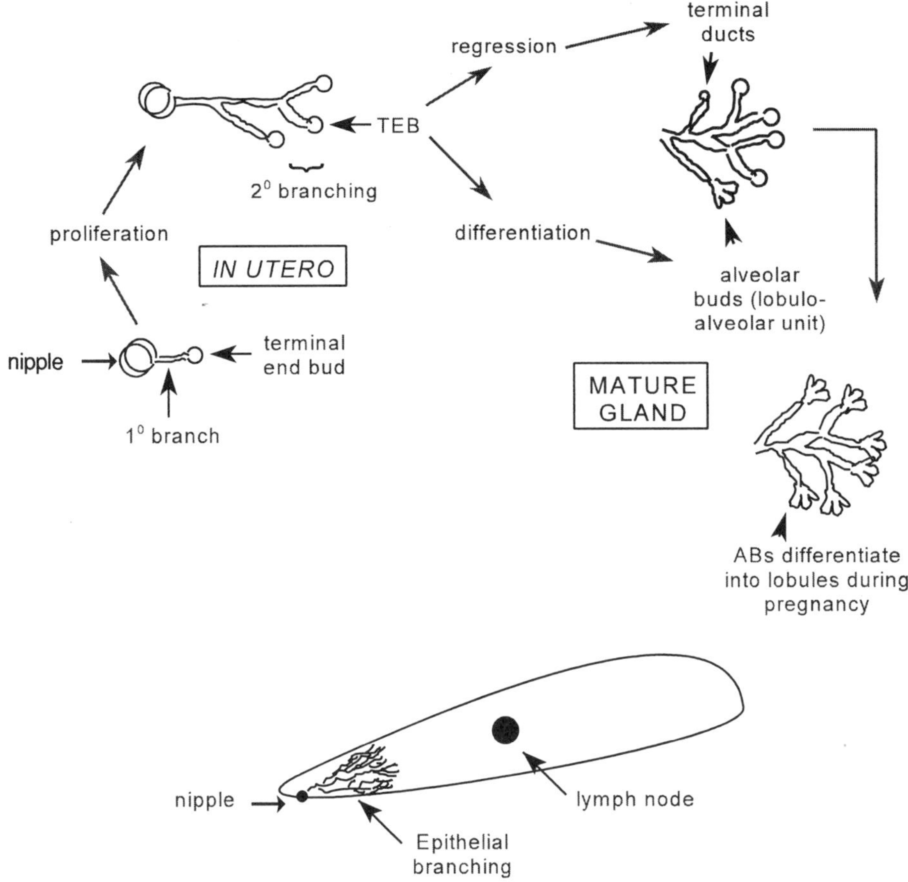

Figure 4. Representation of the development of the rodent mammary gland. The lower panel represents a "low-power" view of the rodent mammary fat pad.

Estrogenic Exposures throughout Life

There are various sources of estrogenic exposures that vary throughout life (Table 3). These exposures also vary with respect to the potency of the estrogen, the duration of exposure, and the susceptibility of the target tissues. The nature, duration and potency of the estrogenic response will, at least in part, reflect the sum of the interactions among these various parameters. The ability to elicit an estrogenic response in a particular tissue, and perhaps also both the nature and magnitude of this response, will be highly time-dependent. The time dependency will be both durational, *i.e.,* how long the tissue is exposed, and developmental, *i.e.,* the point in the life cycle when exposure occurs.

Sources of Estrogens

There are several major sources of estrogenic compounds (Table 1). Some of these provide potential estrogenic stimuli throughout life, *e.g.,* dietary and environmental estrogens. The likely importance of each source will reflect the potency of an estrogen relative to the estrogenic stimuli from other sources. For example, estrogens from adipose tissues

Table 3. Affinity for ER and SHBG, and relative potency of several phytoestrogenic compounds. CAT induction is a marker of biological potency determined by the ability of a compound to regulate expression of a marker gene linked to a promoter construct under the influence of an ERE

Compound	Relative Affinity[#]	[**]Kd	SHBG[*]	CAT induction[†]
17β-Estradiol	100.0%	0.2 nM	100%	100%
Zearalenol	25.2%	1.0 nM	21%	ND
Coumestrol	4.9%	2.0 nM	14%	0.06%
Zearalenone	4.0%	6.0 nM	5%	2.5%
Genistein	1.3%	10.0 nM	27%	0.04%

[#]Adapted from reference [140]
[*]adapted from reference [25]
[**]adapted from reference [25]
[†]adapted from reference [141]

probably provide a smaller component of the total estrogenic exposure in post-menarcheal/premenopausal women with functional ovaries when compared with post-menopausal women. Body fat may account for as much as one-third of the circulating estrogens in premenopausal women. [106] However, it should be noted that even in pre-menopausal women, the contribution from the ovaries varies with time throughout each menses (Figure 5). Thus, those times during the follicular and luteal phases when the ovarian estrogen levels are relatively low, provide an endocrinologic window when estrogens from other sources may be able to regulate estrogenic responses.

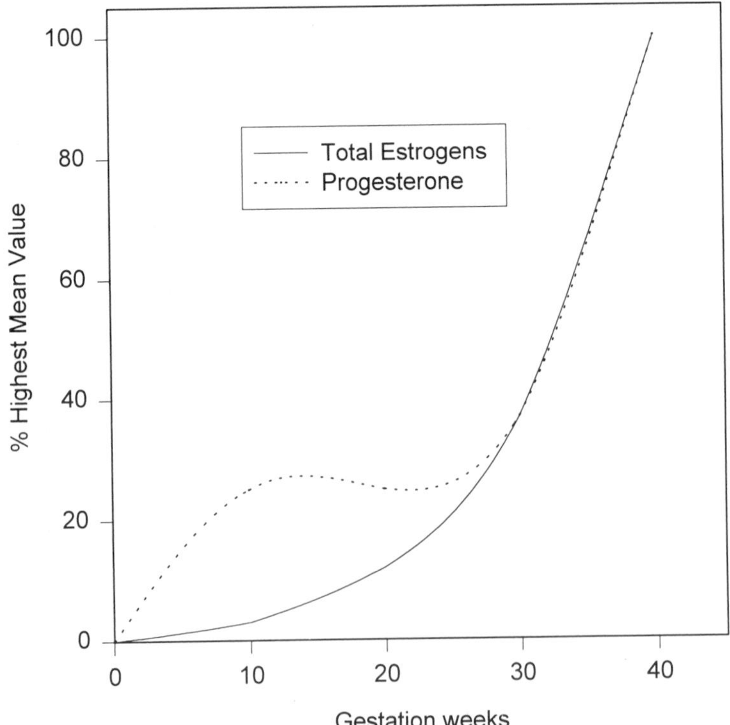

Figure 5. Schematic representation of the endocrine changes that occur during a normal pregnancy. Note that the levels of estrogens rise throughout pregnancy, reaching their peak around birth. Adapted from: Rhoades, R.A. & Tanner, G.A. "Medical Physiology" Little, Brown and Company, Boston, pp785, 1995.

While the primary sources of estrogens for most of a woman's life are the ovaries, this is not the only source for some tissues. The brain appears to make estrogen throughout much of life. Aromatase activity, the enzyme that catalyzes the final step in the biosynthesis of estrogen (testosterone/androstenedione ⇨ estradiol/estrone), is widely distributed in the brain.[63] Extra gonadal sources of estrogens are likely to be most important in postmenopausal women. There is an association between body weight and serum estrogens.[107] In postmenopausal women adipose-derived estrogens are of primary importance. The precise role of dietary and environmental estrogens is unclear.

ESTROGENIC STIMULI IN THE MALIGNANT BREAST

Early Life and Breast Cancer

It is becoming increasingly evident that factors operating during early life play an important role in affecting subsequent risk to develop breast cancer.[108-110] This is not surprising considering the fact that critical events of mammary gland development occur *in utero*. An early estrogen exposure also affects mammary tumorigenesis. C3H mice carrying mouse mammary tumor virus that were treated with either E2 (up to 100 μg daily) or DES during early postnatal life (days 1-5) exhibited an increased incidence of mammary tumors.[111,112] We have shown that female rats exposed to 20 ng E2 *in utero,* through maternal injections, exhibit significantly elevated incidence of carcinogen-induced mammary tumors.[113] Since differentiated epithelial cells are thought to be protected from malignant transformation, early E2 exposure may increase mammary tumorigenesis by preventing mammary cell differentiation.

There is clinical evidence to suggest that fetal life may influence breast cancer risk. The daughters of women who suffered from pre-eclampsia/eclampsia during pregnancy have a significantly lowered risk of developing breast cancer.[109] Since pre-eclampsia/eclampsia is associated with low circulating estrogen, the levels of this hormone in pregnant women may be the mechanism by which early life affects breast cancer risk among the daughters. In accordance with the hypothesis that estrogen is the critical factor *in utero*, daughters whose mothers took DES during pregnancy to prevent miscarriage may have an elevated breast cancer risk.[114]

Estrogen concentrations are high in pregnant women. Total estrogens increase by 100% from the beginning of pregnancy until the end(Figure 5).[115] Estriol, which is a relative weak estrogen, is primarly produced during pregnancy[115] (Table 1). Estrone levels also increase during pregnancy. The low potency component of the total estrogenic exposure during pregnancy could be significantly influenced by phytoestrogenic compounds (see Figure 6).

Pregnancy estrogen concentrations often exhibit a marked, unexplained inter-individual variability.[116] An important factor that could affect maternal estrogen levels during normal pregnancy is diet. Several dietary components alter estrogen levels.[107] We have investigated whether maternal exposure to a diet high in polyunsaturated fatty acids increases the incidence of carcinogen-induced mammary tumorigenesis among female offspring.[113] Pregnant rats were kept either on a low-fat diet containing 12 or 16% calories from fat, or a high-fat diet containing 36 or 46 % of calories from fat. Offspring born to high- and low-fat mothers were kept under standard laboratory conditions from postnatal day 1 onwards, and were treated with 7,12-dimethylbenz(a)anthracene (DMBA) at the age of 55 days. The incidence of DMBA-induced mammary tumors in the female rats exposed to a high-fat feed *in utero* was significantly increased. These results suggest that maternal exposure to high

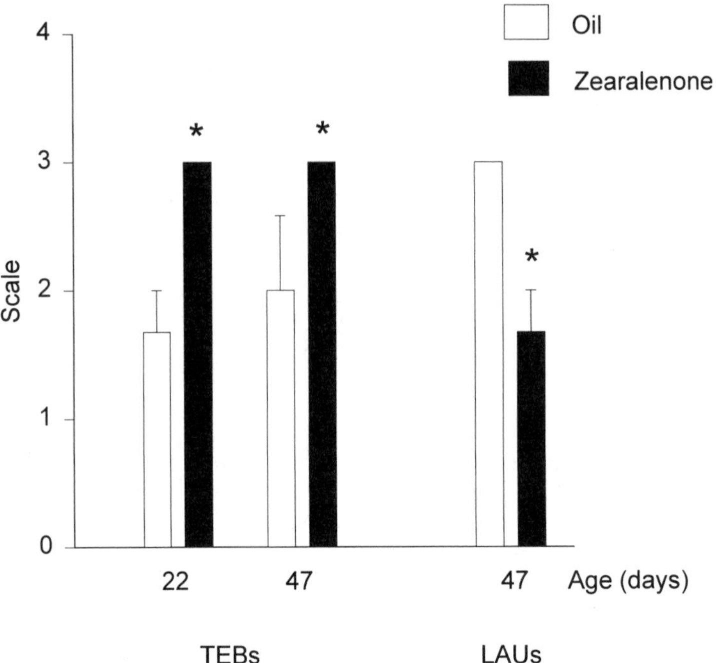

Figure 6. Effect of zearalenone on rat mammary gland development. The number of terminal end buds (TEB), and the degree of differentiation of TEBs to lobulo-alveolar units (LAU) of the 4th abdominal gland in 22- and 47-day old female rats exposed to oil-vehicle or 1 μg zearalenone *in utero* during gestation days 14-20. These parameters were evaluated using a scoring system with scores ranging from 0 (absent) to 3 (high). Means ± SEM obtained from 3 animals per group. *Significantly different from controls $p<0.05$ (Fishers Least Significant test).

dietary fat during pregnancy induces a permanent increase in the events leading to neoplastic transformation of the mammary gland by DMBA among the female offspring.

If these data have implications for development of breast cancer in humans, new dietary guidelines for pregnant women may have to be formulated. The fact that food manufacturing and dietary habits underwent radical changes at the time when mothers of those women diagnosed with breast cancer today were pregnant, makes it very important to determine the role and mechanisms of maternal diet on daughter's breast cancer risk.

Early Phytoestrogen Exposure and Mammary Tumors

In the light of the reported connection among perinatal estrogens, diet and breast cancer risk described above, it is likely that an early exposure to diet-derived phytoestrogens influences breast cancer risk. There already is evidence that neonatal exposure to phytoestrogens affects tumorigenesis. In female mice, treatment with zearalenone during the neonatal period causes ovarian dysfunction and preneoplastic and/or neoplastic changes in the cervicovaginal epithelium.[117] Schoental[30] has investigated the effects of neonatal treatment (postnatal days 7 and 14) with 10 mg/kg zearalenone on spontaneous tumorigenesis in Wistar rats. Her results indicate that early zearalenone exposure induces an increased incidence of mammary tumors.[30] In contrast, neonatal treatment with genistein lengthens the latency to the appearance of DMBA-induced mammary tumors and reduces tumor incidence.[118]

We have studied the effect of *in utero* exposure to zearalenone on mammary gland development in female rats. Pregnant Sprague-Dawley rats were injected with 1 μg zearalenone or oil-vehicle during gestation days 14 and 20. Zearalenone did not have any adverse effects on pregnancy, litter-size or body weight gain in the offspring.

We used the methods of Russo[119] and Engelman *et al.*[120] to determine proliferation and differentiation of mammary epithelial structures in the wholemounts of mammary glands obtained from female rats exposed *in utero* to zearalenone. Wholemounts of the 4th abdominal glands of 22-, and 47-day-old female rats were taken and were stained with carmine aluminum.[121]

The stained mammary wholemounts were examined under an Olympus dissecting scope. The total mammary gland and epithelial cell area were measured using calipers. In addition, the following characteristics of the mammary glands were evaluated using a 4-point scale (0 = Absent, 1 = low, 2 = medium, 3 = high): (i) density of epithelial ducts; (ii) number of TEBs; and (iii) degree of differentiation of TEBs to EBs or LAUs. Detailed characterization of these various epithelial structures in rats is provided by Russo and Russo.[95]

We found a significant interaction between age and treatment with respect to the epithelial cell area. This area was larger in 22-day-old female rats exposed to zearalenone *in utero* than in their controls, but by 47 days, the epithelial cell area was larger in the controls. Similar data have been obtained after early postnatal treatment with E2 in our previous study.[104]

Glandular structures were also markedly altered by early zearalenone exposure (Figure 6). Offspring of mothers treated with zearalenone during pregnancy had more TEBs and their differentiation to LAUs was reduced, when compared with vehicle-treated controls. These observations closely reflect those obtained in female mice treated with E2 during early life [104] and in female rats exposed to E2 or high-fat diet *in utero.*[113] Thus, zearalenone appears to act as an estrogen in influencing the morphology of the developing mammary gland. Whether zearalenone increases carcinogen-induced mammary tumorigenesis as well, remains to be determined.

MECHANISMS OF ACTION OF PHYTOESTROGENS IN BREAST CANCER

It is unclear why some phytoestrogens (*e.g.*, genistein) are antitumorigenic and some (*e.g.*, zearalenone) stimulate tumor cell growth. Several possible explanations are available. Genistein may modulate the effects of estrogen by binding to the ER Experimental evidence to support this assumption is provided by the observation that genistein inhibits the uterovaginal action of estradiol.[122] However, effects mediated through perturbations in tyrosine kinase/topoisomerase II activities[20,21] cannot readily be excluded, and there is no direct evidence indicating that genistein acts as an antiestrogen in the mammary gland. Zearalenone appears to act like an estrogen: it increases the growth of uterine cells,[25,30,123] and also induces the expression of estrogen-regulated genes in a manner similar to E2.[124]

The effects of genistein and zearalenone on tumorigenesis may be due to various other mechanisms besides binding to ER. Phytoestrogens have different effects on SHBG expression. Since estrogens bind to SHBG, increased levels of SHBG reduce the metabolic clearance rate for estradiol and uptake of sex hormones. The converse is true for low SHBG levels. Isoflavonic compounds stimulate synthesis of SHBG (for review, see ref. 22). Therefore, genistein may inhibit tumorigenesis by increasing the levels of SHBG and by reducing the availability of estrogens. Zearalenone has a very low affinity to SHBG (5% vs.

100% for E2),[25] and consequently it may exhibit elevated free (unbound) serum concentrations, increasing its ability to activate estrogenic pathways that lead to malignant growth.

It is also possible that isoflavonoids and mycotoxins have different effects on estrogen metabolism. Isoflavonoids increase the hydroxylation of estrogen to 2-OH metabolites,[22] increasing the formation of catechol estrogens that act as antiestrogens.[125] It has also been reported that low SHBG levels result in increased 16-hydroxylation of estrogens.[126,127]

Genistein inhibits tyrosine-protein kinase[20] and topoisomerase II activities[21] and arrests cell cycle progression at G_2-M.[128] All these effects of genistein may be responsible for inhibition of tumor cell growth. Our preliminary data show that zearalenone increases both MCF-7 proliferation and the proportion of cells in S and G_2/M.[129] Thus, genistein reduces and zearalenone increases cell mitosis. Recent observations further support the antitumorigenic action of genistein. This isoflavone inhibits *in vitro* angiogenesis [130]. Isoflavones also inhibit the human estrogen synthetase (aromatase) that catalyzes the conversion of androgens to estrogens,[131] and reduce the expression of EGFR mRNA.[132] Importantly, clinical studies have revealed that relapse-free survival and total survival are significantly shorter for breast tumors that are positive for epidermal growth factor receptor (EGFR).[133] Further, high levels of EGFR correlate with a failure of endocrine therapy.[134-137] However, the effects of zearalenone on angiogenesis, aromatization of androgens, or growth factors have not been studied.

CONCLUSIONS AND FUTURE PROSPECTS

The precise role of phytoestrogenic compounds in affecting breast cancer risk is likely to be highly complex. It seems unlikely that women are exposed to a single phytoestrogenic compound that is either good or bad. Our diets are complex and generally varied, and the overall exposure will reflect this complexity. Thus, exposure to phytoestrogenic compounds is probably continuous, with the balance of "good" or "bad" exposure constantly changing with the amount and nature of what is consumed, and the time at which it is consumed.

The consumption of a diet that produced an overall agonist/estrogenic exposure in a postmenopausal breast cancer patient, could be considered "bad", with the effects apparent in the short term. A similar exposure in a normal postmenopausal woman might be essentially"good", reducing both her risk of cardiovascular disease and the effects of osteoporosis, *e.g.*, zearalenone has been administered to women as an estrogen therapy.[33,34] In a younger woman this exposure may have few immediate effects, although it could function as a contraceptive if the exposure was sufficient.[32] However, a prolonged exposure could produce an increased breast cancer risk.

It also is possible that a significant exposure to phytoestrogenic stimuli could influence the response of a postmenopausal breast cancer patient to TAM. The affinity of TAM and its major metabolite N-desmethyltamoxifen for ER are low relative to that of 17β-estradiol. Several of the phytoestrogenic compounds could conceivably compete with these compounds and influence the antiestrogenic effects of the therapy.

A prolonged overall antiestrogenic exposure might be "good" for a postmenopausal breast cancer patient, whereas it may increase risk of cardiovascular disease and/or osteoporosis in a normal postmenopausal woman. Both antiestrogenic and agonist/estrogenic responses may have significant effects upon the reproductive capacity of premenopausal women. The triphenylethylene antiestrogens were initially used as fertility agents,[15] and the estrogenic compounds like zearalenone have been known to induce significant reproductive toxicity.[36,117]

The extent to which any of these hypothetical biological interactions could occur will depend upon the duration of exposure, the affinity and potency of the stimulus relative to any endogenous estrogens, and the sensitivity of the target organ/tissue. Since it seems likely that these will vary from one individual to another, determining the role/contribution of an individual's lifetime phytoestrogenic exposure to their breast cancer risk may be very difficult. Our understanding of these compounds remains relatively empirical, and will remain so until we have a better understanding of the true extent and likely potency of exposures. Thus, there are several issues that we need to better understand, including (a) which compounds are "good" or "bad", (b) under what circumstances are compounds "good" or "bad", (c) the complexity of dietary exposure, *e.g.,* how "good" *versus* how "bad" are the dietary phytoestrogenic stimuli, and (d) the pharmacokinetics of phytoestrogenic compounds.

ACKNOWLEDGMENTS

This work was supported in part by grants R01-CA58022, P30-CA51008 (R. Clarke), F31GM16433 (R. Clarke/M.R. James) and P50-CA58185 (L.A. Hilakivi-Clarke) from the Public Health Service, CN-80420 from the American Cancer Society (L.A. Hilakivi-Clarke), 90-BW65 from the American Institute for Cancer Research (R. Clarke), and from the Cancer Research Foundation of America (two awards - R. Clarke and L.A. Hilakivi-Clarke).

REFERENCES

1. G.T. Beatson, On the treatment of inoperable cases of carcinoma of the mamma: suggestions from a new method of treatment, with illustrative cases, *Lancet* ii:104-107 (1896).
2. J.R. Daling, K.E. Malone, L.F. Voigt, E. White, and N.S. Weiss, Risk of breast cancer in young women: relationship to induced abortion, *J Natl Cancer Inst* 86:1584-1592 (1994).
3. B.E. Henderson, R. Ross, and L. Bernstein, Estrogens as a cause of human cancer, *Cancer Res* 48:246-253 (1988).
4. R. Clarke, N. Brünner, B.S. Katzenellenbogen, E.W. Thompson, M.J. Norman, C. Koppi, S. Paik, M.E. Lippman, and R.B. Dickson, Progression from hormone dependent to hormone independent growth in MCF-7 human breast cancer cells, *Proc Natl Acad Sci USA* 86:3649-3653 (1989).
5. K. Seibert, S.M. Shafie, T.J. Triche, J.J. Whang-Peng, S.J. O'Brien, J.H. Toney, K.K. Huff, and M.E. Lippman, Clonal variation of MCF-7 breast cancer cells *in vitro* and in athymic nude mice, *Cancer Res* 43:2223-2239 (1983).
6. R. Clarke, R.B. Dickson, and M.E. Lippman, Hormonal aspects of breast cancer: growth factors, drugs and stromal interactions, *Crit Rev Oncol Hematol* 12:1-23 (1992).
7. D.J.P. Ferguson and T.J. Anderson, Morphological evaluation of cell turnover in relation to the menstrual cycle in the nesting human breast, *Br J Cancer* 44:177-181 (1981).
8. M. Metzler, Biochemical toxicology of diethylstilbestrol, in: "Reviews in Biochemical Toxicology", E. Hodgson et al., Elsevier, NY, pp. 191-220 (1984).
9. T. Tsutsui, H. Maizumi, J.A. McLachlan, and J.C. Barrett, Aneuploidy induction and cell transformation by diethylstilbestrol: a possible chromosomal mechanism in carcinogenesis, *Cancer Res* 43:3814-3821 (1983).
10. J. Fishman, Aromatic hydroxylation of estrogens, *Ann Rev Physiol* 45:61-72 (1983).
11. M. Metzler, Metabolic activation of xenobiotic stilbene estrogens, *Fed Proc* 46:1855-1857 (1987).
12. P. Garcia-Morales, M. Saceda, N. Kenney, N. Kim, D.S. Salomon, M.M. Gottardis, H.B. Solomon, P.F. Sholler, C.V. Jordan, and M.B. Martin, Effect of cadmium on estrogen receptor levels and estrogen-induced responses in human breast cancer cells, *J Biol Chem* 269:16896-16901 (1994).
13. D. Duval, S. Durant, and F. Homo-Delarche, Non-genomic effects of steroids. Interactions of steroid molecules with membrane structures and functions, *Biochim Biophys Acta* 737:409-442 (1983).

14. M. Farhat, S. Abi-Younes, R. Vargas, R.M. Wolfe, R. Clarke, and P.W. Ramwell, Vascular non-genomic effects of estrogen, in: "Sex Steroids and the Cardiovascular System", P.W. Ramwell et al., Springer-Verlag, Berlin, pp. 145-159 (1992).
15. R. Clarke and M.E. Lippman, Antiestrogens resistance: mechanisms and reversal, in: "Drug Resistance in Oncology", B.A. Teicher., ed., Marcel Dekker, Inc, New York, pp. 501-536 (1992).
16. A.E. Wakeling, M. Dukes, and J. Bowler, A potent specific pure antiestrogen with clinical potential, *Cancer Res* 51:3867-3873 (1991).
17. H.-Y.P. Lam, Tamoxifen is a calmodulin antagonist in the activation of cAMP phosphodiesterase, *Biochem Biophys Res Comm* 118:27-32 (1984).
18. C.A. O'Brian, R.M. Liskamp, D.H. Solomon, and I.B. Weinstein, Triphenylethylenes: a new class of protein kinase C inhibitors, *J Natl Cancer Inst* 76:1243-1246 (1986).
19. B.S. Katzenellenbogen, A.M. Miller, A. Mullick, and Y.Y. Sheen, Antiestrogen action in breast cancer cells: modulation of proliferation and protein synthesis, and interaction with estrogen receptors and additional antiestrogen binding sites, *Breast Cancer Res Treat* 5:231-243 (1985).
20. T. Akiyama, J. Ishida, S. Nakagawa, H. Ogawa, S. Watanabe, N. Itou, M. Shibata, and Y. Fukami, Genistein, a specific inhibitor of tyrosine-specific protein kinease, *J Biol Chem* 262:5592-5595 (1987).
21. A. Okura, H. Arakawa, H. Oka, T. Yoshinari, and Y. Monden, Effect of genistein on topoisomerase activity and on the growth of (val 12) Ha-ras-transformed NIH 3T3 cells, *Biochem Biophys Res Commun* 157:183-189 (1988).
22. H. Adlercreutz, Y. Mousavi, J. Clark, K. Hockerstedt, E. Hamalainen, K. Wahala, T. Makela, and T. Hase, Dietary phytoestrogens and cancer: in vitro and in vivo studies, *J Steroid Biochem Mol Biol* 41:331-337 (1992).
23. T. Hirano, K. Oka, and M. Akiba, Antiproliferative effects of synthetic and naturally occurring flavoids on tumor cells of the human breast carcinoma cell line, ZR-75-1, *Res Comm Chem Path Pharm* 64:69-78 (1989).
24. W.V. Welshons, C.S. Murphy, R. Koch, G. Calaf, and V.C. Jordan, Stimulation of breast cancer cells in vitro by environmental estrogens enterolactone and the phytoestrogen equol, *Breast Cancer Res Treat* 10:379-381 (1987).
25. P.M. Martin, K.B. Horwitz, D.S. Ryan, and W.L. McGuire, Phytoestrogen interaction with estrogen receptors in human breast cancer cells, *Endocrinology* 103:1860-1867 (1978).
26. W.M. Hagler, K. Tyczkowska, and P.B. Hamilton, Simultaneous occurrence of deoxynivalenol, zearalenone, and aflatoxin in 1982 scabby wheat from the Midwestern United States, *Appl Environ Microbiol* 47:151-154 (1984).
27. L.W. Burgess, P.E. Nelson, and T.A. Toussoun, Characterization, geographic distribution and ecology of Fusarium crookwellense sp. nov, *Trans Br Mycol Soc* 79:497-505 (1982).
28. Y. Luo, T. Yoshizawa, and T. Katayama, Comparative study on the natural occurrence of fusarium mycotoxins (trichothecenes and zearalenone) in corn and wheat from high- and low-risk areas for human esophageal cancer in China, *Appl Environ Microbiol* 56:3723-3726 (1990).
29. P. Golinski, R.F. Vesonder, D. Latus-Zietkiewicz, and J. Perkowski, Formation of fusarenone X, nivalenol, zearalenone, α-trans-zearalenol, β-trans-zearalenol, and fusarin C by Fusarium crookwellense, *Appl Environ Microbiol* 54:2147-2148 (1988).
30. R. Schoental, Trichothecenes, Zearalenone, and other carcinogenic metabolites of Fusarium and related microfungi, *Adv Cancer Res* 45:217-290 (1985).
31. T. Kuiper-Goodman, Uncertainties in the risk assessment of three mycotoxins: aflatoxin, ochratoxin, and zearalenone, *Can J Physiol Pharm* 68:1017-1024 (1990).
32. P.H. Hidy and R.S. Baldwin, Method of preventing pregnancy with lactone derivates, *U. S. Pat* June 22:3,966,274(1976).
33. W.H. Utian, Comparative trial of P-1496, a new non-steroidal oestrogen analogue, *Br Med J* 1:579-581 (1973).
34. Sandoz Ltd, Pharmacological trials of Frideron P-1496, *Sandoz Ltd* (1980).
35. A.T. Ralston, Effect of zearalanol on weaning weight of male calves, *J Anim Sci* 47:1203-1206 (1978).
36. J.P. Wiggins, H. Rothenbacher, L.L. Wilson, R.J. Martin, P.J. Wangness, and J.H. Ziegler, Growth and endocrine responses of lambs to zearanol implants: effects of preimplant growth rate and breed of sire, *J Anim Sci* 49:291-297 (1979).
37. R. Schoental, Role of podophyllotoxin in the bedding and dietary zearalenone on incidence of "spontaneous" tumors in laboratory animals, *Cancer Res* 34:2419(1974).
38. E.W. Thompson, D. Katz, T.B. Shima, A.E. Wakeling, M.E. Lippman, and R.B. Dickson, ICI 164,384: a pure antagonist of estrogen-stimulated MCF-7 cell proliferation and invasiveness, *Cancer Res* 49:6929-6934 (1989).

39. H.H. Keasling and F.W. Schueler, The relationship between estrogenic activity and chemical constitution in a group of azomethine derivatives, *J Am Pharm Assoc* 39:87-90 (1950).
40. J.P. Raynaud, T. Ojasco, M.M. Bouton, E. Bignon, M. Pons, and A. Crastes de Paulet, Structure-activity relationships of steroid estrogens, in: "Estrogens in the Environment", J.A. McLachlan., ed., Elsevier Science Publishing Company, New York, pp. 24-41 (1985).
41. W.L. Duax and J.F. Griffin, Structure-activity relationships of estrogenic chemicals, in: "Estrogens in the Environment", J.A. McLachlan., ed., Elsevier Science Publishing Company, New York, pp. 15-23 (1985).
42. J.A. Katzenellenbogen, B.S. Katzenellenbogen, T. Tatee, D.W. Robertson, and S.W. Landvatter, The chemistry of estrogens and antiestrogens: relationships between structure, receptor binding and biological activity, in: "Estrogens in the Environment", J.A. McLachlan., ed., Elsevier North Holland, New York, pp. 33-52 (1980).
43. W.L. Duax and C.M. Weeks, Molecular basis of estrogenicity: X-ray crystallographic studies, in: "Estrogens in the Environment", J.A. McLachlan., ed., Elsevier North Holland, New York, pp. 11-32 (1980).
44. J.P. Raynaud, T. Ojasoo, M.M. Bouton, and D. Philibert, Receptor binding as a tool in the development of new bioactive steroids, *Drug Design* viii:169-214 (1979).
45. G.M. Anstead and P.R. Kym, Benz(a)anthracene diols: predicted carcinogenicity and structure-estrogen reecptor binding affinity relationships, *Steroids* 60:383-394 (1995).
46. D. Thouvenot and R.F. Morfin, Interferences of zearalenone, zearalenol or estradiol-17β with the steroid metabolizing enzymes of the human prostate gland, *J Steroid Biochem* 13:1337-1345 (1980).
47. J.J. Li, S.A. Li, J.K. Klicka, and J.A. Heller, Some biological and toxicological studies of various estrogen mycotoxins and phytoestrogens, in: "Estrogens in the Environment", J.A. McLachlan., ed., Elsevier Science Publishing Company, New York, pp. 168-183 (1985).
48. B.M. Weichman and A.C. Notides, Estrogen receptor activation and the dissociation kinetics of estradiol, estriol and estrone, *Endocrinology* 106:434-439 (1980).
49. G. Stack, K. Korach, and J. Gorski, Relative mitogenic activities of various estrogens and antiestrogens, *Steroids* 54:227-243 (1989).
50. D.P. McDonnell, D.L. Clemm, T. Hermann, M.E. Goldman, and J.W. Pike, Analysis of estrogen receptor function *in vitro* reveals three distinct classes of antiestrogens, *Mol Endocrinol* 9:659-669 (1995).
51. R. Clarke and N. Brünner, Cross resistance and molecular mechanisms in antiestrogen resistance, *Endocr Related Cancer* 2:59-72 (1995).
52. R. Clarke, T. Skaar, F. Leonessa, B. Brankin, M.R. James, N. Brünner, and M.E. Lippman, The acquisition of an antiestrogen resistant phenotype in breast cancer: the role of cellular and molecular mechanisms, in: "Advances in Cancer Research", W. Hait., ed., Kluwer, Norwell, pp. in press(1996).
53. R.T. Turner, B.L. Riggs, and T.C. Spelsberg, Skeletal effects of estrogens, *Endocrine Rev* 15:275-300 (1994).
54. J.A. O'Keefe, E.B. Pedersen, A.J. Castro, and R.J. Handa, Estrogen receptors in hippocampal and neocortical transplants during development, *Society for Neuroscience abstracts* 16:(1990).
55. C.J. Coscia, W.T. Bem, G.E. Thomas, J.Y. Mamone, B.K. Levy, M. Szucs, and F.E. Johnson, Sigma and opioid receptors in human cancers, *Society for Neuroscience abstracts* 16:(1990).
56. R.E. Watson, Jr., M.C. Langub, Jr., and J.W. Landis, Further evidence that LHRH neurons are not directly estrogen responsive: LHRH and estrogen receptor immunoreactivity in the guinea pig brain, *Society for Neuroscience abstracts* 16:(1990).
57. T.L. Dellovade, J.D. Blaustein, and E.F. Rissman, Distribution of estrogen receptors in the brain of the female musk shrew: an immunocytochemical study, *Society for Neuroscience abstracts* 16:(1990).
58. D.W. Pfaff, Estrogens and Brain Function, Springer, New York, (1980).
59. N.G. Simon and R.E. Whalen, Hormonal regulation of aggression: evidence for a relationship among genotype, receptor binding, and behavioral sensitivity to androgens and estrogen, *Aggress Behav* 12:255-266 (1986).
60. L.A. Hilakivi-Clarke, P.K. Arora, M.B. Sabol, R. Clarke, R.B. Dickson, and M.E. Lippman, Alterations in behavior, steroid hormones and natural killer cell activity in male transgenic TGF-α mice, *Brain Res* 588:97-103 (1992).
61. G.N. Wade and J.M. Gray, Gonadal effects on food intake and adiposity: a metabolic hypothesis, *Physiol Behav* 22:583-593 (1979).
62. S.G. Beck, W.P. Clarke, and J. Goldfarb, Chronic estrogen effects on 5-hydroxytryptamine-mediated responses in hippocampal pyramidal cells of female rats, *Neurosci Lett* 106:181-187 (1989).
63. K. Sinchak, C.E. Roselli, and L.G. Clemens, Determination of aromatase activity and estrogen receptor levels in discrete brain regions from intact and castrated hybrid B6D2F1 male house mice that copulate and those that do not, *Society for Neuroscience abstracts* 16:(1990).

64. K.-P. Lesch and S. Gross, Estrogen receptor immunoreactivity in meningiomas, *J Neurosurg* 67:237-243 (1987).
65. P. Bouillot, J.F. Pellissier, B. Devictor, N. Graziani, N. Bianco, F. Grisoli, and D. Figarella-Branger, Quantitative imaging of estrogen and progesterone receptors, estrogen-regulated protein, and growth fraction: immunocytochemical assays in 52 meningiomas. correlation with clinical and morphological data, *J. Neurosurg.* 81:765-773 (1994).
66. K. Iwasaki, S.A. Toms, G.H. Barnett, M.L. Estes, M.K. Gupta, and B.P. Barna, Inhibitory effects of tamoxifen and tumor necrosis factor alpha on human glioblastoma cells, *Cancer Immunol. Immunother.* 40:228-234 (1995).
67. G. Baltuch, G. Shenouda, A. Langleben, and J.G. Villemure, High dose tamoxifen in the treatment of recurrent high grade glioma: a report of clinical stabilization and tumour regression, *Can. J. Neurol. Sci.* 20:168-170 (1993).
68. J.P. Geisler, M.C. Wiemann, G.A. Miller, Z. Zhou, and H.E. Geisler, Estrogen and progesterone receptors in malignant mixed mesodermal tumors of the ovary, *J. Surg. Oncol.* 59:45-47 (1995).
69. S. Li, Relationship between cellular DNA synthesis, PCNA expression and sex steroid hormone receptor status in the developing mouse ovary, uterus and oviduct, *Histochemistry* 102:405-413 (1994).
70. G. Scambia, L. Catozzi, P.B. Panici, G. Ferrandina, F. Coronetta, R. Barozzi, G. Baiocchi, L. Uccelli, A. Piffanelli, and S. Mancuso, Expression of ras oncogene p21 protein in normal and neoplastic ovarian tissues: correlation with histopathologic features and receptors for estrogen, progesterone, and epidermal growth factor, *Am. J. Obstet. Gynecol.* 168:71-78 (1993).
71. Y. Ohno, K. Hosokawa, T. Tamura, Y. Fujimoto, M. Kawashima, K. Koishi, and H. Okada, Endometrial oestrogen and progesterone receptors and their relationship to sonographic endometrial appearance, *Hum. Reprod.* 10:708-711 (1995).
72. P.C. Morris, J.R. Anderson, B. Anderson, and R.E. Buller, Steroid hormone receptor content and lymph node status in endometrial cancer, *Gynecol. Oncol.* 56:406-411 (1995).
73. I. Cohen, A. Shulman, M. Altaras, R. Tepper, M. Cordoba, and Y. Beyth, Estrogen and progesterone receptor expression of decidual endometrium in a postmenopausal woman treated with tamoxifen and megestrol acetate, *Gynecol. Obstet. Invest.* 38:127-129 (1994).
74. T. Paavonen, L.C. Andersson, and H. Adlercreutz, Sex hormone regulation of in vitro immune response. Estradiol enhances human B cell maturation via inhibition of suppressor T cells in pokeweed mitogen stimulated cultures, *J Exp Med* 154:1935-1945 (1981).
75. A.S. Ahmed, M.J. Dauphinee, and N. Talal, Effects of short term administration of sex hormones on normal and autoimmune mice, *J Immunol* 134:204-210 (1985).
76. Z.M. Sthoeger, N. Chiorazzi, and R.G. Lahita, Regulation of the immune response by sex steroids. I. In vitro effects of estradiol and testosterone on pokeweed mitogen-induced human B cell differentiation, *J Immunol* 141:91-98 (1988).
77. I. Screpanti, A. Santoni, A. Gulino, R.B. Herberman, and L. Frati, Estrogen and antiestrogen modulation of mouse natural killer activity and large granular lymphocytes, *Cell Immunol* 106:191-202 (1987).
78. W.E. Seaman, M.A. Blackman, T.D. Gindhart, J.R. Roubinian, J.M. Loeb, and N. Talal, β-Estradiol reduces natural killer cells in mice, *J Immunol* 121:2193-2198 (1978).
79. N. Hanna and M. Schneider, Enhancement of tumor metastases and suppression of natural killer cell activity by β-estradiol treatment, *J Immunol* 130:974-980 (1983).
80. W.E. Seaman and N. Talal, The effect of 17β-estradiol on natural killing in the mouse, in: "Natural Cell-Mediated Immunity Against Tumors", R.B. Herberman., ed., Academic Press, New York, pp. 765-777 (1980).
81. M.M. Gottardis, R.J. Wagner, E.C. Borden, and C.V. Jordan, Differential ability of antiestrogens to stimulate breast cancer cell (MCF-7) growth in vivo and in vitro, *Cancer Res* 49:4765-4769 (1989).
82. L. Boix, J. Bruix, A. Castells, J. Fuster, C. Bru, J. Visa, F. Rivera, and J. Rodes, Sex hormone receptors in hepatocellular carcinoma. Is there a rationale for hormonal treatment?, *J. Hepatol.* 17:187-191 (1993).
83. S. Masood, A.B. West, and K.W. Barwick, Expression of steroid hormone receptors in benign hepatic tumors. An immunocytochemical study, *Arch. Pathol. Lab. Med.* 116:1355-1359 (1992).
84. D.R. Ciocca, A.D. Jorge, O. Jorge, C. Milutin, R. Hosokawa, M. Diaz Lestren, E. Muzzio, S. Schulkin, and R. Schirbu, Estrogen receptors, progesterone receptors and heat-shock 27-kd protein in liver biopsy specimens from patients with hepatitis b virus infection, *Hepatology* 13:838-844 (1991).
85. C.W. Hendrickse, C.E. Jones, I.A. Donovan, J.P. Neoptolemos, and P.R. Baker, Oestrogen and progesterone receptors in colorectal cancer and human colonic cancer cell lines, *Br. J. Surg.* 80:636-640 (1993).
86. S. Galandiuk, S. Miseljic, A.R. Yang, M. Early, M.D. McCoy, and J.L. Wittliff, Expression of hormone receptors, cathepsin d, and her-2/neu oncoprotein in normal colon and colonic disease, *Arch. Surg.* 128:637-642 (1993).

87. F. Meggouh, P. Lointier, and S. Saez, Sex steroid and 1,25-dihydroxyvitamin d3 receptors in human colorectal adenocarcinoma and normal mucosa, *Cancer Res* 51:1227-1233 (1991).
88. G. Viale, C. Doglioni, M. Gambacorta, G. Zamboni, G. Coggi, and C. Bordi, Progesterone receptor immunoreactivity in pancreatic endocrine tumors. An immunocytochemical study of 156 neuroendocrine tumors of the pancreas, gastrointestinal and respiratory tracts, and skin, *Cancer* 70:2268-2277 (1992).
89. A. Andren-Sandberg and J. Johansson, Influence of sex hormones on pancreatic cancer, *Int. J. Pancreatol.* 7:167-176 (1990).
90. T.O. Siu and W.B. Kwan, Hormones in chemotherapy for pancreatic cancer, chemoagents or carriers?, *In Vivo.* 3:255-258 (1989).
91. L.A. Hilakivi-Clarke, J. Rowland, R. Clarke, and M.E. Lippman, Psychosocial factors in the development and progression of breast cancer, *Breast Cancer Res Treat* 29:141-160 (1993).
92. R. Clarke, N. Brünner, D. Katz, P. Glanz, R.B. Dickson, M.E. Lippman, and F. Kern, The effects of a constitutive production of TGF-α on the growth of MCF-7 human breast cancer cells in vitro and in vivo, *Mol Endocrinol* 3:372-380 (1989).
93. W. Imagawa, G.K. Bandyopadhyay, and S. Nandi, Regulation of mammary epithelial cell growth in mice and rats, *Endocrine Rev* 11:494-523 (1990).
94. S. Coleman, G.B. Silberstein, and C.W. Daniel, Ductal morphogenesis in the mouse mammary gland: evidence supporting a role for epidermal growth factor, *Dev Biology* 127:304-315 (1988).
95. J. Russo and I.H. Russo, Biological and molecular bases of mammary carcinogenesis, *Lab Invest* 57:112-137 (1987).
96. B. Heuberger, I. Fitzka, G. Wasner, and K. Kratochwil, Induction of androgen receptor formation by epithelium-mesenchyme interaction in embryonic mouse mammary gland, *Proc Natl Acad Sci USA* 79:2957-2961 (1982).
97. K. Kratochwil, Organ specifity in mesenchymal induction demonstrated in the embryonic development of the mammary gland of the mouse, *Dev Biology* 20:46-71 (1969).
98. K. Kratochwil, Development and loss of androgen responsiveness in the embryonic rudiment of the mouse mammary gland, *Dev Biology* 61:358-365 (1977).
99. H. Wasner and C.W. Turner, Ontogeny of mesenchymal androgen receptors in the embryonic mouse mammary gland, *Endocrinology* 113:1771-1780 (1983).
100. A. Raynaud, Observations sur les modifications provoquees pas les hormones oestrogenes, du mode de developpement des mamelons des foetus de souris, *C R Acad Sci* 240:674-676 (1955).
101. I.H. Russo, J. Medado, and J. Russo, Endocrine influences on the mammary gland, in: "Integument and Mammary Glands", T.C. Jones et al., Springer-Verlag, Berlin, pp. 252-266 (1994).
102. M.R. Warner, Effect of various doses of estrogen to BALB/cCrgl neonatal female mice on mammary growth and branching at 5 weeks of age, *Cell Tissue Kinet* 9:429-438 (1976).
103. T. Tamooka and H.A. Bern, Growth of mouse mammary glands after neonatal sex hormone treatment, *J Natl Cancer Inst* 69:1347-1352 (1982).
104. L. Hilakivi-Clarke, E. Cho, M. Raygada, and N. Kenney, Alterations in mammary gland development following neonatal exposure to estradiol, transforming growth factor alpha, and estrogen receptor antagonist ICI 182,780, *Endocrinology* Submitted:(1995).
105. L.A. Jones and H.A. Bern, Cervicovaginal and mammary gland abnormalities in BALB/cCrgl mice treated neonatally with progesterone and estrogen, alone or in combination, *Cancer Res* 39:2560-2567 (1979).
106. R.E. Frisch, Adipose Tissue and Reproduction, Karger, Basel, (1990).
107. H. Adlercreutz, Diet and sex hormone metabolism, in: "Nutrition, Toxicity, and Cancer", I.R. Rowland., ed., CRC Press, Boca Raton, pp. 137-195 (1991).
108. D. Trichopoulos, Hypothesis: does breast cancer originate in utero?, *Lancet* 335:939-940 (1990).
109. A. Ekbom, D. Trichopoulos, H.O. Adami, C.C. Hsieh, and S.J. Lan, Evidence of prenatal influences on breast cancer risk, *Lancet* 340:1015-1018 (1992).
110. R. Anbazhagan, B. Nathan, and B.A. Gusterson, Prenatal influences and breast cancer, *Lancet* 340:1477-1478 (1992).
111. J. Lopez, L. Ogren, R. Verjan, and F. Talamantes, Effects of perinatal exposure to a synthetic estrogen and progestin on mammary tumorigenesis in mice, *Teratology* 38:129-134 (1988).
112. B.E. Walker, Tumors of female offspring of mice exposed prenatally to diethylstilbestrol, *J Natl Cancer Inst* 73:133-140 (1984).
113. L. Hilakivi-Clarke, R. Clarke, I. Onojafe, M. Raygada, E. Cho, and M.E. Lippman, Maternal dietary fat increases breast cancer risk among female offspring, *submitted* (1995).
114. T. Colton, R. Greenberg, K. Noller, L. Resseguie, C... Bennekom, T. Heeren, and Y. Zhang, Breast cancer in mothers prescribed diethylstilbestrol in pregnancy, *J Am Med Assoc* 269:2096-2100 (1993).

115. R.A. Rhoades and G.A. Tanner, Medical Physiology, Little, Brown and Company, Boston, (1995).
116. M.L. Casey, P.C. MacDonald, and E.R. Simpson, Textbook of Endocrinology, W.B. Saunders Company, New York, (1992).
117. B.A. Williams, K.T. Mills, C.D. Burroughs, and H.A. Bern, Reproductive alterations in female C57BL/Crgl mice exposed neonatally to zearalenone, an estrogenic mycotoxin, *Cancer Lett* 46:225-230 (1989).
118. C.A. Lamartiniere, J.B. Moore, J.B. Holland, and S. Barnes, Chemoprevention of mammary cancer from neonatal genistein treatment, *Proc. Am. Assoc. Cancer Res.* 35:Abstract 3689(1994).
119. J. Russo, G. Wilgus, and I.H. Russo, Susceptibility of the mammary gland to carcinogenesis: Differentiation of the mammary gland as determinant of tumor incidence and type of lesion, *Am J Pathol* 96:721-736 (1979).
120. R.W. Engelman, N.K. Day, and R.A. Good, Calorie intake during mammary development influences cancer risk: lasting inhibition of C3H/HeOu mammary tumorigenesis by prepubertal calorie restriction, *Cancer Res* 54:5724-5730 (1994).
121. S.Z. Haslam, Progesterone effects on deoxyribonucleic acid synthesis in normal mouse mammary glands, *Endocrinology* 122:464-470 (1988).
122. Y. Folman and G.S. Pope, The interaction in the immature mouse of potent oestrogens with coumestrol, genistein and other utero-vaginotrophic compounds of low potency, *J Endocrinol* 34:215-225 (1966).
123. D.T. Kiang, B.J. Kennedy, S.V. Pathre, and C.J. Mirocha, Binding characteristics of zearalenone analogs to estrogen receptors, *Cancer Res* 38:3611-3615 (1978).
124. U.E. Mayr, Estrogen-controlled gene expression in tissue culture cells by zearalenone, *FEBS Lett* 239:223-226 (1988).
125. J.J. Michnovicz and H.L. Bradlow, Induction of estradiol metabolism by dietary indole-3-carbinol in humans, *J Natl Cancer Inst* 82:947-949 (1990).
126. H.L. Bradlow, R.E. Hershcopf, and J. Fishman, Oestradiol 16α-hydroxylase: a risk marker for breast cancer, *Cancer Surv* 5:573-583 (1986).
127. A.L. Harris, The epidermal growth factor receptor as a target for therapy, *Cancer Cells* 2:321-323 (1990).
128. Y. Matsukawa, N. Marui, T. Sakai, Y. Satomi, M. Yoshida, K. Matsumoto, H. Nishino, and A. Aoike, Genistein arrests cell cycle progression, *Cancer Res* 53:1328-1331 (1993).
129. F. Leonessa, W.-Y. Lim, V. Boulay, M.E. Lippman, and R. Clarke, Effect of the phytoestrogen zearalenone on human breast cancer cells, in: "Exercise, Calories, Fat and Cancer", M.M. Jacobs., ed., Plenum Press, New York, pp. 286(1992).
130. T. Fotsis, M. Pepper, H. Adlercreutz, G. Fleischmann, T. Hase, R. Montesano, and L. Schweigerer, Genistein, a dietary-derived inhibitor of in vitro angiogenesis, *Proc Natl Acad Sci USA* 90:2690-2694 (1993).
131. J.T.J. Kellis and L.E. Vickery, Inhibition of human estrogen synthetase (aromatase) by flavones, *Science* 225:1032-1034 (1984).
132. H. Adlercreutz, K. Hockerstedt, C. Bannwart, S. Bloigu, E. Hamalainen, T. Fotsis, and A. Ollus, Effect of dietary components, including lignans and phytoestrogens, on enterohepatic circulation and liver metablism of estrogens and on sex hormone binding globulin (SHBG), *J Steroid Biochem* 27:1135-1144 (1987).
133. J.R.C. Sainsbury, A.J. Malcolm, D.R. Appleton, J.R. Farndon, and A.L. Harris, Presence of epidermal growth factor receptor as an indicator of poor prognosis in patients with breast cancer, *J Clin Pathol* 38:1225-1228 (1985).
134. S. Nicholson, J.R.C. Sainsbury, P. Halcrow, P. Chambers, J.R. Farndon, and A.L. Harris, Expression of epidermal growth factor receptors associated with lack of response to endocrine therapy in recurrent breast cancer, *Lancet* i:182-184 (1989).
135. S. Nicholson, J. Richard, C. Sainsbury, P. Halcrow, P. Kelly, B. Angus, C. Wright, J. Henry, J.R. Farndon, and A.L. Harris, Epidermal growth factor receptors: results of a 6 year follow-up study in operable breast cancer with emphasis on the node negative subgroup, *Br J Cancer* 63:146-150 (1991).
136. P. Bolufer, F. Miralles, A. Rodriguez, C. Vasquez, A. Lluch, J. Garcia-Conde, and T. Olmos, Epidermal growth factor receptor in human breast cancer: correlation with cytosolic and nuclear estrogen receptors and with biological and histological tumor characteristics, *Eur J Cancer* 26:283-290 (1990).
137. S. Nicholson, P. Halcrow, J.R.C. Sainsbury, B. Angus, P. Chambers, J.R. Farndon, and A.L. Harris, Epidermal growth factor receptor (EGFr) status associated with failure of primary endocrine therapy in elderly postmenopausal patients with breast cancer, *Br J Cancer* 58:810-814 (1988).
138. K.D.R. Setchell, Naturally occurring non-steroidal estrogens of dietary origin, in: "Estrogens in the environment", J.A. McLachlan., ed., Elsevier, pp. 69-85 (1985).

139. W.S. Pierpoint, Flavonoids in the human diet, in: "Plant Flavonoids in Biology and Medicine: Biochemical, Pharmacological and Structure-Activity Relationships", Alan R. Liss, Inc, New York, pp. 125-140 (1986).
140. K. Verdeal, R.R. Brown, T. Richardson, and D.S. Ryan, Affinity of phytoestrogens for estradiol-binding proteins and effect of coumesterol on growth of 7,12-dimethylbenz(a)anthracene-induced rat mammary tumors, *J Natl Cancer Inst* 64:285-290 (1980).
141. U. Mayr, A. Butsch, and S. Schneider, Validation of two in vitro test systems for estrogenic activities with zearalenone, phytoestrogens, and cereal extracts, *Toxicology* 74:135-149 (1992).

7

SOY ISOFLAVONOIDS AND CANCER PREVENTION

Underlying Biochemical and Pharmacological Issues

Stephen Barnes,[1,2,3] Jeff Sfakianos,[1] Lori Coward,[1,3] and Marion Kirk[3]

[1] Departments of Pharmacology and Toxicology
[2] Departments of Biochemistry and Molecular Genetics
[3] Comprehensive Cancer Center Mass Spectrometry Shared Facility
University of Alabama at Birmingham
Birmingham, Alabama 35294

ABSTRACT

The isoflavonoids in soy, genistein and daidzein, have been proposed to contribute an important part of the anti-cancer effect of soy. Although there have been many interesting studies on the effects of isoflavones on biochemical targets in tissue culture experiments, in most cases the concentrations used by investigators have exceeded 10 μM. However, based on simple pharmacokinetic calculations involving daily intake of isoflavones, absorption from the gut, distribution to peripheral tissues, and excretion, it is unlikely that blood isoflavone concentrations even in high soy consumers could be greater than 1-5 μM. Experiments designed to evaluate these pharmacological principles were carried out in anesthetized rats with indwelling biliary catheters and in human volunteers consuming soy beverages. The data from these experiments indicate that genistein is efficiently absorbed from the gut, taken up by the liver and excreted in the bile as its 7-O-β-glucuronide. Re-infused genistein 7-O-β-glucuronide was also well absorbed from the gut, although this occurred in the distal small intestine. In human subjects fed a soy beverage for a period of two weeks, plasma levels of genistein and daidzein, determined by HPLC-mass spectrometry, ranged from 0.55-0.86 μM, mostly as glucuronide and sulfate conjugates. In summary, genistein is well absorbed from the small intestine and undergoes an enterohepatic circulation. Although the plasma genistein levels achievable with soy food feeding are unlikely to be sufficient to inhibit the growth of mature, established breast cancer cells by chemotherapeutic-like mechanisms, these levels are sufficient to regulate the proliferation of epithelial cells in the breast and thereby may cause a chemopreventive effect.

Dietary Phytochemicals in Cancer Prevention and Treatment
Edited under the auspices of the American Institute for Cancer Research, Plenum Press, New York, 1996

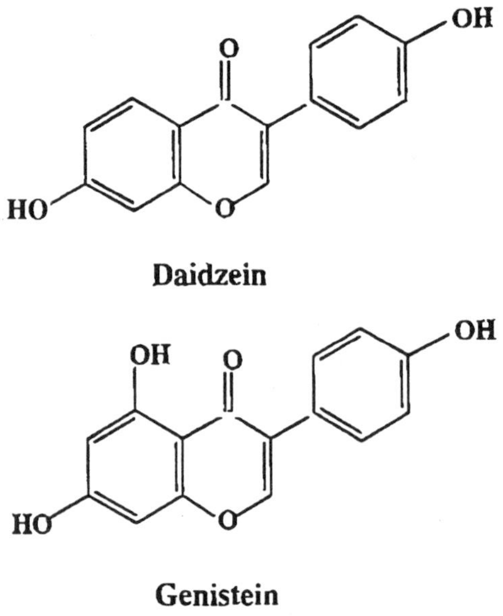

Daidzein

Genistein

Figure 1. Chemical structures and numbering scheme of genistein and daidzein.

INTRODUCTION

Soy foods appear to have an important role in the prevention of cancer (Messina et al., 1994), cardiovascular disease (Anderson et al., 1995) and possibly osteoporosis (Kalu et al., 1988; Blair et al., 1995; Kao & P'eng, 1995; Messina, 1995). Several phytochemical components of soy have been reported to have a role in causing these preventive effects (see review by Messina & Barnes, 1991).

Over the past ten years we have been investigating the importance of the isoflavonoids in soy, particularly genistein (4',5,7-trihydroxyisoflavone) (Fig. 1), in animal models of cancer. Although in our original hypothesis we viewed genistein as a phytoestrogen (Barnes et al., 1990), acting as an estrogen antagonist in a similar manner to the anti-cancer drug tamoxifen, Akiyama et al. (1987) showed that genistein is an excellent and specific inhibitor of protein tyrosine kinases (PTKs), thereby considerably expanding the potential mode of action of genistein. Many steps of the cellular biochemical pathways activated by extracellular growth factors involve tyrosine phosphorylation. This is especially important for transformed cells where the protein products of 50% of the known cellular oncogenes are subject to uncontrolled tyrosine phosphorylation (Cantley et al., 1991). As a result, genistein has been extensively used to explore a wide variety of biochemical and biological events which involve an important PTK activity. Indeed, since 1992 over 800 reports in which genistein has been used have been published, mostly by investigators who are unaware of the food origins of genistein.

Evidence for the role of genistein as a chemopreventive agent has come from experiments in which it has been shown that genistein inhibits the proliferative growth of many cancer cells in tissue culture, irrespective of whether the cells contain estrogen receptors (Peterson et al., 1991; Peterson, 1995). Further investigations have shown that genistein inhibits the appearance of tumors in animal models of breast cancer (Lamartiniere et al., 1995) and skin cancer (Wei et al. 1993, 1995), and it inhibits the formation of aberrant crypts in models of colonic cancer (Pereira et al., 1994; Helms & Gallaher, 1995).

When considering whether to ingest genistein either by eating soy foods or genistein itself, the key questions from a pharmacologist's viewpoint are the traditional ones of absorption, distribution, delivery to the intended target, metabolism and excretion of genistein. In the present study we examine the pharmacological principles that affect the use of genistein. To demonstrate these principles, experiments have been carried out in bile fistula rats given genistein and in human volunteers in soy feeding studies.

MATERIALS AND METHODS

Materials

Genistin was isolated from soy molasses as previously described (Peterson & Barnes, 1991, 1993). 4-^{14}C-Genistein was custom synthesized (Moravek Biochemicals, Brea, CA) and had a radiochemical purity of >98% by HPLC and a specific activity of 16 mCi/mmol. Daidzein was purchased from LC Labs, MA.

β-Glucuronidase and sulfatase were purchased from Sigma Chemical Co., St. Louis, MO. Acetonitrile and trifluoroacetic acid were sequencing grades and were obtained from Fisher Chemical Co., Norcross, GA, and Pierce Chemical Co., Milwaukee, WS, respectively. All other chemicals were of the highest grades obtainable. Sep-Pak C_{18} cartridges were purchased from Waters (Milford, MA).

Animals

Female Sprague-Dawley rats (225-275 g) were bred at UAB. Both fathers and dams (purchased from Harlan Sprague-Dawley, Indianapolis, IN) were fed an isoflavone- and soy-free diet (AIN-76A, Harland Teklad, Madison, WS) prior to impregnation of the dams. The pregnant dams were maintained on this diet throughout pregnancy and weaning. The pups were also placed on the AIN-76A diet after weaning.

Rats used in these experiments were anesthetized with ketamine/xylocaine (0.1 ml/100 g body wt.) and their body temperature was maintained at 38°C (monitored rectally) by placing them on a heating pad. Following a midline incision, the bile duct was exposed and was cannulated with PE-10 tubing, tied and secured with 5-0 silk (Ethicon, Somerville, NJ).

To assess intestinal recovery of genistein, 10 μCi (0.17 mg) of 4-^{14}C-genistein was infused in a basic infusate (10 mM sodium taurocholate, 10 mM glucose, 154 mM NaCl and 10 mM KCl) at 70 μl/min over a one hour period into the duodenum of a bile duct cannulated rat. After this, the basic perfusate was infused for the remainder of the experimental period (4 h). Biles were collected over 20 min intervals into tared vials.

In a second experiment to assess whether the biliary genistein metabolites also underwent an enterohepatic circulation, the biles collected over the first 60-100 min from the first experiment were pooled and diluted using the basic perfusate. This mixture was re-infused into the duodenum or the proximal ileum for 60 min, followed by infusion of the basic perfusate for a further 4 h. Biles were collected through a biliary cannula at 20 min intervals throughout the experiment.

Soy Feeding Studies

Blood samples were obtained at 2:30 pm from normal healthy volunteers (3 males, 1 female) consuming two 8-oz soy protein beverages containing 20 g of isolated soy protein (Altima, Protein Technologies International, St. Louis, MO) each day (at 8 am and 12 noon) for 14 days. They were on a soy-free diet for seven days prior to the study. Blood samples were drawn prior to the study, and then at 2:30 pm (6.5 h after drinking the beverage), on days 1, 2, 3, 7, 10 and 14 after starting the study Plasma from each sample was separated by centrifugation and stored at -20°C prior to use. The protocol for this study was approved by the Institutional Review Board for Human Use.

HPLC Analysis of Genistein Metabolites in Bile

^{14}C-Radioactivity was recovered from biles by absorption onto activated Sep-Pak C_{18} cartridges and elution with methanol. After evaporation of the methanol, the residues were reconstituted in 100 µl of 80% aqueous methanol. Aliquots (1-10 µl) were used for HPLC analysis. To ascertain whether the individual peaks observed were glucuronides or sulfates, aliquots of the extracts were evaporated to dryness and reconstituted in 50 mM Tris-HCl buffer, pH 7, containing 500 units of β-glucuronidase or 0.5 units of sulfatase and incubated overnight at 37°C. The pH was lowered to 5.0 by the the addition of 0.75 ml of 1.0 M ammonium acetate, pH 5.0. The sample was diluted with 5 ml water and the isoflavones were extracted by passage over an activated Sep-Pak C_{18} cartridge.

HPLC analysis was carried out on a Hewlett Packard model 1050 instrument with a 25 cm x 0.46 cm i.d. Brownlee Aquapore C_8 reversed-phase column, using a linear 0-50% gradient (5% per min) of acetonitrile in a background of 10 mM ammonium acetate at a flow rate of 1.5 ml/min. Isoflavones in the eluate were detected by their absorbance at 262 nm. Radioactivity was determined by liquid scintillation counting of fractions collected every 30 sec during the analysis.

Analysis of Isoflavones in Plasma

Plasma samples (1 ml) were treated with β-glucuronidase/sulfatase to hydrolyze the isoflavone and metabolite conjugates. Incubates were extracted with hexane (4 ml), acidified with acetic acid and the isoflavones were extracted with diethyl ether. Biochanin A (100 pmol) was added as an internal standard. Ether extracts were evaporated to dryness and reconstituted in 100 µl of 80% aqueous methanol.

HPLC-Mass Spectrometry

Extracted samples of bile were separated by reversed-phase HPLC on a 15 cm x 0.21 cm i.d.Brownlee Aquapore C_8 column using a linear 0-50% gradient (5% per min) of acetonitrile in 10 mM ammonium acetate at a flow rate of 0.2 ml/min. The column eluate was split 1:1 and one stream passed into the IonSpray™ interface of a PE-Sciex (Concord, Ontario, Canada) API III triple quadrupole mass spectrometer operating in the negative ion mode, with an orifice potential of -60V.

Plasma isoflavone extracts were analyzed on a 15 cm x 4.6 mm i.d. 300 Å pore size, Aquapore C_{18} reversed-phase HPLC column. Isocratic conditions (30% acetonitrile in 10 mM ammonium acetate) at a flow rate of 1 ml/min were employed. Plasma extracts (20 µl) were injected onto the HPLC column using a model 7125 Rheodyne injector. After chromatographic separation, the eluate stream was diluted with 13 µl/min of ammonium hydroxide provided by a Harvard infusion pump and passed into the heated nebulizer-atmospheric pressure chemical ionization interface. The orifice potential of the mass spectrometer was set at -60V.

In the MS-MS experiments, daughter ion spectra were obtained by selecting parent ions in the first quadrupole, which were then collided with argon-10% nitrogen gas in the second quadrupole and analyzed in the third quadrupole. Multiple reaction ion monitoring (MRM) was carried out in a similar manner to MS-MS by selection of specific ions not only in the first quadrupole, but also in the third quadrupole (Coward et al., 1996). The operation of the mass spectrometer and analysis of data were carried out using two MacIntosh Quadra 950 computers interfaced with an Ethernet link.

RESULTS

Intestinal Uptake and Biliary Recovery of Genistein

Following administration of 4-^{14}C-genistein into the upper small intestine of anesthetized rats, ^{14}C-radioactivity promptly appeared in bile. The total recovery over the next four hours was 70-80% of the administered dose (Fig. 2).

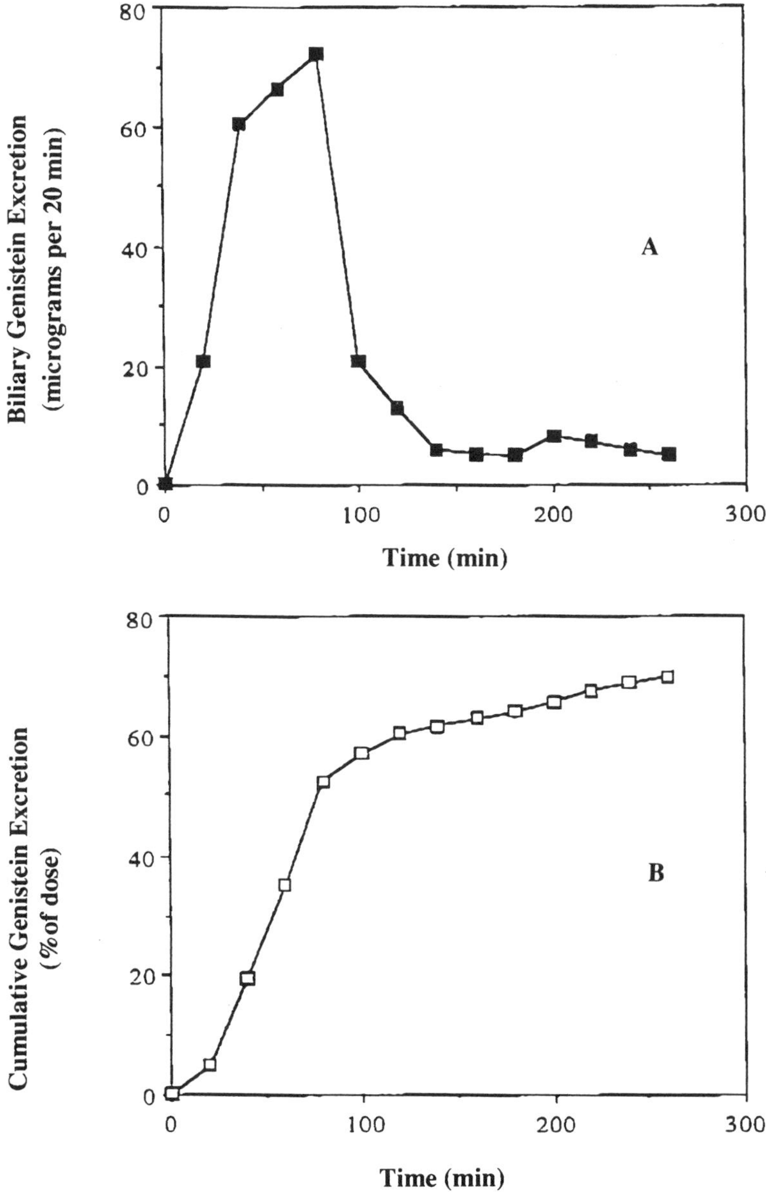

Figure 2. Recovery of ^{14}C-radioactivity (expressed as μg of genistein) in bile following duodenal infusion of 4-^{14}C-genistein for 1 h followed by a perfusate for 3 h in an anesthesized rat fitted with an indwelling biliary cannula. A: biliary radioactivity excreted in 20-min intervals throughout the experiment. B: cumulative biliary recovery of radioactivity as a percentage of the dose infused.

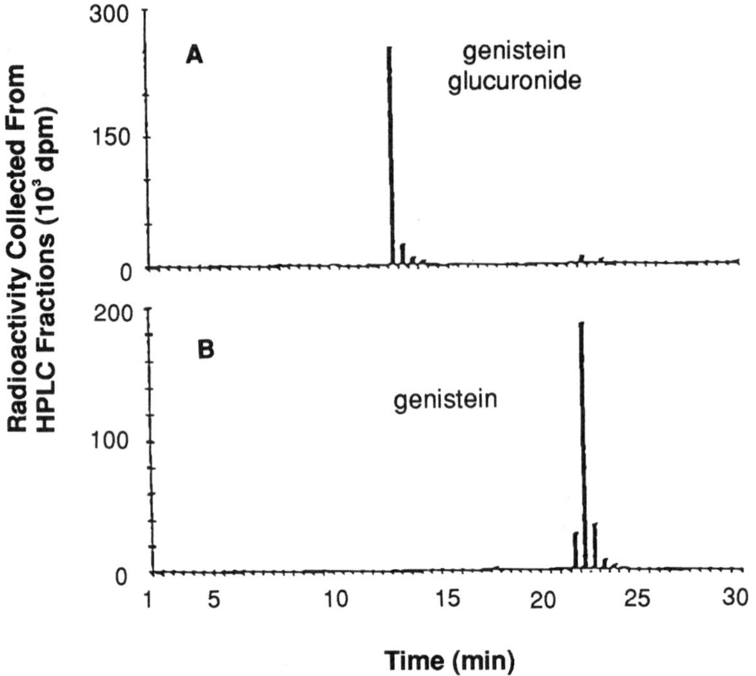

Figure 3. Radio-HPLC analysis of biliary genistein ^{14}C-metabolite with (A) and without (B) treatment with β-glucuronidase.

HPLC analysis of the biliary ^{14}C-radioactivity revealed that it consisted of a single peak, substantially more hydrophilic than genistein (Fig. 3A). Treatment of this peak with β-glucuronidase demonstrated that it was a β-glucuronide of genistein (Fig. 3B).

Confirmation of the identification of the genistein metabolite was made by HPLC-MS; the radioactive peak gave rise to a molecular [M-H]- ion with a m/z value of 445. The MS-MS spectrum of this ion showed the expected loss of the glucuronide group (-176) to give the m/z 269 aglucone ion (Fig. 4). Proton NMR data indicated that it was the 7-O-β-glucuronide (Coward et al., 1996).

Re-infusion of the ^{14}C-labeled genistein 7-O-β-glucuronide into the intestine of another set of rats also resulted in the appearance of ^{14}C-radioactivity in the bile. When it was infused into the upper small intestine, recovery into bile was noticeably delayed, but when infused into the mid-small intestine recovery was as rapid and complete as with ^{14}C-genistein (Fig. 5). Once again, in both experiments the biliary metabolite was genistein 7-O-β-glucuronide.

Urinary output of ^{14}C-radioactivity was low in these experiments, being 2.4% of the administered dose for infusions with ^{14}C-genistein and 7.4% for infusions with genistein 7-O-β-glucuronide. In experiments in which the rats were placed in metabolic cages and urine and feces collected over dry ice for the next 120 h, the total ^{14}C-radioactivity excreted in the urine and feces was 21% and 2%, respectively.

Plasma Isoflavones following a Soy Beverage

The total intake of genistein and daidzein with each serving of the soy beverage was 21.02 mg (78 μmol) and 13.55 mg (53 μmol), respectively, a mole ratio of 1.46. Blood

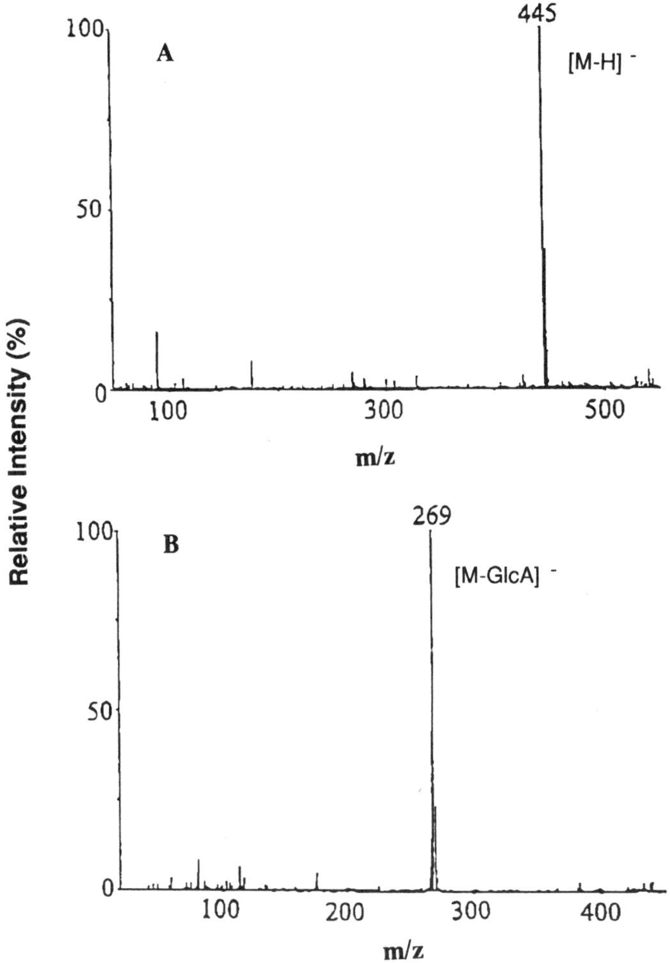

Figure 4. Electrospray ionization mass spectrometry of the biliary genistein metabolite. A: Negative ionization mass spectrum of the metabolite showing that it has a molecular weight of 446. B: Daughter ion spectrum of the m/z 445 ion showing the loss of 176, indicative of a glucuronide group.

samples were taken at 2:30 pm and the isoflavones in 1 ml aliquots of plasma were determined (Fig. 6). Plasma concentrations of genistein and daidzein on the first day of the study were 496-644 nM and 289-424 nM, respectively, the genistein/(daidzein + metabolites) mole ratio varying from 1.30 to 1.54. Over the next 14 days there was a gradual fall in the plasma concentrations of daidzein and genistein.

DISCUSSION

Genistein has many of the pharmacological characteristics of a typical xenobiotic. Xenobiotics taken orally enter the blood stream by a process of *absorption* from the gut. Absorption, which occurs by passage of the xenobiotic through the mucosal cells lining the gut wall, depends on whether the xenobiotic is sufficiently *hydrophilic* to maintain solubility

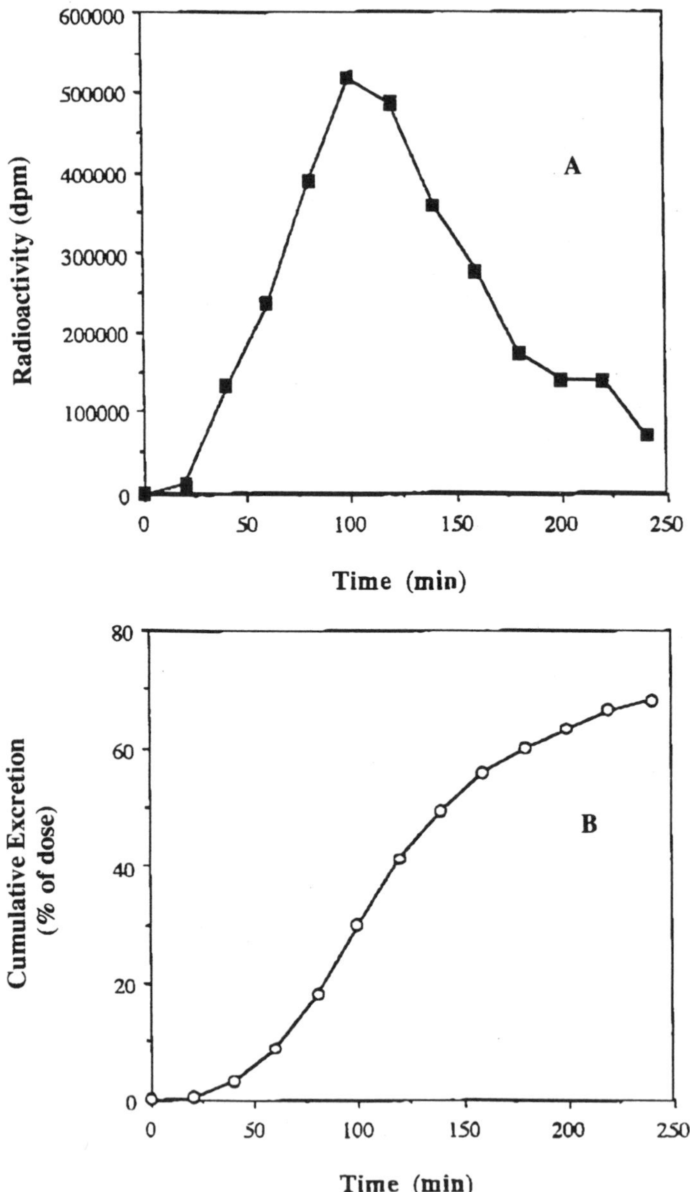

Figure 5. Recovery of ^{14}C-radioactivity in bile following mid small intestine infusion of 4- ^{14}C-labeled genistein 7-O-β-glucuronide for 1 h followed by a perfusate for 3 h in an anesthesized rat fitted with an indwelling biliary cannula. A: biliary radioactivity excreted in 20-min intervals throughout the experiment. B: cumulative biliary recovery of radioactivity as a percentage of the dose infused.

in the aqueous milieu on the luminal side of the mucosal cell and sufficiently *hydrophobic* to be soluble in the membrane. The concentration gradient between the lumen of the gut and the interior of the mucosal cell therefore is dissipated by the entry of the xenobiotic into the cell. On the serosal side of the cell, a similar process occurs in which blood sweeps away xenobiotic diffusing through the basolateral membrane. For a few compounds such as bile

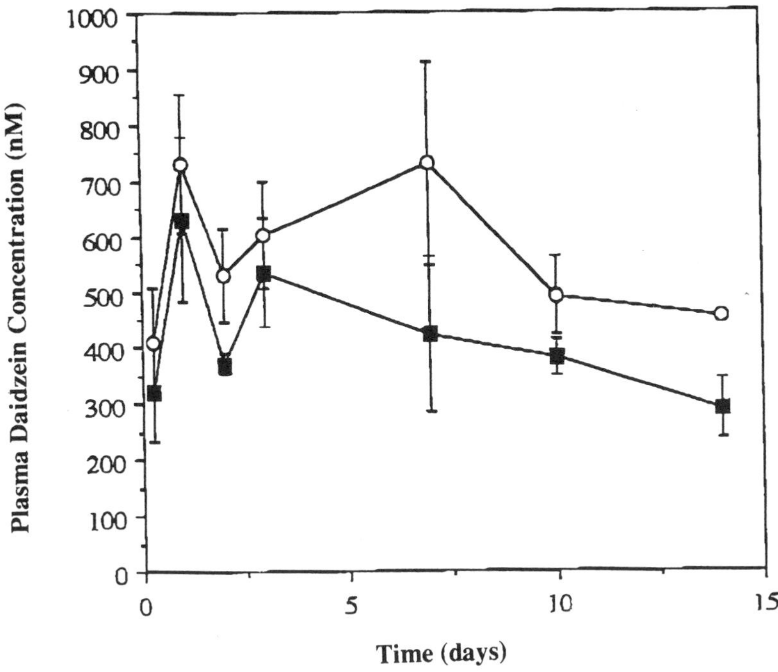

Figure 6. Plasma concentrations of the isoflavones genistein and daidzein as determined by HPLC- mass spectrometry (see Methods) from subjects consuming two soy beverages a day for 14 days. Data are presented as means ± SEM.

acids, there are specific transporters which allow for active transport against a concentration gradient (Weiner et al., 1964), but for most xenobiotics uptake occurs by simple passive diffusion. In some cases, metabolism of the xenobiotic within the gut mucosal cells may prevent its back diffusion to the gut lumen.

The hydrophobicity of xenobiotics is influenced by the charge they carry, which in turn is influenced by the luminal pH. Negatively charged organic acids can be protonated in the upper gut due to gastric acid secretion and become uncharged (and more hydrophobic) for example, this is where aspirin (acetylsalicylic acid) absorption is the greatest from the intestine.

Genistein, by having weakly basic phenolic hydroxyl groups, is fully protonated under the pH conditions of the upper gut and therefore should be easily absorbed. On the other hand, it has a low aqueous solubility at acid pH. Its solubility may be further reduced by its binding to proteins in the food.

In non-fermented soy foods, genistein is present in several chemical forms (Barnes et al., 1994; Wang & Murphy, 1994), namely 6"-O-malonylglucosides (6OMalGlc) (Kudou et al., 1991), 6"-O-acetylglucosides (6OAcGlc) (Farmakalidis & Murphy, 1985) and β-glucosides (Fig. 7). These are more soluble in water than genistein, particularly the 6OMalGlc conjugates (G-P. Ji & S. Barnes, unpublished data), but they would not absorbed intact. Instead, they must be first hydrolyzed by β-glucosidases to genistein. Since the complex glucosides (6OMalGlc and 6OAcGlc) are poorly hydrolyzed by β-glucosidases (Farmakalidis & Murphy, 1985), they may not be converted to genistein until they reach the more distal regions of the gut where bacterial concentrations are very high (Barnes et al., 1995). In this scenario, genistein may also undergo *first pass bacterial metabolism*, i.e., genistein

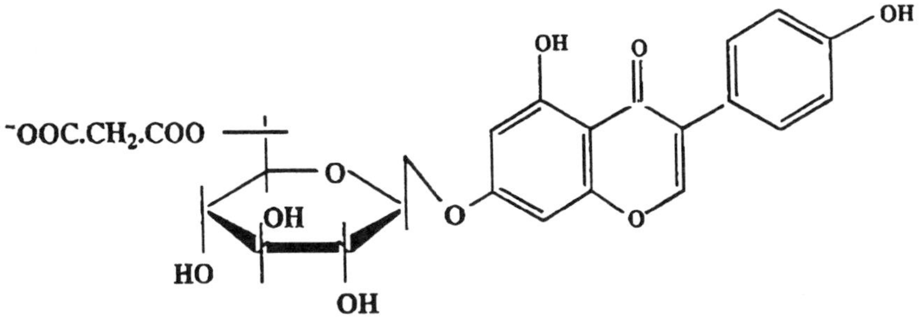

6''-O-Malonylgenistin

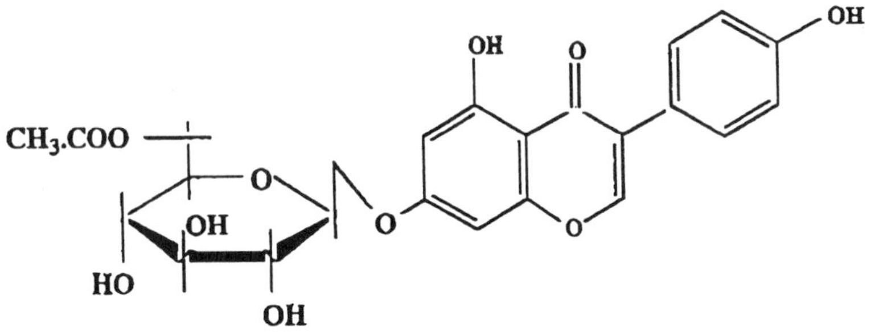

6''-O-Acetylgenistin

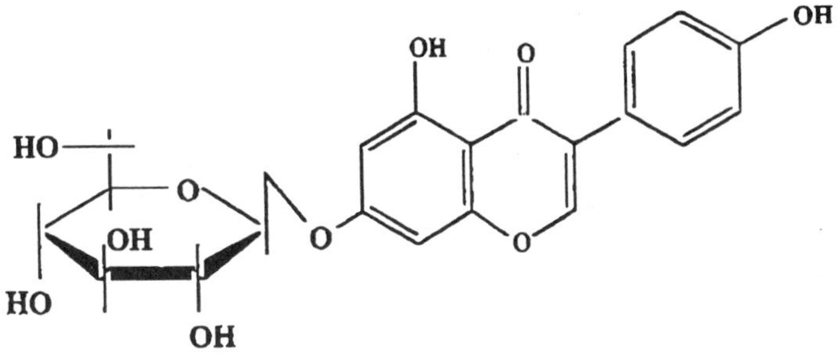

Genistin

Figure 7. Structures of glucosides of genistein found in soy foods.

liberated from the glucoside conjugates may be converted into bacterial metabolites before it is absorbed for the first time.

As shown in the present study in rats with indwelling biliary cannulas, genistein was well absorbed from the intestine and was excreted rapidly as a metabolite into bile. These data do not allow description of the individual steps of this process. However, it can be reasonably assumed that once absorbed from the gut, genistein would be carried in the mesenteric venous drainage to the portal vein and then the liver. There, genistein would be taken up by the hepatocytes, converted into one or more polar metabolites, and excreted through the bile canalicular membrane into bile. The transport processes for both hepatic uptake of genistein and canalicular excretion of its metabolite(s) may be saturable at higher doses. It should be noted that the intestinal wall is not necessarily a passive organ for selective

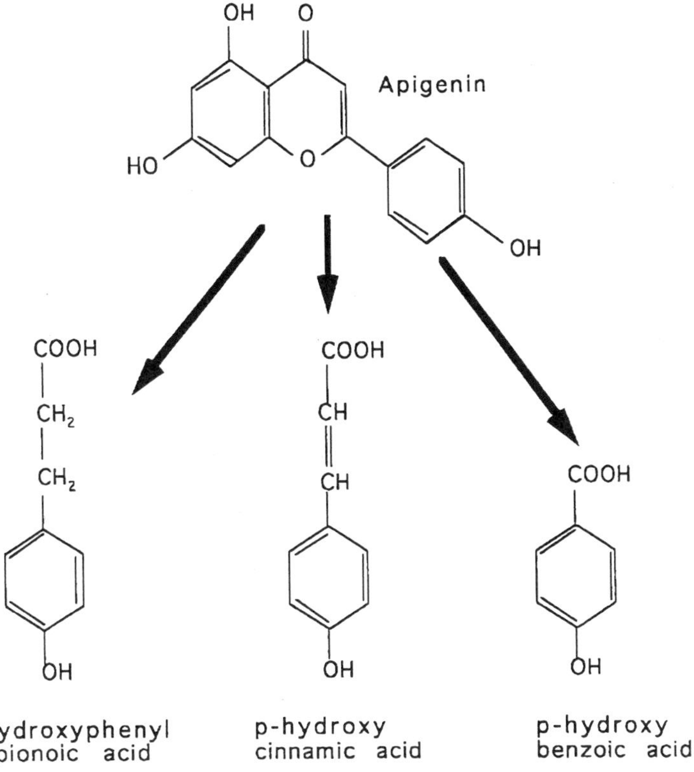

Figure 8. Metabolites of apigenin, the flavone isomer of genistein.

expression of both phase I and phase II enzymes occurs there, permitting a type of *first pass clearance.*

The rapid biliary excretion also allowed the genistein metabolite (shown in the present study to be its 7-O-glucuronide) to re-enter the small intestine. There it was hydrolyzed, allowing genistein to be absorbed for a second time and/or further metabolized to other metabolites. It is likely that genistein has such an enterohepatic circulation since this has been described to differing extents for several flavonoids (Hackett, 1986).

Although *metabolism* of xenobiotics occurs mostly in the liver and kidney, other organs may also be involved. Bacterial metabolism of slowly absorbed xenobiotics is a distinct possibility. Metabolism of flavonoids by bacteria is well described and usually involves ring fission (Booth et al., 1958; Griffiths, 1982). Genistein and daidzein have been shown to be converted to p-ethylphenol in sheep (Batterham et al., 1965) - however, metabolites derived from the A-ring of these isoflavonoids have not yet been identified. Apigenin, the flavone isomer of genistein, is converted to p-hydroxyphenylpropionic acid, p-hydroxycinnamic acid and p-hydroxybenzoic acid (Fig. 8) which appear in the urine as glucuronide and sulfate conjugates (Griffiths & Smith, 1972).

Metabolism of many drugs to their β-glucuronides and sulfates is not only a signal for them to be excreted into the bile, but also to be excreted in the urine. The processes involved in urinary excretion have close analogy to those in the gastrointestinal tract. Extensive reabsorption of metabolites which enter the tubules either by filtration at the glomerulus or by active secretion occurs in the proximal tubules, and conjugation reduces the efficiency of reabsorption by exactly the same physicochemical principles as in the gut.

This results in the more rapid urinary clearance of the more hydrophilic compounds. Many investigators have reported that daidzein is more rapidly excreted in the urine than genistein (Kelly et al., 1993; Xu et al., 1994), from this data it has been suggested that genistein has a lower bioavailability than daidzein (Xu et al., 1994), but instead it may be because daidzein is more hydrophilic than genistein, as observed by their relative mobilities during reversed-phase HPLC analysis.

Excretion of genistein or its metabolites can also occur by the fecal route which is a result of incomplete intestinal reabsorption. Studies of fecal output of genistein are limited in number (Xu et al., 1994; Adlercreutz et al., 1994). Nonetheless, only 1-2% of the genistein in the foods used in these studies was excreted in the feces. Genistein is rapidly degraded in fecal culture (Xu et al., 1995).

Following intestinal absorption, a proportion of the genistein in portal blood may escape uptake by the liver and will enter the peripheral blood circulation. The liver therefore has an important role in regulating the level of genistein in the blood, both qualitatively and quantitatively. Extensive removal of genistein from portal blood is termed the *hepatic first pass clearance* and diminishes the amount of genistein distributed to the rest of the body.

For breast cancer or prostate cancer, delivery of genistein via the blood supply is the only route available physiologically. The effective concentration of genistein in blood will depend on *plasma protein binding*. Physiologic steroids are 98% bound to plasma proteins, i.e., less than 2% is available for uptake into tissues. This effect has not been examined yet for genistein, but it was reported that 40% of daidzein in rat plasma was protein bound.

The concentration of genistein at its tissue target is crucial. Some xenobiotics become concentrated in certain cells due to specific transporters. Although steroids have been long considered to cross the membranes of target cells by simple diffusion, recent evidence suggests that this may not be always the case. Kralli et al. (1994) reported the discovery of a transporter (LEM1) for glucocorticoids which belongs to the multidrug resistance gene super family. The transporter is ATP-dependent and pumps some, but not all, of the glucocorticoids out of the cell. Data on tissue concentrations of genistein have not been reported and may await development of radioimmunoassay procedures for genistein, but the concentration of daidzein in tissues did not exceed the plasma level. However, the dose used by these investigators was 500 times that available by dietary means, possibly leading to saturation of a specific transporter.

Studies in human volunteers consuming a soy beverage (delivering approximately 42 mg of genistein and 27 mg of daidzein per subject per day) revealed that genistein is present in peripheral blood at concentrations from 0.5 - 1.0 μM, mostly as the glucuronide and sulfate conjugates. These concentrations are much lower than those required to inhibit the growth of most cultured cancer cells (Barnes & Peterson, 1995). However, we have recently shown that non-transformed human mammary epithelial cells are far more sensitive to genistein, with an IC_{50} for EGF-stimulated growth of 1-2 μM (Peterson & Barnes, 1995). This implies that genistein may have its greatest chemopreventive effect before the tumor cell phenotype is acquired. Furthermore, these cells were also inhibited by genistein 7-sulfate, with an IC_{50} for EGF-stimulated growth of 0.5-1.0 μM, suggesting that conjugation of genistein does not necessarily reduce its biological activity.

Finally, the steady fall in the plasma concentrations of daidzein and genistein over a two-week period indicate that these isoflavonoids may induce their own metabolism. This may vary from individual to individual and may explain the relatively low plasma concentrations of genistein and daidzein reported in a study of Japanese men who were habitual soy eaters (Adlercreutz et al., 1991).

ACKNOWLEDGMENTS

These studies were supported in part by grants from the American Institute for Cancer Research (91B47R and 92), the National Cancer Institute (5R01 CA-61668) and the United Soybean Board. A genistin concentrate, used in the isolation and purification of genistin and genistein, and the soy protein beverage were kindly donated by Protein Technologies International, a St. Louis, MO-based Company. The mass spectrometer was purchased by funds from a NIH Instrumentation Grant (S10RR06487) and from this institution. Operation of the UAB Comprehensive Cancer Center Mass Spectrometry Shared Facility has been supported in part by a NCI Core Research Support Grant to the UAB Comprehensive Cancer (P30 CA13148). We are indebted to Ms. Laura Whitaker and Dr. Clinton J. Grubbs, Department of Nutritional Sciences, UAB, for the breeding of rats on isoflavone-free AIN-76A diets.

REFERENCES

Adlercreutz H, H. Honjo, A. Higashi, T. Fotsis, E. Hämäläinen, T. Hasegawa and H. Okada. Urinary excretion of lignans and isoflavonoid phytoestrogens in Japanese men and women consuming a traditional Japanese diet. *Am J Clin Nutr* 54: 1093-1100 (1991).

Adlercreutz, H., T. Fotsis, K. Wähälä, T. Mäkelä, and T. Hase. Isotope dilution gas chromatographic-mass spectrometric method for the determination of unconjugated lignans and isoflavonoids in human feces, with preliminary results in omnivorous and vegetarian women. *Anal Biochem* 225: 101-108 (1995).

Akiyama, T., J. Ishida, S. Nakagawa, H. Ogawara, S. Watanabe, N.M. Itoh, M. Shibuya, and Y. Fukami. Genistein, a specific inhibitor of tyrosine-specific protein kinases. *J Biol Chem* 262: 5592-5595 (1987).

Anderson J.W., B.M. Johnstone, and M.E. Cook-Newell. Meta-analysis of the effects of soy protein intake on serum lipids. *New Engl J Med* 333: 276-282 (1995).

Barnes, S., C. Grubbs, K.D.R. Setchell, and J. Carlson. Soybeans inhibit mammary tumors in models of breast cancer, *in*: "Mutagens and carcinogens in the diet", ed. M. Pariza, Alan R. Liss, New York, pp. 239-253 (1990).

Barnes, S., M. Kirk, and L. Coward. Isoflavones and their conjugates in soy foods: extraction conditions and analysis by HPLC-mass spectrometry. *J Agric Food Chem* 42: 2466-2474 (1994).

Barnes, S., M. Kirk, and L. Coward. Isoflavonoids in soy - all is not what it seems, *in*: "Physiologically functional foods", A. Caregay, ed., AOCS Press, in press (1995).

Barnes, S., M. Kirk, and L. Coward. Soy isoflavonoids: the key to good health, *in*: "Hypernutritious foods", J.W. Finley & D.J. Armstrong, eds., American Chemical Society, Washington, DC, in press (1995).

Barnes, S. and T.G. Peterson. Biochemical targets of the isoflavone genistein in tumor cell lines. *Proc Soc Exptl Biol Med* 208: 103-108 (1995).

Batterham, T.J., N.K. Hart, and J.A. Lamberton. Metabolism of oestrogenic isoflavones in sheep. *Nature* 206: 509, 1965.

Blair, H.C., S.E. Jordan, T.G. Peterson, and S. Barnes. Variable effect of tyrosine kinase inhibitors on avian osteoclastic activity and reduction of bone loss in ovariectomized rats. *J Cell Biochem*, accepted for publication, 1996.

Booth, A.N., F.T. Jones & F. DeEds. Metabolic fate of hesperidin, eriodictyol, homoeriodictyol, and diosmin. *J Biol Chem* 230:661-668 (1958).

Cantley, L.C., K.R. Auger, C. Carpenter, B. Duckworth, R. Kapeller, and S. Soltoff. Oncogenes and signal transduction. *Cell* 64: 281-302 (1991).

Coward, L., M. Kirk, and S. Barnes. Analysis of Plasma Isoflavones by Reversed-Phase HPLC-Multiple Reaction Ion Monitoring-Mass Spectrometry. *Clin Chim Acta*, accepted for publication, 1996.

Farmakalidis E. and P.A. Murphy. Isolation of 6"-O-acetyldaidzein and 6"-O-acetylgenistein from toasted defatted soy flakes. *J Agric Food Chem* 33: 385-389 (1985).

Griffiths, L.A. and G.E. Smith. Metabolism of apigenin and related compounds in the rat. Metabolite formation *in vivo* and by the intestinal microflora *in vitro*. Biochem J, 128: 901-911 (1972).

Griffiths, L.A. *In*: "The Flavonoids: Advances in Research", J. Harborne & T.J. Mabry, eds, Chapman and Hall, London (1982).

Hackett, A.M. The metabolism of flavonoid compounds in mammals, *in*: "Plant Flavonoids in Biology and Medicine: Biochemical, Pharmacological, and Structure-Activity Relationships", A.R. Liss, pp 177-194 (1986).

Helms J.R. and J.J. Gallaher. The effect of dietary soy protein isolate and genistein on the development of preneoplastic lesions (aberrant crypts) in rats. *J Nutr* 125: 802S (1995).

Kalu, D.N., E.J. Masoro, B.P. Yu, R.R. Hardin, and B.W. Hollis. Modulation of age-related hyperparathyroidism and senile bone loss in Fischer rats by soy protein and food restriction. *Endocrinol* 122: 1847-1854 (1988).

Kao, P.C. and F.K. P'eng. How to reduce the risk factors of osteoporosis in Asia? *Chinese Med J* 55: 209-213 (1995).

Kelly, G.E., C. Nelson, M.A. Waring, G.E. Joannu, and A.Y. Reeder. Metabolites of dietary (soya) isoflavones in human urine. *Clin Chim Acta* 223: 9-22, 1993.

Kralli, A., S.P. Bohen, and K.R. Yamamoto. LEM1, an ATP-binding cassette transporter selectively modulates the biological efficacy of steroid hormones. *Mol Biol Cell* 5: (suppl) 18A (1994).

Kudou, S., Y. Fleury, D. Welti, D. Magnolato, T. Uchida, K. Kitamura, and K. Okubo. Malonyl isoflavone glycosides in soybean seeds (*Glycine max* MERRILL). *Agric Biol Chem* 55: 2227-2233, 1991.

Lamartiniere, C.A., J. Moore, M. Holland, and S. Barnes. Genistein and chemoprevention of breast cancer. *Proc Soc Exptl Biol Med* 208: 120-123 (1995).

Messina, M. and S. Barnes. Workshop report from the Division of Cancer Etiology, National Cancer Institute, National Institutes of Health. The role of soy products in reducing risks of certain cancers. *J Natl Cancer Inst* 83: 541-546 (1991).

Messina, M., V. Persky, K.D.R. Setchell, and S. Barnes. Soy intake and cancer risk: A review of *in vitro* and *in vivo* data. *Nutr Cancer* 21: 113-131 (1994).

Messina, M. Modern applications for an ancient bean: soybeans and the prevention and treatment of chronic disease. *J Nutr* 125: 567S-569S (1995).

Pereira, M.A., L.H. Barnes, V.L. Rassman, G.V. Kelloff, and V.E. Steele. Use of azoxymethane-induced foci of aberrant crypts in rat colon to identify potential cancer chemopreventive agents. *Carcinogenesis* 15: 1049-1054 (1994).

Peterson, T.G. and S. Barnes. Genistein inhibition of the growth of human breast cancer cells: independence from estrogen receptors and the multi-drug resistance gene. *Biochem Biophys Res Commun* 179, 661-667 (1991).

Peterson, T.P. and S. Barnes. Genistein potently inhibits the growth of human primary breast epithelial cells: correlation with a lack of genistein metabolism. *Mol Biol Cell* 5: 384a (1994).

Peterson, T.G. Evaluation of the biochemical targets of genistein in tumor cells. *J Nutr* 125: 784S-789S (1995).

Wang H-J. and P.A. Murphy. Isoflavone content in commercial soybean foods. *J Agric Food Chem* 42: 1666-1673 (1994).

Wei, H., L. Wei, K. Frankel, R. Bowen, and S. Barnes. Inhibition of tumor promoter-induced hydrogen peroxide formation *in vitro* and *in vivo* by genistein. *Nutr Cancer* 20: 1-12 (1993).

Wei, H., R. Bowen, Q. Cai, S. Barnes, and Y. Wang. Antioxidant and antipromotional effects of the soybean isoflavone genistein. *Proc Soc Exptl Biol Med* 208: 124-130 (1995).

Weiner, I.M., J.E. Glasser, and L. Lack. *Am J Physiol* 207: 964-970 (1964).

Xu, X., H-J. Wang, P.A. Murphy, L. Cook, and S. Hendrich. Daidzein is a more bioavailable soymilk isoflavone than is genistein in adult women. *J Nutr* 124: 825-832 (1994).

Xu, X., K.S. Harris, H.J. Wang, P.A. Murphy & S. Hendrich. Bioavailability of soybean isoflavones depends upon gut microflora in women. J Nutr *125*: 2307-2315 (1995).

Yueh T-L. and H-Y. Chu. The metabolic fate of daidzein. *Sci Sin* 20: 513-521 (1977).

8

QUERCETIN AS A MODULATOR OF THE CELLULAR NEOPLASTIC PHENOTYPE

Effects on the Expression of Mutated H-*ras* and *p53* in Rodent and Human Cells

Matías A. Avila, José Cansado, K. William Harter, Juan A. Velasco, and Vicente Notario

Division of Experimental Carcinogenesis
Department of Radiation Medicine
Georgetown University Medical Center, Washington, DC 20007

INTRODUCTION

Quercetin (3,3′,4′,5,7-pentahydroxyflavone) is a widely distributed plant-derived flavonoid present in most vegetables and fruits, and it is therefore a common component of the human diet.[1] Quercetin has been shown to exert multiple biochemical effects in mammalian cells, including the increase of cAMP levels,[2] the inhibition of enzymatic activities such as protein kinase C,[3,4] protein tyrosine kinases,[5-7] and cAMP and cGMP phosphodiesterases, [8,9] as well as the interaction with estrogen type II binding sites.[10] These biological actions of quercetin may explain its predominantly inhibitory effect of tumor-derived cell lines,[10-14] and its ability to arrest tumor cells in the G_1 phase[13,15] or, less frequently, in the G_2-M phase[16] of the cell cycle. Although some reports[17,18] indicate that quercetin could be carcinogenic under certain experimental conditions, the overwhelming evidence demonstrating its ability to inhibit the growth of tumor cells identifies quercetin as an anticarcinogenic phytochemical. Furthermore, the dosage of quercetin could be easily modified in the human diet to attain the appropriate levels for optimum cancer chemopreventive action.

In the present report we describe the effect of quercetin on the expression of (*i*) the H-*ras* oncogene in *ras*-transformed mouse NIH/3T3 fibroblasts, and (*ii*) the *p53* tumor suppressor gene in human mammary carcinoma cells. In these systems, the continuous presence of either the activated H-*ras* oncogene or the mutated *p53* gene, respectively, is required for the maintenance of the cellular neoplastic phenotype.[19,20] Our results indicate that growth inhibitory concentrations of quercetin interfere with the functional expression of both mutated genes, thus explaining its antineoplastic action. These data suggest that quercetin may be used efficiently to down- modulate the neoplastic properties of mammalian tumor cells.

MATERIALS AND METHODS

Cell Lines, Culture Conditions, and Quercetin Treatment

Studies on the H-*ras* oncogene were performed using murine NIH/3T3 fibroblasts, obtained from the American Type Culture Collection (A.T.C.C., Rockville, MD), as the recipient cells in transfection assays carried out by the calcium phosphate co-precipitation method.[21] Studies on mutant *p53* were performed using the human breast carcinoma MDA-MB468 cells, an estrogen receptor negative cell line which overexpress a transcriptionally active mutant p53 protein,[22,23] also obtained from the A.T.C.C. NIH/3T3 and derived cell lines were grown in Dulbecco's modified Eagle's (DMEM) medium, and MDA-MB468 cells were grown in minimum essential Eagle's (MEM) medium. Both media were supplemented with 2 mM glutamine, 10% fetal bovine serum, 100 units/ml penicillin, and 100 μg/ml streptomycin. Cultures were incubated at 37 °C in a 5% CO_2 humidified atmosphere. Quercetin was purchased from Sigma (St. Louis, MO). The quercetin analogues, quercetrin and quercetin-3-rutinoside, were generously provided in purified form by Dr. Terrance Leighton. All flavonoids were dissolved in ethanol at concentrations such that the final ethanol content in the culture medium never exceeded 0.9% (v/v). The same concentration of ethanol was added to the medium in control experiments.

Growth Rate, Clonogenic Assays, and FACS Analysis

In experiments designed to study the effect of quercetin on cell growth, cells ($5x10^6$ per plate) were seeded in 60 mm dishes. The culture medium was changed 24 h (NIH/3T3 cells) or 48 h (MDA-MB468 cells) after inoculation. Various amounts of quercetin were added at this time, and the number of viable cells was determined at different time intervals after quercetin addition by the Trypan blue dye exclusion test. The potential toxicity of quercetin was studied using a clonogenic assay endpoint. Cells ($0.5\text{-}1x10^3$ per plate) were seeded into 10 cm plates, and various amounts of quercetin were added 24 h later. The medium was changed every three days, and after two weeks the surviving colonies were stained with 1% crystal violet in 70% ethanol, dried and counted. Cell cycle effects were studied on exponentially growing cells at various times after quercetin addition. Cell cycle analysis was performed as described[24] using a Becton Dickinson FACStar Plus flow cytometer.

Northern Blot Hybridization Analysis

Total RNA from control and quercetin-treated cells was extracted[25] and electrophoretically resolved in 1% formaldehyde/agarose gels as described.[16] Prehybridizations and hybridizations were carried out using the Wahl buffer[26] with full-length human H-*ras* or *p53* cDNA probes, under high stringency conditions.[16] After appropriate autoradiographic exposures, blots were stripped and rehybridized to a β-actin probe (Clontech, Palo Alto, CA) to ensure equal loading of the samples.

Protein Analyses

Total cell lysates from control and quercetin-treated cultures were analysed by western blotting procedures after 12.5% SDS-polyacrylamide gel electrophoresis. Procedures for lysate preparation, protein electrophoresis and immunodetection were as described.[16] Analysis of the H-*ras* p21 product was performed using the Y259 anti-*ras* monoclonal antibody,[27] and p53 was detected using the Ab-1 anti-p53 monoclonal antibody from Oncogene Science (Uniondale, NY). De novo protein synthesis was followed by immunoprecipitation and autoradiography of total extracts of cells metabolically labeled with L-[^{35}S]-methionine, after 12.5% SDS-polyacrylamide gel electrophoresis. In this case, p53 synthesis was monitored with the Ab-2 anti-p53 monoclonal antibody from Oncogene Science, which works better when p53 protein molecules are not attached to a solid support.

Other Experimental Procedures

Protein kinase C activity was determined as described[28] in cytosolic and membrane fractions prepared from murine cells, using a commercial N-acetylated peptide substrate derived from myelin basic protein (Boehringer Mannheim Biochemicals, Indianapolis, IN). The content of GDP- and GTP-bound p21^{ras} proteins were determined in total cell extracts as described previously.[29]

RESULTS

Effect of Quercetin on *ras*-Transformed NIH/3T3 Cells

NIH/3T3 cells transformed by transfection with activated human H-, K-, or N-*ras* oncogenes were tested in clonogenic assays for their sensitivity to quercetin in comparison with the untransfected murine fibroblasts. Results demonstrated that the LD_{50}, that is, the dose of quercetin that would allow the growth of only 50% of the cells plated relative to untreated controls, was not significantly affected in K-*ras*-transformed cells ($LD_{50} \approx$ 8.7μg/ml vs.8.5 μg/ml for NIH/3T3 cells), whereas cells transformed with either H-*ras* (LD_{50} $\approx$ 6 μg/ml) or N-*ras* ($LD_{50} \approx$ 4 μg/ml) were more sensitive to quercetin exposure. These results showed a perfect correlation with the observed effect of equivalent concentrations of quercetin on the growth rate of *ras*-transformed cells. Figure 1 shows the inhibitory effect of increasing concentrations of quercetin on the growth of H-*ras* transformed NIH/3T3 cells (Fig. 1B) relative to untransfected fibroblasts (Fig. 1A). At any given concentration, quercetin inhibited more efficiently the growth of *ras*-transformed cells than the growth of untransformed cells.

On the basis of these results, it could be predicted that the presence of quercetin during the transfection of NIH/3T3 cells with *ras* oncogenes would prevent the development of foci of transformed cells over the background of normal cells, and that this effect would be particularly pronounced in the case of H- or N-*ras*. We tested this prediction using an activated human H-*ras* oncogene[30] in transfection assays on NIH/3T3 cells carried out by the calcium phosphate co-precipitation method[21] during which quercetin was added at the LD_{50} for H-*ras*-transformed cells (6 μg/ml). Quercetin was added in all possible combinations, at each and every phase of the transfection process (before DNA addition, during DNA/gene transfer and/or during foci development for two weeks post-transfection). Regardless of the timing and pattern of addition, quercetin nearly completely blocked the appearance of foci. Under our experimental conditions, transfection assays in the absence of quercetin yielded about 45 to 50 foci of transformed cells per plate, whereas they resulted in 1 or 2 foci when the cells were exposed to quercetin. Interestingly, this effect could be reversed by removal of quercetin from the culture medium two weeks after transfection. However, the foci which developed after quercetin removal were phenotypically different from those generated in transfection assays performed in the absence of quercetin, showing a more relaxed piling-up of less refractile cells. Parallel experiments in which the H-*ras* oncogene was transfected as a recombinant construct in a vector carrying the NEO gene

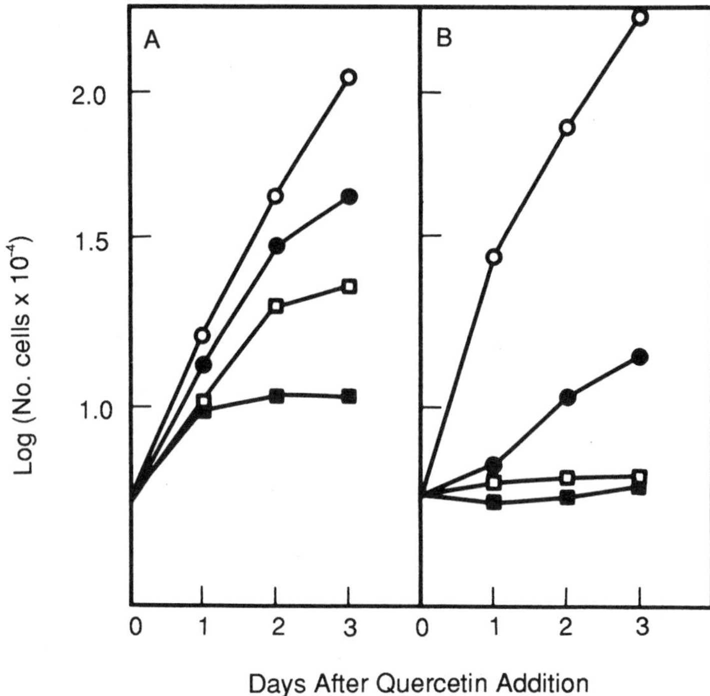

Figure 1. Effect of quercetin on the growth of mouse NIH/3T3 (A) and *ras*-transformed NIH/3T3 fibroblasts (B). Cells were cultured for several days in the presence of 4 μg/ml (●), 8 μg/ml (□) or 12 μg/ml (■) of quercetin, or were grown in the absence of the bioflavonoid (O).

demonstrated that the failure of the transformed foci to develop in the presence of quercetin was not due to the flavonoid preventing the entry of the *ras* DNA into cells. In several replicated experiments, above 90% of the foci which developed after removal of quercetin from the culture medium were resistant to G-418.

As an initial approximation to understand the mechanism of quercetin action we examined H-*ras*-transformed NIH/3T3 cells grown in the presence of increasing concentrations of quercetin for (*i*) their level of $p21^{ras}$ protein, (*ii*) their levels of protein kinase C (PKC) activity, and (*iii*) their content of GDP- and GTP-bound $p21^{ras}$ protein relative to untransfected NIH/3T3 cells. Although, as shown in Table 1, both cytosolic and membrane-bound PKC activities were inhibited effectively by increasing concentrations of quercetin, the extent and dose-dependence of this inhibition were not significantly different from those observed in parallel experiments with untransfected NIH/3T3 cells, suggesting that PKC is not directly involved in the response to quercetin in this cell system. On the contrary, preliminary experiments indicated that, despite the fact that the total content of $p21^{ras}$ protein in *ras*-transformed cells was not altered by quercetin treatment (Fig. 2A), the predominant form of the p21 protein which is isolated from quercetin-treated cells was in the GDP-bound, rather than the GTP-bound, active state (Fig. 2B). Although these data await further confirmation, they may be indicative of quercetin mediating the functional inactivation of the p21 protein in *ras*-transformed cells through a mechanism which does not affect the normal murine $p21^{ras}$ present in untransfected NIH/3T3 cells. Whether this is caused by a direct preferential effect of quercetin on the oncogenic p21 itself, or by interfering with its sorting to the cell surface or its membrane attachment remains to be elucidated. However,

Table 1. Quercetin effect on PKC activity in H-*ras*-NIH/3T3 cells

Treatment	Cytosol	Membrane
Control	760.00 ± 16.00	213.11 ± 2.22
Quercetin		
10 μg/ml	614.38 ± 18.62	158.61 ± 4.22
30 "	600.03 ± 6.49	151.95 ± 5.86
50 "	537.49 ± 10.35	148.08 ± 3.73
75 "	480.63 ± 9.69	113.56 ± 8.99
100 "	415.60 ± 15.33	95.82 ± 4.52

Activity is expressed as pmol ^{32}P/min/mg protein, determined as the mean ± SEM from five experimental replicas.

the specificity of the effect described above was demonstrated by the fact that quercetin glycoside analogues such as quercetrin and quercetin-3-rutinoside did not show any differential biological activity between untransfected and *ras*-transformed NIH/3T3 cells.

Effect of Quercetin on Human Mammary Carcinoma Cells Carrying Mutant p53

Quercetin induced a dose-dependent inhibition of the growth of MDA-MB468 cells, for which the LD_{50} was estimated to be about 7 μg/ml. Addition of 5, 10, 15 or 30 μg/ml of quercetin to exponentially growing cultures resulted in the reduction of the viable cell number to 62, 38, 27, and 20%, respectively, of the control value. These effects were

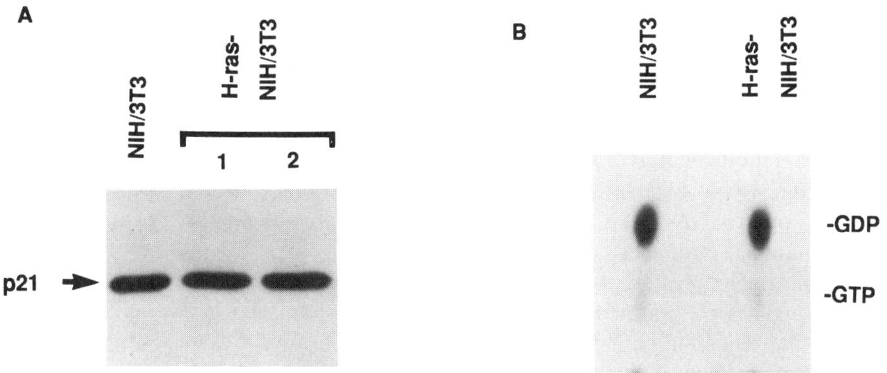

Figure 2. Effect of quercetin on the steady-state levels (A) and GTP-/GDP-bound ratio (B) of p21^{ras} in NIH/3T3 cells and *ras*-transformed NIH/3T3 cells. The ras protein was immunoprecipitated from total extracts of untreated cells or from two independent H-*ras*-transformed clones (1 and 2). Nucleotides were eluted and separated by thin-layer chromatography. The positions of GDP and GTP standards are indicated.

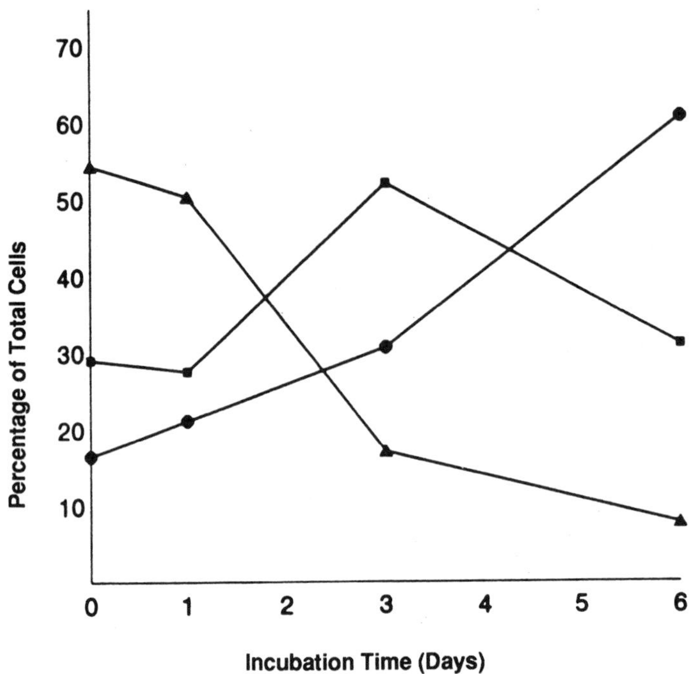

Figure 3. Effect of quercetin on the cell cycle progression of MDA-MB468 human mammary carcinoma cells. Two days after inoculation, cells were exposed to 30 μg/ml of quercetin, and the cultures were monitored for up to 6 days. The percentage of cells in G_1 (▲), S (■) or the G_2-M (●) phase of the cycle was determined by quantifying the DNA histograms obtained from FACS analysis. Each point corresponds to the mean of duplicated experiments.

correlated with the progressive accumulation of cells in the G_2-M phase of the cell cycle (Fig. 3) in a time- and dose-dependent fashion.

Because the mutant p53 present in MDA-MB468 cells has been directly implicated in the maintenance of their neoplastic phenotype,[20,31] we monitored the level of *p53* expression after exposure to quercetin. As shown in Figure 4A, the addition of increasing concentrations of quercetin, up to 75 μg/ml, had no effect on the steady-state levels of *p53* mRNA. However, quercetin induced a significant decrease in the level of p53 protein (Fig. 4B). This effect was dose-dependent, and at high concentrations (75 μg/ml) quercetin brought the total cellular p53 content to nearly undetectable levels. For any given dose of quercetin, its inhibitory effect on p53 levels was also time-dependent, with observable p53 inhibition after 4 h of exposure. These results suggested that p53 is a primary target for quercetin action in MDA-MB468 cells and, perhaps other human mammary carcinoma cells carrying mutant p53 protein.

In an attempt to gain some insight into the mechanism underlying the action of quercetin on p53, we examined whether the decrease in total p53 in the cells was due to quercetin either preventing p53 synthesis, or stimulating its degradation. Cells were labeled with L-[^{35}S]-methionine in the presence or absence of various concentrations of quercetin. Labeled cells were lysed after 5 h of quercetin exposure, and the de novo synthesized p53 protein was analyzed by immunoprecipitation, SDS-polyacrylamide gel electrophoresis and autoradiography. As shown in Figure 4C, quercetin treatment resulted in a dose-dependent inhibition of the synthesis of p53, strongly suggesting that the decrease in the steady-state

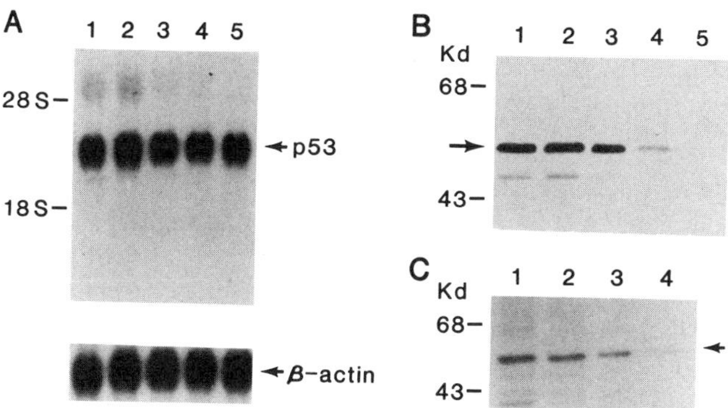

Figure 4. Effect of quercetin on the steady-state levels of *p53* mRNA (A) and protein (B), and the de novo synthesis of the p53 protein (C) in MDA-MB468 cells. Cells were not treated (lane 1 in all panels) or exposed to increasing concentrations (lanes 2-5 in panels A and B, lanes 2-4, in panel C) of quercetin. Total RNA was analyzed by northern hybridization with a full-length human *p53* cDNA probe (A, top panel) and rehybridized with a β-actin probe (A, bottom panel). Total p53 protein was immunodetected by western analysis (B), and the de novo synthesis of p53 was analyzed by immunoprecipitation from extracts of L-[^{35}S]- methionine. The positions of the *p53* mRNA and the p53 protein are indicated by name or by arrows. The migration of ribosomal RNAs and protein molecular size standards is also indicated. This figure includes, with permission from Cancer Research, a combination of data taken from reference 16.

levels of p53 protein observed after quercetin exposure was mediated by the inhibition of the translation of the p53 protein.

To determine the specificity of the quercetin effect on p53, and to rule out the possibility that it may be due to an overall inhibition of protein synthesis by quercetin, we studied the synthesis of other proteins in MDA-MB468 cells after quercetin exposure under identical experimental conditions. We included in our studies P-glycoprotein, which is known to be expressed constitutively at high levels in these cells, and the EGF-receptor, which can be induced efficiently in these cells by EGF treatment. The fact that quercetin treatment had no detectable effect on the synthesis of P-glycoprotein or the inducibility of the EGF-receptor[16] demonstrated the specificity for p53 of its translational inhibitory action. Nevertheless, the possibility that quercetin activated a mechanism by which the nascent p53 polypeptide was actively degraded as it was synthesized, or soon thereafter, can not be ruled out at this point, although it appears to be a rather unlikely explanation.

The specificity of the quercetin effect was further demonstrated by the fact that the quercetin glycoside analogues quercetrin and quercetin-3-rutinoside did not show any effect on the steady-state levels of p53 in MDA-MB468 cells under similar experimental conditions.

DISCUSSION

Screening increasing numbers of tumor-derived cell lines for their sensitivity to quercetin treatment continues to provide strong support for the notion that quercetin exerts its inhibitory effects on cell proliferation preferentially on transformed cells, and therefore it may be useful for the differential down-modulation of preneoplastic and neoplastic cells over a background of normal cells during the onset and/or development of tumors. Because quercetin is a component of the human diet, the possibility of modifying the diet to reach effective doses, or times of quercetin exposure becomes more of a reality. Unfortunately, little is known on the molecular basis of quercetin action, and consequently the selective enhancement of metabolic pathways which could potentiate even further the antineoplastic action of quercetin remains a distant objective.

In this regard, it is important to mention at this point that quercetin has been reported to inhibit the growth of the human breast cancer cell line MCF-7,[10] but its effect was much less dramatic than the action observed in our laboratory on MDA-MB468 cells. In the case of MCF-7, the inhibitory effect of quercetin was related to its ability to interact with type II estradiol binding sites.[10] In our case, because MDA-MB468 cells are estrogen receptor negative, this mechanism of action could not account for quercetin action. We performed experiments to determine the effect of quercetin on the expression of wild-type p53 using MCF-7 cells. Results (data not shown) indicated that, although normal p53 was somewhat down-regulated, much higher doses of quercetin were needed to attain the same magnitude of inhibition observed on mutant p53 in MDA-MB468 cells. Ongoing experiments are designed to elucidate whether this differential effect is due to a reduced quercetin uptake in MCF-7 cells, or to actual differences in the interaction of quercetin, or intermediates of quercetin, with wild-type and mutant p53 proteins. The p53 in MDA-MB468 cells contains a mutation at codon 273 which results in the substitution of the normal arginine for a histidine residue. This is a frequent mutation in a variety of human tumors.[32] Further studies will examine the usefulness of quercetin to control the proliferation of tumors from different tissue/organ origin containing the same p53 mutant species.

Results presented in this report point to the fact that, in two different systems, quercetin appears to target the function of gene products ($p21^{ras}$ and p53) presumed to be the primary determinants of the neoplastic properties of the cells. It is not immediately obvious what could be the common pathways through which quercetin causes the functional inactivation of both p21 and p53. However, it is possible to speculate that they might be either (*i*) a pathway(s) upstream of p53 which, by regulating its synthesis, is involved in the control of the transduction of signals received through members of the *ras* family of proteins, or (*ii*) a generalized response affecting a cellular cofactor required for efficient synthesis/stabilization of proteins, in the case of p53, and for the regulation of the activity of small G-proteins, such as $p21^{ras}$. A general regulatory role of that nature has been described for the nucleotide pool in the cells, in particular for the GTP/GDP ratio in different subcellular compartments.[33] In fact, the GTP/GDP ratio and the activity of enzymes which participate in its maintenance at physiological levels have been reported to be involved in the expression of the malignant phenotype of tumor cells.[34] Indeed, a novel oncogene, termed *cph*, isolated in our laboratory from neoplastic Syrian hamster embryo cells initiated with the carcinogen 3-methylcholanthrene, appears to be a new member of the family of GDP-exchange factors.[35] This is a quickly expanding group of proteins involved in the regulation of ras and other small G-proteins by controlling the local GTP-GDP ratio around their membrane attachment locations. However, further investigation is required in this area because there is no available information on the effect of quercetin or any other antineoplastic phytochemicals on the level and/or activity of these factors and enzymes. Our data clearly contribute to delineate two systems which could be used to begin answering some of the most urgent mechanistic questions on quercetin action.

ACKNOWLEDGMENTS

The authors wish to thank Dr. Terrance Leighton, for providing purified quercetin glycoside analogues, Dr. Owen C. Blair, for cell cycle analysis, and Dr. Anatoly Dritschilo, for support during the development of this project. This work was supported in part by the

Lombardi Cancer Research Center Flow Cytometry Core Facility, U.S. Public Health Service Grant 2P30-CA-51008.

REFERENCES

1. Kuhnau, J. The flavonoids. A class of semi-essential food components: their role in human nutrition. Wld. Rev. Nutr. Diet, *24:* 117-191, 1976.
2. Graziani, Y., and Chayoth, R. Regulation of cyclic AMP level and synthesis of DNA, RNA and protein by quercetin in Ehrlich ascites tumor cells. Biochem. Pharmacol., *28:* 397-403, 1979.
3. Gschwendt, M., Horn, F., Kittestein, W., and Marks, I. Inhibition of the calcium and phospholipid-dependent protein kinase activity from mouse brain cytosol by quercetin. Biochem. Biophys. Res. Commun, *117:* 444-447, 1983.
4. Grunicke, H., Hofmann, J., Maly, K., Uberall, F., Posch, L., Oberhuber, H., and Fiebig, H. The phospolipid- and calcium-dependent protein kinase as a target in tumor chemotherapy. Adv. Enzyme Regul., *28:* 201-216, 1989.
5. Graziani, Y., Erikson, E., and Erikson, R.L. The effect of quercetin on the phosphorylation activity of the Rous sarcoma virus transforming gene product in vitro and in vivo. Eur. J. Biochem., *135:* 583-589, 1983.
6. Levi, J., Teuerstein, I., Marbach, M., Radians, S., and Sharoni, Y. Tyrosine protein kinase activity in the DMBA-induced rat mammary tumor: inhibition by quercetin. Biochem. Biophys. Res. Commun., *123:* 1227-1233, 1984.
7. Akiyama, T., Ishida, J., Nakagawa, S., Ogawara, H., Watanaba, S., Itoh, N., Shibuya, M., and Fukami, Y. Genistein, a specific inhibitor of tyrosine-specific protein kinases. J. Biol. Chem., *262:* 5592-5595, 1987.
8. Beretz, A., Anton, R., and Stoclet, J.C. Flavonoid compounds are potent inhibitors of cyclic AMP phosphodiesterase. Experientia, *34:* 1054-1055, 1978.
9. Ruckstuhl, M., Beretz, A., Anton, R., and Landry, Y. Flavonoids are selective cyclic GMP phosphodiesterase inhibitors. Biochem. Pharmacol., *28:* 535-538, 1979.
10. Markaverich, B.M., Roberts, R.R., Alejandro, M.A., Johnson, G.A., Middledich, B.S., and Clarke, J.H. Bioflavonoid interaction with rat uterine type II binding sites and cell growth inhibition. J. Steroid Biochem., *30:* 71-78, 1988.
11. Suolinna, E.M., Buchsbaum, R.N., and Racker, E. The effect of flavonoids on aerobic glycolysis and growth of tumor cells. Cancer Res., *35:* 1865-1872, 1975.
12. Castillo, M.H., Perkins, E., Cambell, J.H., Doerr, R., Hassett, J. M.,Kandaswami, C., and Middleton, E. The effect of the bioflavonoid quercetin on squamous cell carcinoma of head and neck origin. Am. J. Surg., *158:* 351-355, 1989.
13. Yoshida, M., Sakai, T., Hosowa, N., Marvi, N., Matsumoto, K., Fujioka, A., Nishino, H., and Aoike, A. The effect of quercetin on cell cycle progression and growth of human gastric cancer cells. FEBS lett., *260:* 10-13, 1990.
14. Hosokawa, N., Hosokawa, Y., Sakai, T., Yoshida, M., Marvi, N., Nishino, H., Kawai, K., and Aoike, A. Inhibitory effect of quercetin on the synthesis of a possible cell-cycle-related 17 KDa protein in human colon cancer cells. Int. J. Cancer, *45:* 1119-1124, 1990.
15. Yoshida, M., Yamamoto, M., and Nikaido, T. Quercetin arrests human leukemic T-cells in late G_1 phase of the cell cycle. Cancer Res., *52:* 6676-6681, 1992.
16. Avila, M.A., Velasco, J.A., Cansado, J., and Notario, V. Quercetin mediates the down-regulation of mutant p53 in the human breast cancer cell line MDA-MB468. Cancer Res., *54*: 2424-2428, 1994,
17. Saito, D., Shirai, A., Matsushima, T., Sugimura, T., and Hirono, I. Test of carcinogenicity of quercetin, a widely distributed mutagen in food. Teratogenesis Carcinog. Mutagen., *1*: 213-221, 1980.
18. Tanaka, K., Ono, T., and Umeda, M. Pleiotropic effects of quercetin on the transformation of BALB 3T3 cells. Jpn. J. Cancer Res., *78*: 819-825, 1987.
19. Barbacid., M. ras genes. Annu. Rev. Biochem., *56*: 779-827, 1987.
20. Wang, N.P., To, H., Lee, W., and Lee, E.Y. Tumor suppressor activity of RB and *p53* genes in human breast carcinoma cells. Oncogene, *8:* 279-288, 1993.
21. Wigler, M., Silverstein, S., Lee, L.S., Pellicer, A., Chen, Y., Axel, R. Transfer of purified herpes virus thymidine kinase gene to cultured mouse cells. Cell, *11*: 223-232, 1977.
22. Nigro, J.M., Baker, S.J., Preisinger, A.C., Jessup, J.M., Hosteter, R., Cleary, K., Bigner, S.H., Davidson, N., Baylin, S., Devilez, P., Glover, T., Collins, F.S., Weston, A., Modali, R., Harris, C.C., And Vogelstein, B. Mutations in the *p53* gene occur in diverse human tumor types. Nature (London), *342:* 705-708, 1989.

23. Chen, J.Y., Funk, W.D., Wright, W.E., Shay, J.W., and Minna, J.D. Heterogeneity of transcriptional activity of mutant p53 proteins and *p53* DNA target sequences. Oncogene, *8:* 2159-2166, 1993.
24. Vindelov, L.L., Christensen I.J., and Nissen, N.I. A detergent-trypsin method for preparation of nuclei for flow cytometric DNA analysis. Cytometry, *3:* 323-327, 1983.
25. Chomczynski, P., and Sacchi, N. Single-step method of RNA isolation by acid guanidium thiocyanate-phenol-chloroform extraction. Anal. Biochem., *162:* 156-159, 1987.
26. Wahl, G.M., Stern, M., and Stark, G.R. Efficient transfer of large DNA fragments from agarose gels to diazobenzyloxymethyl-paper and rapid hybridization by using dextran sulfate. Proc. Natl. Acad. Sci. USA, *76:* 3683-3687, 1979.
27. Furth, M.E., Davis, L.J., Fleurdelys, B., and Scolnick, E.M. Monoclonal antibodies to the p21 product of the transforming gene of Harvey murine sarcoma virus and of the celular *ras* gene family. J. Virol., *43*: 294-304, 1982.
28. Hofmann, J., Doppler, W., Jakob, A., Maly, K., Posch, L., Überall, F., and Grunicke, H.H. Enhancement of the antiproliferative effect of *cis*-diamminedichloroplatinum(II) and nitrogen mustard by inhibitors of protein kinase C. Int. J. Cancer *42*: 382-388, 1988.
29. Burgering, B.M.Th., Medema, R.H., Maasen, J.A., van de Wetering, M.L., van der Eb, A.J., NcCormick, F., and Bos, J.L. Insulin stimulation of gene expression mediated by p21 ras activation. EMBO J., *10*: 1103-1109, 1991.
30. Santos, E., Tronick, S.R., Aaronson, S.A., Pulciani, S., and Barbacid, M. T24 human bladder carcinoma oncogene is an activated form of the normal human homologue of BALB- and Harvey-MSV transforming genes. Nature, *298*: 343-347, 1982.
31. Casey, G., Lo-Hsueh, M., Lopez, M.E., Vogelstein, B., and Stanbridge, E.J. Growth suppression of human breast cancer cells by the introduction of a wild type *p53* gene. Oncogene, *6:* 1791-1797, 1991.
32. Hollstein, M., Sidransky, D., Vogelstein, B., and Harris, C.C. p53 mutations in human cancers. Science (Washington DC), *253:* 49-53, 1991.
33. Pall, M.L. GTP: a central regulator of cellular anabolism. Curr. Top. Cell. Regul., *25*: 1-20, 1985.
34. Collart, F.R., Chubb, C.B., Mirkin, B.L., and Huberman, E. Increased inosine-5′-phosphate dehydrogenase gene expression in solid tumor tissues and tumor cell lines. Cancer Res., *52*: 5826-5828, 1992.
35. Velasco, J.A., Castro, R., Avila, M.A., Laborda, J., DiPaolo, J.A., Cansado, J., and Notario, V. *cph*, a novel oncogene which cooperates with H-*ras* in the transformation of NIH3T3 fibroblasts. Oncogene, *9*: 2065-2069, 1994.

9

EFFECTS OF MONOTERPENES AND MEVINOLIN ON MURINE COLON TUMOR CT-26 *IN VITRO* AND ITS HEPATIC "METASTASES" *IN VIVO*[*]

Selwyn A. Broitman, John Wilkinson IV, Sonia Cerda, and Steven K. Branch

Departments of Microbiology, and Pathology and Laboratory Medicine
Boston University School of Medicine and the Mallory Institute of Pathology
80 East Concord Street
Boston, Massachusetts 02118

INTRODUCTION

d-Limonene is a monocyclic monoterpene which has been used for many years as a flavoring, perfuming, or degreasing agent. Recently, this product and its biologic metabolites have been highlighted for their chemotherapeutic potential in the management of certain common malignant diseases. Some of its metabolic actions interfere with certain aspects of cholesterol metabolism which are vital to the malignant cell and hopefully less critical to the normal cell. The mammalian cell has, as a major component of its plasma membrane, cholesterol which helps to maintain the integrity of the cell, and modulates the fluidity of the cellular lipid membrane. It thus serves as a major regulatory agent for chemical signals sent from other cells, transport of critical metabolites in and out of the cell, and for the membrane association of receptors, enzymes, and a wide variety of proteins which serve to regulate the intracellular processes of the cell.[1]

A simplified scheme of intracellular cholesterol dynamics redrawn from the work of Brown and Goldstein is depicted in Figure 1.[2] Cholesterol homeostasis is primarily the result of three processes in the cell which are coordinately regulated. The first is the activity of the rate limiting enzymatic step of cholesterol synthesis, 3-hydroxy-3-methylglutaryl coenzyme A (HMGCoA) reductase. This enzyme produces mevalonate in the endoplasmic reticulum

[*] *Abbreviations used*: ACAT, acylCoA:cholesterol-acyl-transferase, FBS, fetal bovine serum; HMG CoA reductase, 3-hydroxy-3-methylglutaryl Coenzyme A reductase; 25-OH-CH, 25-hydroxycholesterol; LDL, low density lipoprotein; LDLR, LDL receptor; LPDS, lipoprotein deficient serum; MEV, Mevinolin.

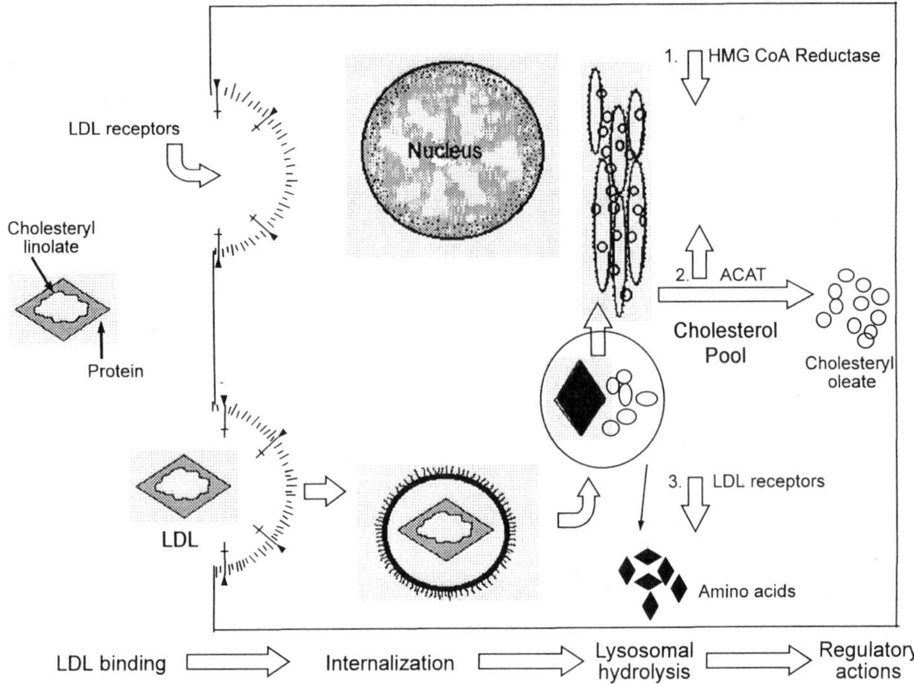

Figure 1. Cholesterol Homeostasis: A simplified scheme of intracellular cholesterol dynamics redrawn from the work of Brown and Goldstein[2] is depicted.

from HMGCoA. The synthesis via transcription and translation, and degradation of this enzyme are under strict control.[3] A second part of cholesterol homeostasis regulation is concerned with the storage and therefore removal of cholesterol which is not immediately required from the intracellular pool. Such excess cholesterol is esterified with oleic acid via the acyl CoA: cholesterol acyl transferase (ACAT) enzyme to form cholesteryloleate. The third element of intracellular cholesterol regulation occurs by controlling the transport of cholesterol into the cell. This is accomplished via alterations in the expression of LDL receptors on the cell surface (low density lipoproteins). These receptors pick up and internalize LDL from circulating plasma or from serum products in tissue culture. The LDL particle comprises a core of cholesterol esterified primarily to linoleic acid, and triglycerides within a phospholipid and free cholesterol monolayer that carries the apolipoprotein Apo-B-100 which is recognized by the LDL receptor. Upon internalization, the receptor-particle complex(s) form an endocytic vesicle in which the receptor is recycled and the particle is degraded to its cholesterol, amino acid, and fatty acid components. Ultimately the cholesterol is liberated thus contributing to the intracellular pool. When cellular demand for cholesterol is high and intracellular pools are low, signals are sent to increase both the level of HMG CoA reductase and the quantity of LDL receptors; conversely, at times when intracellular needs are satisfied these proteins are downregulated and the excess cholesterol will be stored via ACAT. This general schema was determined using human skin fibroblasts such as AG1519 and remains the classic paradigm for cholesterol metabolic regulation. However, significant variations on this theme are seen in hepatocytes, small intestinal cells, and in human and murine colon tumor cells.

In many individuals who are hypercholesterolemic, effective control of their serum cholesterol levels has been achieved using the drug mevinolin (Mevacor™, lovastatin). This competitive inhibitor of HMG CoA reductase effectively reduces the intracellular synthesis of cholesterol in hepatocytes. Diminished intracellular stores signal for the generation of increased LDL receptors, which internalize more plasma LDL to replenish the intracellular pools of cholesterol. The net result is that, over a period of time, plasma LDL levels progressively decrease and intracellular cholesterol pools ultimately achieve homeostasis with the lowered plasma cholesterol levels.

While cholesterol is the bulk product of the pathway controlled by HMG CoA reductase, the direct product of this enzyme, mevalonate, is precursor to a number of non-sterol compounds that are vital to a variety of cellular functions (see Figure 2).[3] The isoprene farnesol, in particular, has gained recent attention as it can become post-translationally incorporated into proteins in its native form or, after gaining five carbons, as the compound geranylgeraniol. This modification facilitates the membrane anchorage of such key proteins as nuclear lamins, G proteins, and perhaps the most intriguing from an oncologic chemotherapeutic perspective, the growth regulatory protein p21^{ras}. Strong inhibition of cholesterol biosynthesis by such drugs as MEV can therefore affect a wide variety of cellular

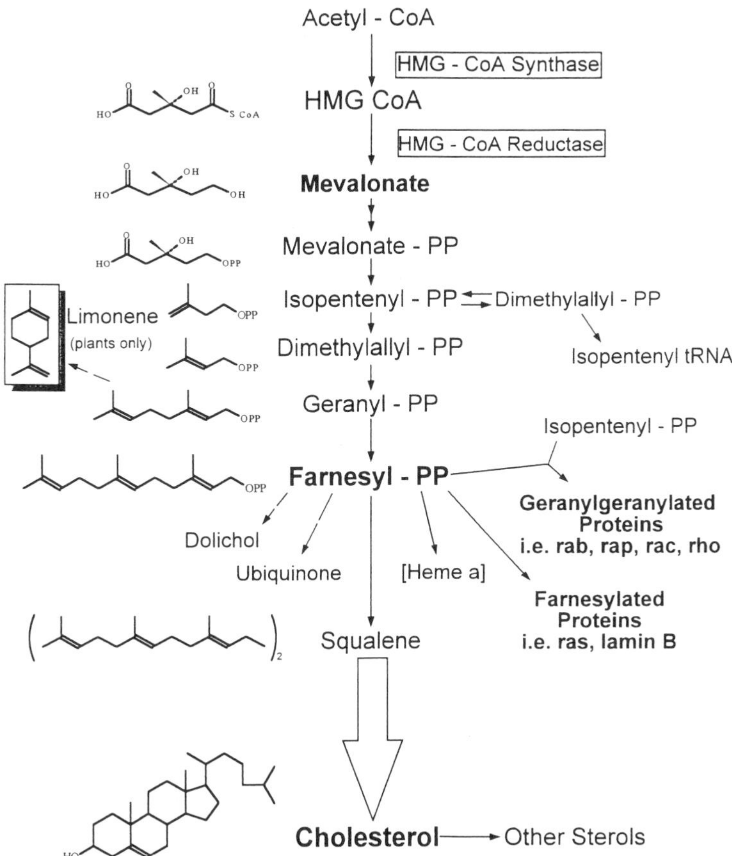

Figure 2. Biosynthesis of isoprenes and cholesterol from mevalonate: the cholesterol biosynthetic pathway and its connections to plant monoterpene generation is depicted. This figure has been adapted from the work of Brown and Goldstein.[3]

functions required for growth of the cell, which are not involved with sterol depletion *per se*.

Mevinolin is relatively nontoxic, and approximately 95% of the absorbed compound accumulates in the liver on the first pass. The possibility that this drug might be helpful in managing hepatic metastasis from colon tumors was derived from a study of a series of human colon tumor cell lines in which it was demonstrated that colon tumors have different dynamics in cholesterol metabolism from fibroblasts.[4] Figure 3 illustrates these differences with colon tumor cell line SW480, a colonic adenocarcinoma that is moderately well differentiated and is compared to the fibroblast AG1519. In these earlier studies, both cell lines were grown in a medium deficient in lipoproteins. This caused each respective cell to synthesize its own cholesterol for cellular needs because cholesterol cannot enter the cell from the suspending medium. With AG1519, the addition of MEV to the cell culture system altered the growth by the second day, illustrating the growth inhibitory effects of cholesterol synthesis inhibition. Addition of LDL caused a reversal of this process by the fourth day, and by the sixth day LDL totally negated the effect of MEV, indicating that LDL has entered the cell and that the cholesterol derived from the LDL particle had been used to replace the deficit in cholesterol synthesis resulting from MEV inhibition. A different picture emerged in the colon tumor cell line SW480 in which MEV treatment inhibited cellular growth, and the addition of LDL, even in high concentrations, was inadequate to counter this inhibition. It was determined that a) there were fewer LDL receptors on the SW480 cell and that even though some LDL gained access to the cell, it was incapable of reversing growth. This

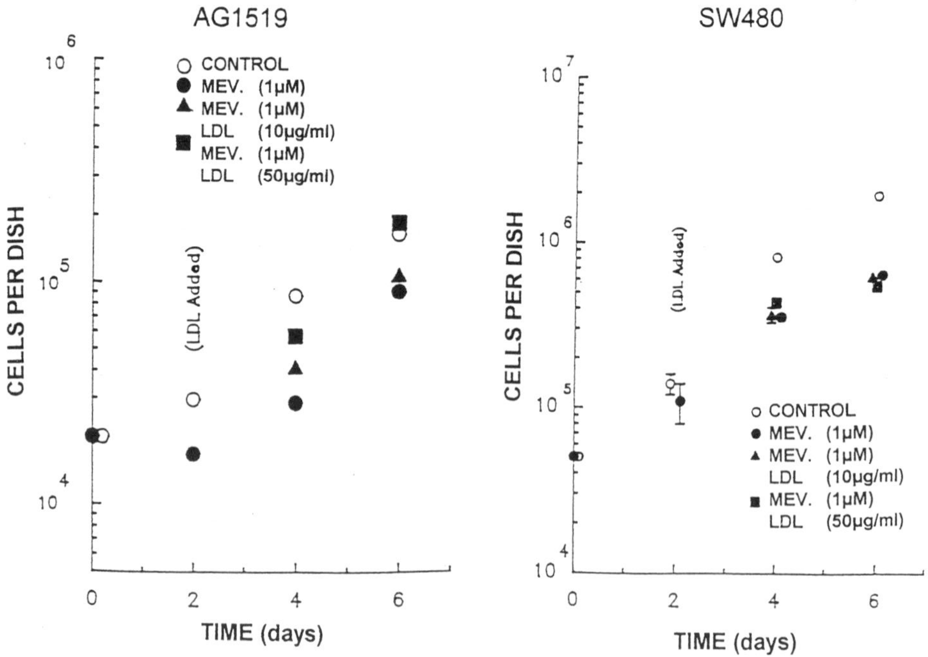

Figure 3. Effect of MEV and LDL on growth of AG1519 fibroblast and SW480 colonic adenocarcinoma cell line[4]: cells were seeded in 35 mm dishes at a density of $1x10^5$ cells/dish in culture medium. MEV (MEV) was added at the concentrations described above to cultures at day 2. Low density lipoprotein (LDL) was added to certain groups as depicted in the concentrations described on day 4. Medium was changed on days two and four. Cell growth was determined by trypan blue staining and viable cells counted on days two, four, and six in a hemocytometer.

suggested that cholesterol dynamics in colon tumor cells were subject to different regulations from those in fibroblasts/hepatocytes. Other studies showed that mevalonate could rescue colon tumor cells from growth inhibition by MEV (see), implying that such inhibition was based on depletion of mevalonate derivatives which occur prior to cholesterol synthesis rather than depletion of cholesterol as seen with the fibroblast. Among these derivatives are isoprene compounds which in addition to serving as the skeletal structure of sterols, were shown to play a critical role in post-translational protein modification.

Isoprenylation of proteins was first noted by Kamiya *et al.* in 1978[5] and later by Sakagami in 1981,[6] for their role in fungal mating tube formation. Schmidt *et al.*[7] indicated that isoprenylation similarly occurred in mammalian cells and this phenomenon was further characterized by Beck *et al.* 1988.[8] Isoprenylation or generically lipidation facilitates protein insertion into a membrane bilayer enabling it to anchor to the membrane.[9] Approximately 0.5% of animal cell proteins are prenylated.[10] Of these it is believed that geranylgeranylation accounts for about 80-90% of isoprenylated protein, while a few, such as prelamin A, lamin B, and p21ras are farnesylated.[11-14]

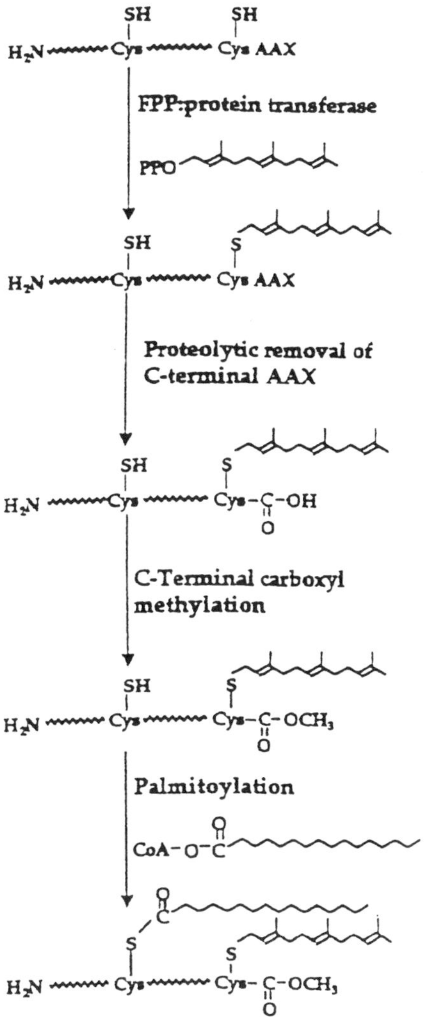

Figure 4. Isoprenylation and further processing of ras proteins: membrane localization of ras involves a series of post-translational modifications increasing its hydrophobicity and enabling its anchorage to the plasma membrane (from Grunler *et al.*).[15]

The processing of the ras gene product p21^{ras} is illustrated in Figure 4 (from Grunler *et al.*).[15] This protein functions normally in signal transduction transferring a regulated intermittent cellular growth signal. A mutation in the pattern encoding for certain strategically placed amino acids impairs the regulation in transmitting its signal, committing it to a constant transmission of its growth signal. For this protein to function normally it requires a series of post-translational modifications that will increase its lipophilicity and thereby enable it to anchor into the membrane. This process of prenylation, whether it involves a 15-carbon farnesyl or the 20-carbon geranylgeranyl portion of these isoprenoid moieties, depends upon the specific sequence of the terminal amino acids.[16,17] In p21 ras the terminal signal amino acids are commonly called CAAX box, in which cysteine is depicted as C, aliphatic amino acids are designated as A, and the carboxy terminal amino acid is designated as X, which can be one of various amino acids. This CAAX motif is found on all proteins that are farnesylated,[16,17] and other proteins with different motifs may be geranylgeranylated.[18] Figure 4 illustrates the enzymatic steps in this process. Initially a thioether linkage between farnesyl pyrophosphate and the C terminal cysteine of the CAAX ras motif is formed via a farnesyl transferase enzyme. The C terminal amino acids AAX are removed by proteolysis and the carboxy group of the now terminal cysteine is methylated. In addition, certain species of ras as N and H but not K undergo an additional post-translation modification, palmitoylation of a cysteine adjacent to the C terminal region. In turn, the avidity of the membrane association is enhanced by this process. A number of investigators have targeted ras prenylation as a focal point in cancer intervention.[3,18,19]

Recently, Crowell *et al.* demonstrated that d-limonene and such related compounds as perillyl alcohol may selectively inhibit the isoprenylation of 21-26-kDa proteins.[20] d-Limonene, a monoterpene plant product, is synthesized from isoprenoids as shown in Figure 2. d-Limonene ($C_{10}H_{16}$molecular weight 136.2) occurs in nature in a variety of foods as well as non-food products. It is a skin irritant and sensitizer of moderate toxicity. The lethal dosage in humans is between 0.5 to 5.0 grams per kilo. It si particularly abundant in the peel and pulp of citrus fruits; after juice production it is a product of the deterpenation of citrus oils. In orange juice, its average concentration is around 100 ppm. In 1984 approximately 254,000 kg of this compound was utilized in the United States in such products as a lemon fragrance in a number of household cleaners and soaps, perfumes, and as flavoring agents for various food products, including beverages, baked goods, puddings, etc. It has usage in a variety of lubricating oils and as an industrial solvent replacing chlorinated hydrocarbons.[21]

Because of its widespread use in the United States, a toxicology and carcinogenesis study was conducted by the U.S. National Toxicology Program.[21] An excerpt of that report is depicted in Figure 5. d-Limonene was administered by gavage in corn oil to rats and mice for periods of 13 weeks or 2 years. As depicted, male and female rats and male and female mice could tolerate 1,200 and 1,000 mg/k for 13 weeks. A major difficulty over the short term experiment was weight loss which ranged from 2% to 12% compared to controls. Female rodents tended to tolerate the d-limonene to a somewhat better extent than males. In the 2 year study, somewhat less than 20% of male rats experienced renal tubular adenomas and adenocarcinomas. Information since this report indicates that this is peculiar to the F344/N rat and not other strains. The decreased survival of female rats fed 600 mg/kg resulted principally from the killing of moribund animals and second to a relatively large percentage of accidental deaths due to gavage in this group (approximately 16%). Once again, with the exception of the male mice, a constant finding appeared to be a slight weight loss relative to control animals. The genetic toxicology testing indicated that it was not mutagenic, either in the Ames assay, in the mouse L5178Y lymphoma cell assay with or without activation by S9, nor were there chromosomal aberrations or sister chromatid exchanges in cultured Chinese hamster ovary (CHO) cells. Since these reports, it has been possible to feed higher

d-Limonene Given By Gavage Five Days per Week To F344/N Rats and B6C3F1 Mice

13 Weeks Rodents	Dosage (mg/kg)	Survival % Of Control	Weight Change % of Control
Male Rats	1200	100	-12.0
Female Rats	1200	100	-2.0
Male Mice	1000	90	-11.0
Female Mice	1000	100	-2.3
2 Years Rodents	**Dosage (mg/kg)**	**Survival % Of Control**	**Weight Change % of Control**
Male Rats	150	100 *	-2.0
Female Rats	600	57	-5.0
Male Mice	500	100	+4.0
Female Mice	1000	100	-10.0

* Renal Tubular Adenomas And Adenocarcinoma in less than 20%

Genotoxicity Testing Of d-Limonene - Negative

Figure 5. Excerpt from the national toxicology report on d-Limonene (from Jameson).[21]

levels of d-limonene for short periods of time to rodents, particularly if an allowance is made for the animals to accommodate to the diet. It is irritating to rats and mice, consequently rodents will avoid the diet initially, but, after a brief period of time, start to consume the d-limonene when mixed in a chow or semi-synthetic diet.

Homburger *et al.*[22] first called attention to the inhibitory properties of d-limonene on the induction of subcutaneous tumors in mice by benzo(*rst*)pentaphene. Wattenberg noted that orange oil served as an "antipromoter" of DMBA (7,12-dimethylbenz[*a*]anthracene) induced mammary carcinomas.[23] Others noted promotional effects or no effects of d-limonene in other tumor systems[21]. Elgebede *et al.*[24] demonstrated shortly thereafter that female Sprague-Dawley rats had a significant reduction in DMBA-induced mammary gland tumors when fed diets containing 1000 or 10,000 ppm d-limonene from 1 week before to 27 weeks after a single oral dose of DMBA. Subsequently, they fed rats a 10% d-limonene diet and noticed regression of carcinogen induced mammary gland tumors.[25] Other work has focused attention on the feasibility of using d-limonene and related monoterpenes as chemopreventive and chemotherapeutic agents for the prevention or treatment of mammary cancer.[26]

This laboratory has focused similarly on the ability of another protein isoprenylation inhibitor, MEV, to inhibit growth of colon tumor cells both *in vitro*[4,27,28] and *in vivo*.[29] While MEV inhibits colon tumor growth *in vitro* it was unclear whether the drug was inhibiting the cells through cholesterol or mevalonate deprivation or both. Subsequent experiments indicated that when mevalonate was added *in vitro* to colon tumor cells subjected to growth inhibition with MEV, the cells were rescued, implying a different demand for isoprenoids in tumor cells than in fibroblasts. This suggestion is supported by an increased rate of synthesis evidenced by colonic tumor cell lines, which is not repressible by lipoprotein (cholesterol) addition to culture media.[30] This dependency upon isoprenoids in these colonic adenocarcinomas led to the concept that in these cells MEV might be interfering with cellular functions that require these intermediates namely protein isoprenylation. Inhibitors of HMG CoA reductase have been shown to possess protein isoprenylation.[31] Thus, it was reasoned that the use of d-limonene (*in vivo*) and its hydroxyl derivative, perillyl alcohol (*in vitro*)[20] in conjunction with MEV could provide an additional growth inhibitory effect since the monoterpenes had been demonstrated to be isoprenylation inhibitors.

MATERIALS AND METHODS

Cell Culture

The murine colonic cancer cell line, CT-26, a gift from Dr. Michael Brattain, was grown in monolayer in plastic 75 cm^2 flasks and incubated at 37°C in a 5% CO_2/95% air atmosphere. Culture medium consisted of RPMI 1640 containing antibiotics (penicillin, 200 units/ml; streptomycin, 0.2 mg/ml), glutamine (0.2 mg/ml) and nonessential amino acids and nonessential vitamins both at 10 ml of 100X solution (GIBCO, Grand Island, NY), per 500 ml and supplemented with 10% fetal bovine serum (FBS) (SIGMA, St. Louis, MO). Cells were plated at a density of 10^6 cells per flask, and cultures divided weekly (1:19 or 1:20) using 0.05% trypsin/0.02% disodium EDTA. After removal of media, cells were rinsed with isotonic phosphate buffered saline (PBS) (pH 7.4), and 4 ml of trypsin was added. Flasks were incubated at 37°C until cells began to detach, after which 4-6ml of fresh medium was added. Cells used in experiments were grown in plastic 35 mm^2 dishes.

Preparation of Lipoprotein-Deficient Serum (LPDS)

Lipoproteins were removed from FBS by density gradient centrifugation.[32] FBS was adjusted to a density of 1.21 g/ml with solid potassium bromide (64.4 g KBr per 200 ml FBS) and centrifuged for 48 hours at 5°C at 244,000 x g_{max} (45,000 rpm in a 50.2ti rotor in a Beckman L8-70 ultracentrifuge). The top lipoprotein fraction ($d<1.21$ g/ml) was removed and the bottom lipoprotein-deficient fraction dialyzed extensively against 0.01% EDTA/double distilled water (pH 8) and then PBS. The LPDS was sterilized by passage through a 0.22 μ filter.

Preparation of MEV

This reagent was kindly provided by A. Alberts of Merck Sharp & Dohme (Rahway, NJ) in the lactone form. Saponification was carried out by adding 0.1 ml absolute ethanol and 0.1 ml of 0.1 N NaOH to 4 mg. of MEV and heating at 50°C for two hours; the solution was neutralized with 5% HCl to pH 7.3. Stock solutions (4 mg/ml) were prepared by bringing the volume to 1 ml with dimethyl sulfoxide (DMSO) and were stored at -20°C. At the concentrations used, DMSO was found to have no effect on cell growth. MEV was diluted in MEM to the concentrations used in the experiments. All the studies regarding the use of MEV were conducted at a 1 uM concentration as published previously.[4]

Protein Determinations

Proteins were determined by the BCA reagent method (Pierce Chemical Company). A standard curve was prepared using a stock BSA standard in sample diluent. Samples and standard were mixed with the protein reagent and absorbance read at 562 nm by spectrophotometry.

Inhibition of Cell Growth with MEV and Perillyl Alcohol Combinations

Cells were seeded in triplicate in 12 well dishes (Costar, Cambridge, MA) at a density of $1x10^4$ cells/well in culture medium. Perillyl alcohol (Fluka, Ronkonkoma, NY) and MEV were added at varying concentrations, as described in the figure legend, to cultures at day 1. The final volume per dish was 0.5 ml. Medium was replaced again on day three. Cell growth was determined by trypsinizing the wells with trypsin/EDTA and resuspending with medium containing serum. Aliquots were stained with trypan blue and viable cells were counted in a hemocytometer on days 1, 2, 3, and 4.

Cholesterol Synthesis Assay

Cells were seeded in 6 well (35 mm^2) dishes at 20,000 cells/well in triplicate per group studied per experiment, in culture media and were grown to a subconfluent-confluent density where approximately 50% of the plate was covered after day 4. Cells were then rinsed with Dulbecco's PBS and switched to a lipoprotein deficient serum (LPDS) based medium (1 ml/well). After 24 to 48 hours, the media was replaced with media containing 5 μCi/ml of 2-^{14}C-acetate μ(51 mCi/mmole NEN), and the following treatments: 1) FBS (fetal bovine serum), 2) LPDS, the basal media for all subsequent treatments (each of which contained 10% LPDS), 3) LDL (50 μg/ml), 4) 25-OH-CH (25-OH-CH at 0.5 μg/ml + cholesterol at 16 μg/ml), 5) MEV at a 1μM concentration, and 6) MEV + LDL. After a 24 hour incubation, the medium was removed and saved; monolayers were harvested in 1 ml 0.1 N NaOH and pooled with their respective saved media to be saponified following the methodology of Brown and Goldstein.[32] Briefly, these mixtures were treated with 0.5 ml 50% KOH, 3 ml ethanol, and (1,2) ^{3}H-cholesterol (10^5 CPM, 50 Ci/mmole) as internal standard. After saponification at 80°C for two hours, unsaponifiable lipids were extracted three times with petroleum ether followed by two washes with NaOH. Samples were dried with nitrogen gas, resuspended in chloroform, spotted on silica gel "G" thin layer chromatography plates, and separated using chloroform as a running solvent to isolate the ^{14}C-cholesterol with appropriate markers. This methodology discriminated 27-C-sterols from lanosterol and squalene, which had higher Rf values. Although this running system does not resolve cholesterol from other 27-C-sterols, GC analysis of the derivatized lipids from the scrapings showed a predominance of cholesterol with some desmosterol and 7α-dehydro-cholesterol (data not shown). Cholesterol synthesis was evaluated as nanomoles of acetate incorporated into cholesterol per mg cellular protein, and these values were normalized to the LPDS control within each experiment. Results of challenge with various treatments were expressed as percentage of LPDS control. Data were analyzed by general linear model ANOVA to determine significant differences between groups.

Protein Isoprenylation Assay

Protein isoprenylation was measured as described.[20,31] Briefly, cells were seeded in MEM + 10% FBS at a density of 5x105/ 35 mm2 dish (Costar, Cambridge, MA). When cells reached 70% confluence, they were treated with 30 μM MEV for 24 hrs. Following this incubation period, cells were treated for 3 additional hrs in MEM + 10% FBS with 30 μM MEV, 15 μCi of ^{14}C-mevalonate at 54.1 mCi/mmol (NEN), and perillyl alcohol at 1 mM. Control cells received the same supplemented media without perillyl alcohol. Following treatments, cells were harvested with trypsin, centrifuged at 3,000 x g, and washed twice with PBS. Pellets were then suspended in 150 μl of electrophoresis sample buffer and equal amounts of protein were subjected to SDS-PAGE as described by Laemmli[33] using 12% separating and 4% stacking polyacrylamide gels. Instruments were from Hoeffer Scientific Instruments (San Francisco, CA) and reagents from Biorad Life Science (Hercules, CA). Gels were then stained with Coomassie blue, permeated with fluorographic enhancer (En3Hance, New England Nuclear, Boston, MA), and dried with a Biorad Gel Dryer for 2.5 hrs at 70°C. They were then left to cool for an additional 30 minutes, followed by exposure to preflashed X-OMAT AR film (Eastman-Kodak Co., Rochester, NY) at -70°C for 14 days. Film was then developed using an x-ray film processor (Fuji, Photofilm USA Inc.) and fluorograms were scanned with a phosphorimager:SI (Molecular Dynamics, Sunnyvale, CA).

Tissue Culture for Mouse Implantation

At confluence, cells were removed from the flask with trypsin/EDTA, washed with media, pelleted at 250 x g, and resuspended in serum-free media to 10^6 cells/0.05 ml for splenic and hepatic administration. Cell viability was monitored by trypan blue exclusion; we have found cellular survival after nine hours on ice to commonly exceed 95%.[34]

Diets Used for Animal Studies

A laboratory powdered chow from Purina-Ralston especially suitable for mixing the liquid (d-limonene) and powdered (MEV) ingredients was utilized in the preparation of various MEV and d-limonene treatment diets. These diets were supplemented in-house using our own diet kitchen. d-Limonene based diets were prepared weekly, since 58% of d-limonene remains intact after two weeks storage in diet at 5°C.[21]

Direct Hepatic Implantation

Mice were randomly assigned to one of four treatment groups in which they received 1) no treatment, 2) MEV administered in the diet (400 mg/kg diet), 3) d-limonene (5% w/w) 4) both MEV and d-limonene. Mice were anesthetized by injection of phenobarbital (50 mg/kg body weight, *i.p.*) and the liver was visualized via laparotomy. Cells of the syngeneic murine colonic tumor CT-26 were inoculated subcapsularly (10^6 in 0.05 ml serum free media) into the anterior right hepatic lobe at an approximate depth of 5-10mm using a 30 gauge needle. Animals were killed at 14, or 28 days and necropsied to determine primary tumor volume, metastases, and the state of uninvolved organs. Tumor volumes were calculated and plotted for each treatment group. Tumors were removed and sections excised and used for routine histology and in the biochemical assays described herein. Growth rates were calculated based on tumor measurements obtained from animals killed at 14, and 28 days following implantation.

Hepatic Colonization from Intrasplenic Inoculation

The ability of CT-26 tumor cells to form liver "metastases" was assessed in 6-8 week old Balb/cbyj mice (Jackson Laboratories) following intrasplenic injection. Mice were anesthetized as described for the direct hepatic assay. A small incision was made exposing the medial aspect of the spleen and the spleen gently externalized. Viable tumor cells (1×10^6 in 0.05 ml) were slowly injected into the medial splenic tip using a 27 gauge needle raising a visible pale welt. The spleen was gently replaced and the abdominal wall was closed with sutures and the skin with metal wound clips. The entire procedure was performed using aseptic techniques. At autopsy after 21 days the organs were removed and examined grossly under a dissecting microscope. Microscopic examinations of H&E sections of the liver, lungs, spleen, pancreas and other organs were carried out routinely. Tumor volumes of individual colonies were measured with vernier calipers, and total tumor volume per liver calculated.

Measurement of HMG CoA Reductase Activity in Liver and CT-26 Implanted in the Liver of Balb/C Host

Washed microsomal fractions from mouse liver and implanted CT-26 tumors were isolated according to the method of Ness et al.[35] In this process any inhibitors, such as MEV, were removed from the HMG-CoA reductase enzyme preparation. Specific activity measurements were made essentially as described by Shapiro *et al.*[36] and represent the cell's *total* amount of reductase enzyme.[37]

RESULTS AND DISCUSSION

The effects of MEV and perillyl alcohol on the growth of the murine colon tumor cell line CT26 were assessed (Figure 6). Control cells were considered as showing 100% growth and various treatments were taken as a percentage of the growth of controls. Growth curves marked with an asterisk are significantly different from the controls. In cells in which 1 μM or 2 μM MEV were used, cell growth of these murine tumor cells was inhibited to 62% and 7% of control growth at the two concentrations of MEV, respectively. The addition of 0.5 mm of perillyl alcohol had no effect, and 1 mm exerted a slight but insignificant effect. However, when 0.5 mm perillyl alcohol was added to 1 mm MEV, growth inhibition was significantly different from the controls and from cells given only 1 mm of MEV alone. Thus, there appeared to be a modest additive effect when MEV and perillyl alcohol were used together.

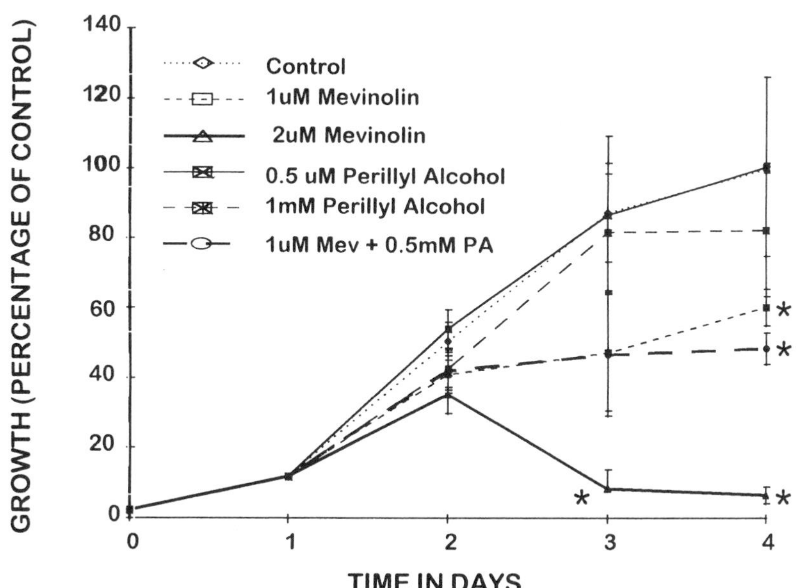

Figure 6. Effect of MEV and perillyl alcohol on growth of CT-26 cells: cells were seeded in triplicate in 12 well dishes at a density of 1×10^4 cells/well in culture medium. Perillyl alcohol and MEV were added at varying concentrations as described above to cultures at day 1. Medium was replaced again on day three. Cell growth was determined by trypan blue staining and viable cells counted daily in a hemocytometer.

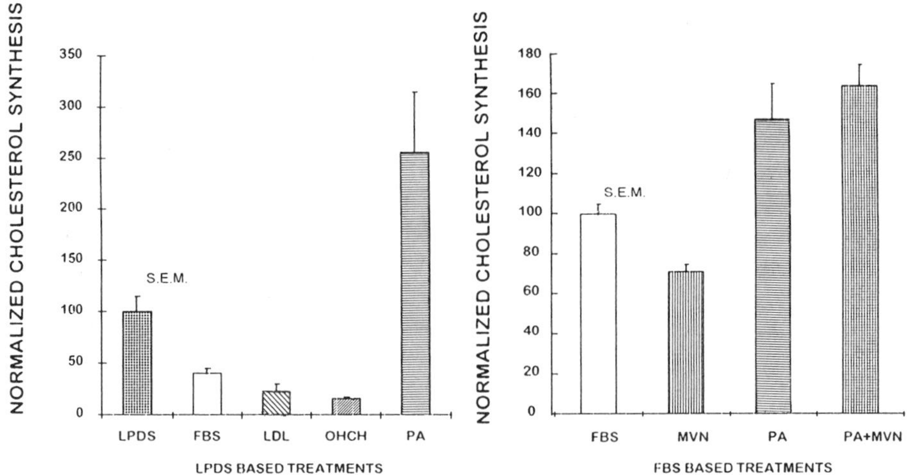

Figure 7. Cholesterol synthesis in CT-26 cells: murine colonic CT-26 tumor cells were grown as described under materials and methods to a state of subconfluency in normal media. Media was then changed to provide a lipid poor (LPDS based) or lipid rich (FBS based) environment. Treatments as described above were added along with ^{14}C-Acetate precursor and cells were incubated for 48 hours. Cells were then harvested and cholesterol production determined as incorporation of ^{14}C-acetate into non-saponifiable sterols.

Figure 7 illustrates cholesterol synthesis in CT26 cells in the presence of MEV and perillyl alcohol. On the left, the series of experiments was carried out in lipoprotein deficient serum (LPDS) which maximizes cholesterol synthesis, to determine the effects of LDL in a media devoid of lipoproteins. Synthesis of cholesterol by these cells was set as a control and considered 100%. Addition of fetal bovine serum (FBS), which is rich in lipoprotein and cholesterol, signals the rate limiting step of HMG CoA reductase to reduce activity and consequently cholesterol synthesis was less than 50% of controls in cells treated in this manner. When LDL was added to these tumor cells, it also regulated cholesterol synthesis causing a reduction. The addition of 25-hydroxycholesterol which regulates the sterol responsive pathways in a non-receptor dependant manner, also inhibited cholesterol synthesis. Conversely, the addition of perillyl alcohol *increased* cholesterol synthesis to 250% of controls.

On the right hand graph, CT26 cells were grown in the presence of fetal bovine serum to ascertain cellular responses in a complete media. Here the FBS treated values obtained were used as a control and set at 100% for cholesterol synthesis. The addition of MEV (1mM) is shown to inhibit cholesterol synthesis to approximately 70%; once again the addition of perillyl alcohol (1mM) significantly increased cholesterol synthesis in these cells. Furthermore, when added with MEV it appeared to override the effects of MEV and raise cholesterol synthesis two-fold over the addition of MEV alone.

Perillyl alcohol and MEV were previously shown to inhibit protein isoprenylation in a number of tumor cell lines.[31,38-40] The effects of these inhibitors alone and in combination upon protein isoprenylation pools in the colonic carcinoma cell line CT-26 were evaluated. Protein isoprenylation was assessed by measuring the incorporation of ^{14}C-mevalonate into proteins as described in the materials and methods section. Figure 8 shows an autoradiogram of ^{14}C-mevalonate labeled proteins under the following treatments: FBS, perillyl alcohol 1 mM, 2 μM MEV, and perillyl alcohol + MEV. Densitometric analysis of ^{14}C-mevalonate

Mevinolin (2μM)	-	+	-	+	-	+	-	+
Perillyl Alcohol (1mM)	-	-	+	+	-	-	+	+
Cycloheximide (20ug/ml)	-	-	-	-	+	+	+	+
Lane Number	1	2	3	4	5	6	7	8

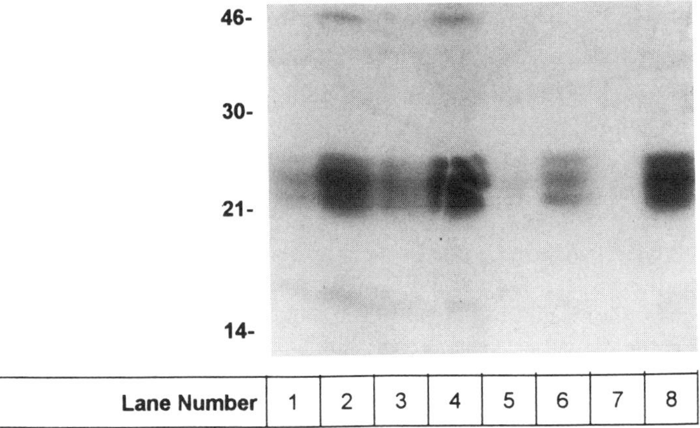

Figure 8. Effect of MEV and perillyl alcohol treatment on protein isoprenylation in CT-26 cells cultured *in vitro*: The effects of these inhibitors alone and in combination upon protein isoprenylation pools in the colonic carcinoma cell line CT-26 were evaluated. Depicted is an autoradiogram of ^{14}C-mevalonate labeled proteins under the following treatments: FBS (lane 1), 2 μM MEV (lane 2), perillyl alcohol 1 mM (lane 3), and perillyl alcohol + MEV (lane 4) Protein isoprenylation was assessed by measuring the incorporation of ^{14}C-mevalonate into proteins as described in the materials and methods section. CA). Cycloheximide (CHX) treatment concurrent with addition of labeled mevalonate precursor prevents labeling of newly synthesized proteins, permitting measurement of pro-proteins that have accumulated as a result of the 18 hour pre-treatment with the tested compounds. Therefore, an increased label under conditions of CHX addition generally indicates an inhibition of isoprenylation, which was necessary to cause such accumulation. These treatments are depicted as: FBS+ CHX (lane 5), 2 μM MEV+ CHX (lane 6), perillyl alcohol 1 mM+ CHX (lane 7), and perillyl alcohol + MEV + CHX (lane 8). Analysis was conducted using densitometry.

incorporation into the 21-28 kDa proteins showed an increase in label incorporation of 1.3 fold in 1 mM perillyl alcohol treated cells as compared to the FBS control (lane 3 vs. lane 1). 2 μM MEV treatment resulted in a 2.1 fold increase in labeling (lane 2 vs. lane 1). Combined treatment generated a 2.1 fold increase similar to MEV alone (lane 4 vs. lane 1). Cycloheximide (CHX) treatment concurrent with addition of labeled mevalonate precursor prevents labeling of newly synthesized proteins, permitting measurement of pro-proteins that have accumulated as a result of the 18 hour pre-treatment with the tested compounds. Therefore, an increased label under conditions of CHX addition generally indicates an inhibition of isoprenylation, which was necessary to cause such accumulation. Treatment with MEV left a pool of unisoprenylated proteins which was two-fold greater than the untreated control cells (lane 6 vs. lane 5). Perillyl alcohol at 1mM caused a decrease in label incorporation into pro-proteins in the presence of cycloheximide (lane 7 vs. lane 5). Concurrent addition of perillyl alcohol and MEV to cells resulted in a large increase in label incorporation into pro-proteins of 4.4 fold as compared to control cells (lane 8 vs. lane 5).

The effects of MEV on protein isoprenylation are predictable, namely generation of a decreased endogenous cold mevalonate pool and therefore an increase in labeled mevalonate incorporation as the tracer is less dilute. Cycloheximide treatment concurrent with labeled precursor addition generated an increased signal. This suggests that the mevalonate inhibition imposed by MEV is sufficient to restrict protein isoprenylation to the extent that pro-unisoprenylated proteins accumulate, waiting for the isoprenoids required for their maturation.

Perillyl alcohol treatment causes a slight increase in label incorporation possibly indicative of 1) a slight inhibitory effect upon HMG CoA reductase similar to that seen with MEV treatment (which also results in increased label incorporation), 2) an enhancement of protein isoprenylation possibly due to restriction of mevalonate derived pathway intermediates into such side products as ubiquinones and dolichols, thus shunting the label towards protein isoprenylation in these cells, 3) an initial inhibition of isoprenylation so pronounced that the enzymes responsible for isoprene transfer may be upregulated to compensate. Cycloheximide treatment concurrent with labeled precursor addition resulted in a decrease in label incorporation into pro-proteins as compared to the control (lane 7 vs. lane 5). This result, when compared to the potential explanations of the non cycloheximide perillyl alcohol treated results are consistent with an increase in protein isoprenylation resulting in a decrease in the pro-protein pool. This could result from either a direct effect of perillyl alcohol or as a compensation for another effect such as an initial decrease in HMG CoA reductase activity. The other alternative is that perillyl alcohol causes an inhibition in protein isoprenylation at a late step in the process in a manner which does not affect precursor or isoprene pools. This latter possibility is not supported by the non-cycloheximide treated group results (which demonstrated an increase in label vs. control). Additionally the monoterpene may be affecting the protein's expression exclusive of its effects on its isoprenylation.[41]

Concurrent treatment with MEV and perillyl alcohol caused a 2.1 fold increase in label incorporation identical to that seen with MEV treatment alone. Cycloheximide treatment concurrent with labeled precursor addition in cells pretreated with MEV and perillyl alcohol generated a 4.4 fold increase in label incorporation into pro-unisoprenylated proteins versus the control group. This implies that MEV and perillyl alcohol are affecting the isoprenylation process through two discrete mechanisms which result in an actual increase in the pro-unisoprenylated protein pool due to MEV. This is further increased due to a distal inhibitory effect of perillyl alcohol on this process at a step in the pathway subsequent to mevalonate synthesis. Another factor that needs to be considered is the potential increase in the specific activity of the labeled precursor due to MEV's inhibition of endogenous mevalonate. This finding is consistent with the observation that perillyl alcohol can inhibit the enzymatic action of protein:farnesylpyrophosphate transferase.[42]

While the actions of these compounds in concert are not entirely consistent with their actions alone, it must be remembered that the entire spectrum of effects of general isoprenoid depletion (MEV) and protein isoprenylation inhibition (perillyl alcohol) are not known, making their effects in combination even less predictable. Additionally, it is possible that isoprenylated proteins themselves may be involved in the regulation of protein isoprenylation and thus general inhibition imposed by these compounds may disrupt the very elements which are responsible for its control of this process.

Thus, *in vitro* it appeared that the effects of the two compounds perillyl alcohol and MEV vary depending on the parameter assessed, as perillyl alcohol is dominant in cholesterol synthesis, while MEV seems to be as strong an isoprenylation inhibitor. Therefore, the following two types of studies were conducted to ascertain if such activities would cause tumor growth inhibition *in vivo*. The premise of these studies is that colon tumors, which metastasize to the liver, may be subject to control by agents which affect cholesterol metabolism and its intermediates differently in tumor cells than in hepatocytes. For this reason, colon tumors were implanted into the liver and mice were fed diets containing d-limonene (5%) or MEV (400 mg/kg) or the combination of both. Two routes of hepatic inoculation were used. In the first, a direct hepatic implant was conducted in which the liver was exposed and a constant site chosen in which 10^5 CT26 cells were implanted. This methodology enables tumors to implant directly into the liver and avoids, to a great extent, the contribution of various immune cells, including the Kupfer cells lining the hepatic sinusoids. In the second method (splenic implantation) 10^5 CT26 cells were implanted directly into the spleen, which was exteriorized following surgery and then replaced. Over a period of time, tumor cells were meted out from the spleen and ultimately implanted in the liver. The size of these tumors was then estimated by measurement and volumes calculated.

In the direct hepatic implantation assay shown in Figure 9, it may be seen that at 14 days there were no differences and by 28 days there was a suggestive effect in animals fed the d-limonene diet. However, because of the wide variations in tumor size and therefore the non-parametric nature of the data, none of these findings were significantly different from the controls. Figure 10 illustrates the effect of d-limonene and MEV and the combination in the splenic colonization assay. There was an approximate 70% to 80% reduction in tumor volumes by d-limonene and MEV added to the diet. This represented a significant inhibition of resultant tumor volumes versus control, ($p < 0.05$, Kruskal Wallis non-parametric ANOVA). Of interest was the finding that the combination of both d-limonene and MEV in the diet was equally effective as either agent alone. However, at this level of tumor growth inhibition, discrimination between additive or synergistic effects was not possible.

To determine if the dietary treatments were capable of altering hepatic or tumor HMG CoA reductase activity *in vivo* the following study was accomplished. Tissues taken at necropsy from mice fed the various diets above which received hepatic or splenic CT-26 tumor implantations were used to isolate microsomal fractions for HMG CoA reductase determinations by the method of Ness *et al.*[35] Results are depicted in Figure 11 and emphasize the differences in cholesterol dynamics which can occur between tumor and normal liver

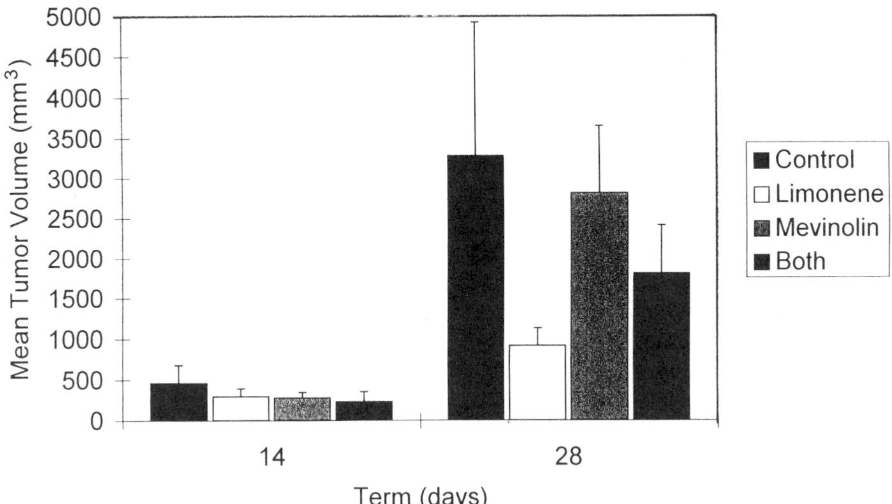

Figure 9. Effect of d-Limonene and MEV on growth of CT-26 in direct hepatic implantation assay in Balb/c host. Mice were randomly assigned to one of four treatment groups in which they received: 1) no treatment, 2) MVN administered in the diet (400mg/kg diet), 3) d-limonene (5% w/w) 4) both MVN and d-limonene for two weeks prior to surgical implantation of 10^6 CT-26 tumor cells directly into the right major hepatic lobe. At 14 or 28 days following implantation, mice were killed and tumor sizes measured.

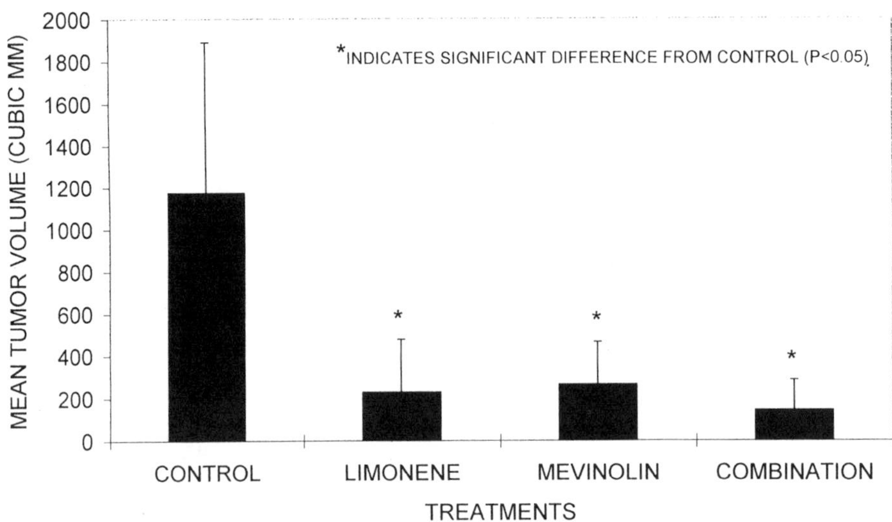

Figure 10. Effect of d-Limonene and MEV on growth of CT-26 in splenic colonization (35 Day) assay in Balb/c host. Thirty-two mice were randomly assigned to one of four treatment groups in which they receive: 1) no treatment, 2) MVN administered in the diet (400mg/kg diet), 3) d-limonene (5% w/w) 4) both MVN and d-limonene for two weeks prior to surgical implantation of 10^6 CT-26 tumor cells into the body of the spleen. At 35 days following implantation, mice were killed and tumor sizes measured.

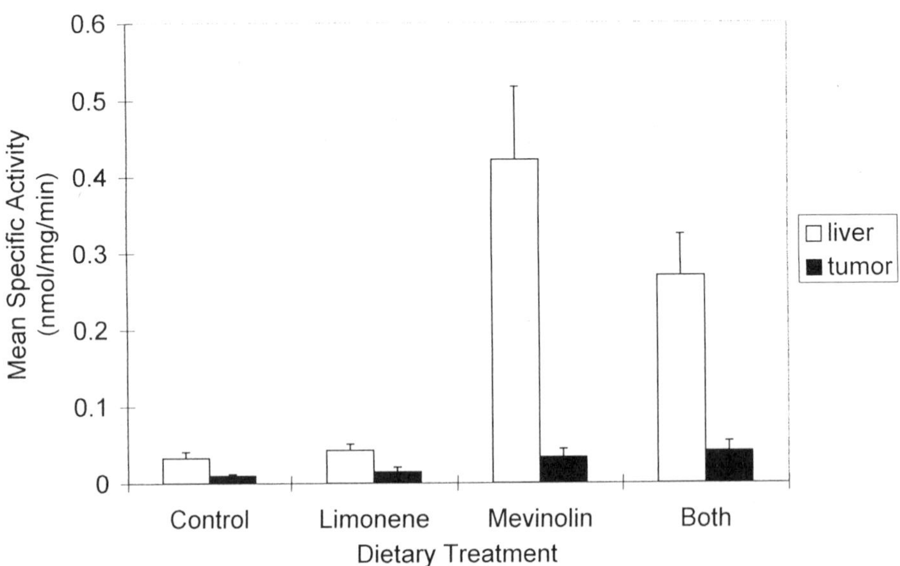

Figure 11. Effect of MEV and d-Limonene on HMGCoA reductase activity in livers and tumors from Balb/c host. Mice received diets containing: 1) no treatment, 2) MVN administered in the diet (400 mg/kg diet), 3) d-limonene (5% w/w) 4) both MVN and d-limonene for 4-6 weeks prior to assessment. Mice were killed and tissue microsomes harvested at 7am for each experiment. Numbers of animals in each group depicted are: control-fed liver, n = 17; control-fed tumor, n = 6; d-limonene-fed liver, n = 16; d-limonene-fed tumor, n = 5; MEV-fed liver, n = 21; MEV-fed tumor, n = 5; MEV and d-limonene-fed liver, n = 18; MEV and d-limonene-fed tumor, n = 7.

tissue. MEV treatment typically caused a chronic inhibition of HMG CoA reductase activity which resulted in a compensatory increase in enzyme protein levels over time. Liver reductase activity followed a predictable response pattern to dietary MEV treatment, by significantly (ANOVA, $p<0.05$; Kruskal-Wallis z-value test for multiple comparisons) increasing total (washed microsomal) enzyme activity 12.8 fold. Reductase activity in livers from mice fed d-limonene treated diets did not differ significantly from the control animals. Liver reductase activity in mice fed diets with both supplements demonstrated a significant 8.2 fold increase versus control. While HMG CoA reductase activity of implanted CT-26 tumors from mice fed untreated diet was 30% of that in the control livers, this was not a statistically significant difference. MEV and the combination treatments resulted in a 3.4 and 4.2 fold increase versus control tumors respectively. However these responses to dietary treatments were not significant.

The responses of the liver HMG CoA reductase activity indicate that treatments were successful in reaching this site. The ability of tumors to respond to a HMG CoA reductase inhibitor was of interest because in normal hepatic/fibroblast cells a compensation typically takes place and this might decrease the impact of the treatment over time. The lessened and therefore insigificant responses seen in tumor tissue may indicate that 1) these treatments do not reach the tumor as successfully, 2) these cell types respond differently to the treatments. Similarities apparent between HMG CoA reductase activity *in vivo* (see Figure 11) and cholesterol synthesis data *in vitro* (see Figure 7) support the second possibility. Cholesterol synthesis in FBS supplemented media *in vitro*, is presumably similar to that in the hepatic site *in vivo* from which the tumors were taken. Under these conditions the cells respond to the related monoterpene perillyl alcohol by increasing their cholesterol synthesis roughly 1.5 fold as compared to a 1.5 fold increase in HMG CoA reductase activity in tumors from mice fed d-limonene treated diets. Taken as a whole, results indicate that while these two agents differentially affect the same pathway, and while it might seem that their individual actions are antagonistic, their overall action is to cause tumor cell growth inhibition both *in vitro* and *in vivo* alone and in combination.

SUMMARY

Tumors derived from the colonic epithelium exhibit cholesterol metabolism which is clearly different from that in fibroblasts, hepatocytes, adrenals, and ovaries. In hepatocytes and fibroblasts MEV inhibition of the rate limiting step in cholesterol synthesis HMG Co A reductase can be overcome by the uptake of LDL . Colon cancer cells however do not overcome MEV inhibition by LDL uptake but rather exhibit further growth suppression Mevinolin (Mevacor™), a drug used to lower serum cholesterol levels has the advantage of accumulating in the liver to approximately 95% with the first pass. A small but variable percentage of non-sterol precursors may escape inhibition and be utilized for other pathways in the isoprenylation of certain proteins, among them members of the ras family. Mutated ras, an oncogene, is found in 40-50% of colon tumors and the expression of a functional gene product is dependent on isoprenylation for anchorage to the tumor cell membrane. d-Limonene, a relatively non-toxic monoterpene found in orange skin oil, selectively inhibits isoprenylation and also accumulates to some extent in the liver. It was hypothesized that the differences in mevalonate metabolism between hepatocytes and colon tumor cells could provide a chemotherapeutic advantage in which MEV and/or d-limonene could effectively inhibit cholesterol synthesis and post-translational modification of proteins with non-sterol cholesterol precursors in colon tumor derived hepatic metastases and thus inhibit their growth. Since each drug affects aspects of mevalonate synthesis at different points, the effects of the combination of their agents on inhibiting tumor metastases was investigated to ascertain if these could be additive. In tissue culture, MEV and d-limonene significantly inhibited the growth of CT-26, a murine transplantable colon tumor. Cholesterol synthesis assessed in these cells indicated that in lipid deficient media the following additions - 25-hydroxycholesterol, and LDL significantly reduced cholesterol synthesis. Conversely, perillyl alcohol increased cholesterol synthesis 2.5 fold. In cells cultured in FBS based medium, which have an FBS control, MEV treatment reduced cholesterol synthesis to 65% of control. Perillyl alcohol increased synthesis 1.4 fold and when given in conjunction with MEV, it abolished the effects of this inhibitor. In isoprenylation studies of 14C-mevalonate incorporation into proteins, MEV impaired isoprenylation by restricting synthesis of mevalonate derived intermediates. Results of CT-26 treatment with perillyl alcohol are inconsistent with its putative role as a protein isoprenylation inhibitor. The combination of these agents indicates an additive action which requires additional investigation to elucidate their mechanism(s). Dietary MEV and d-limonene were evaluated alone and in combination for their chemotherapeutic potential in a hepatic "metastais" model. Using splenic colonization in which CT-26 was implanted into the spleen and ultimately seeded the liver, each of these compounds were found to inhibit the growth of resultant tumors both alone and in combination by approximately 80% versus controls at 35 days post-implantation. Assessment of HMGCoA reductase in liver and tumor indicated that these agents were effective in reaching these target sites. The findings to date indicate that while d-limonene and MEV may differentially affect the same pathway, and their individual actions may appear antagonistic in vitro, their overall action individually or together, appears promising as a chemotherapeutic modality for the possible management of hepatic metastases from colon cancer.

ACKNOWLEDGMENTS

These studies have been supported in part by the research grant 93B34 from the American Institute for Cancer Research, a training grant 5T 32 CA 09423 (JW) and a predoctoral fellowship 1F31CA636226 (SC) from the National Cancer Institute of the National Institutes of Health, and by the Karin Grunebaum Cancer Research Foundation Fellowship (SKB).

The authors would like to acknowledge the excellent techincal assistance of Dr. Sandra Cerda, Marielle Nguyen, Jeff Z. Broitman, and Paul Colon.

REFERENCES

1. Chen H.W. Role of cholesterol metabolism in cell growth. Fed. Proc. 43(1);126-130, (1984).
2. Brown, M.S., and Goldstein, J.L. A receptor mediated pathway for endocytosis. Science, 232:34-47, (1986).
3. Brown, M.S. and Goldstein, J.L. Regulation of the mevalonate pathway. Nature 343:425-430, (1990).
4. Fabricant, M. and Broitman, S.A. Evidence for deficiency of low density lipoprotein receptors on human colonic carcinoma cell lines. Cancer Res. 50, 632-636, (1990).
5. Kamiya, Y., Sakurai, A., Tamura, S., Takahashi, N., Abe, K., Tsuchiga, E., Fakai, S., Kitada, C., and Fujino, M. Structure of rodotorucien A novel lipopeptide inducing mating tube formation in Rhodosporidium toruloides. Biochem. Biophys. Res. Commun. 83:1077-1083, (1978).
6. Sakagami, Y., Yoshida, M., Isogai, A., Suzuki, A. Peptidal sex hormones inducing conjugation tube formation in compatible mating-type cells of Tremella mesenterica. Science 212:1525-1527, (1981).
7. Schmidt, R.A., Schneider, C.J., Glomset, J.A. Evidence for post-translational incorporation of a product of mevalonic acid into Swiss 3T3 cell proteins. J. Biol. Chem. 259:10175-10180, (1984).
8. Beck, L.A., Hosick, T.J. and Sinensky, M. Incorporation of a product of mevalonic acid metabolism into proteins of Chinese hamster ovary cell nuclei. J. Cell Biol. 107:1307-1316, (1988).

9. P.J. Casey. Protein lipidation in cell signaling. Science 268:221-224, (1995)
10. Marshall, C.J. Protein prenylation: a mediator of protein-protein interaction. Science 259:1865-1866, (1993).
11. Sinesky, M., Fantle, K., Trujillo, M.A., McLain, T.M., Kupfer, A., and Dalton, M. The processing pathway of prelamin a. J. Cell. Sci. 107:61-67, (1994).
12. Wolda, S.L. and Glomset, J.A. Evidence for modification of lamin B by a product of mevalonic acid. J. Biol. Chem. 263:5997-6000, (1988).
13. Hancock, J.S., Kadwallader, K. and Marshal, C.J. Methylation and proteolysis are essential for efficient membrane binding of prenylated p25$^{k\text{-}ras(b)}$. E.M.B.O. J. 10:641-646, (1991).
14. Dalton, M.B., Fantle, K.S., Bechtold, H.A., DeMaio, L., Evans, R.M., Krystosek, A. and Sinensky, M. The farnesyl protein transferase inhibitor BZA-5B blocks farnesylation of nuclear lamins and p21^{ras} but does not affect their function or localization. Cancer Res. 55:3295-3304, (1995).
15. Grunler, J., Ericsson, J., Dallner, G. Branch-point reactions in the biosynthesis of cholesterol, dolichol, ubiquinone and prenylated proteins. Biochim. Biophys. Acta 1212:259-277, (1994).
16. Hancock, J.F., Magee, A.I., Childs, J.E., and Marshall, C.J.. All ras proteins are polyisoprenylated but only some are palmitoylated.. Cell 57:11167-11177, (1989).
17. Casey, P.J., Solski, P.A., Der, C.J., and Bus, J.E.. p21ras is modified by a farnesyl isoprenoid. Proc. Natl. Acad. Sci. USA 86:8323-8327, (1989).
18. Sinensky, M. and Lutz, R.J. The prenylation of proteins. Bioessays 14:25-31, (1992).
19. Khosravi-Far, R., Cox, A.D., Kato, K., and Der, C.D. Protein prenylation: key to ras function and cancer intervention?. Cell Growth and Differentiation 3:461-469, (1992).
20. Crowell, P.L., Chang, R.R., Ren, Z., Elson, C.E. and Gould, M.N. Selective inhibition of isoprenylation of 21-26 kDa proteins by the anticarcinogen d-limonene and its metabolites. J. Biol. Chem. 266:17679-17685, (1991).
21. Jameson, C.W. Toxicology and carcinogenesis studies of d-limonene in F344/N rats and B6C3F$_1$ mice. NIH publication #90-2802, (1990).
22. Homburger, F., Trager, A., Boger, E. Inhibition of murine subcutaneous and intravenous benzo(rst)pentaphene carcinogenesis by sweet orange oils in d-limonene. Oncol. 25:1-10, (1971).
23. Wattenberg, L.W. Inhibition of neoplasia by minor dietary constituents. Cancer Res. (suppl) 43:2448s-2453s, (1983).
24. Elgebede, J.A., Elson, C.E., Qureshi, A., Tanner, M.A., Gould, M.N. Inhibition of DMBA-induced mammary cancer by the nonoterpene d-limonene. Carcinogenesis 5:661-664, (1984).
25. Elgebede, J.A., Elson, C.E., Tanner, M.A., Qureshi, A., Gould, M.N. Regression of rat primary mammary tumors following dietary d-limonene. J Natl. Cancer Inst. 76:323-325, (1986).
26. Gould, M.N., Wacker, W.D., and Maltzman, T.H. Chemoprevention and chemotherapy of mammary tumors by monoterpenoids. In: *Mutagens and Carcinogens in the Diet* , Pariza, M.N., Felton, J.S., Aeschbacher, H., and Sato, S., Eds. Wiley-Liss, New York (1990).
27. Broitman, S.A., Wilkinson, J. IV, and Cerda, S. Inhibitors of cholesterol/ isoprenoid pathways as potential chemotherapeutics in colon cancer. Abstract # 1075P, World Congresses Gastroenterology, Abstract #1075P, Los Angeles, CA, (1994).
28. Cerda, S., Wilkinson, IV, J. and Broitman, S.A. Enhanced antitumor activity of lovastatin and perillyl alcohol combination in the colonic adenocarcinoma cell line SW480. Proc. Am. Assoc. Cancer Res., 35:#1996, (1994).
29. Wilkinson, J., Cerda, S., Cerda, S.R., Branch, S.K., and Broitman, S.A. Limonene and MEV affect growth of the human colonic adenocarcinoma LS174T implanted in the livers of nude mice. Proc. Am. Assoc. Cancer Res., 36:1776, (1995).
30. Cerda, S., Wilkinson, J. IV, and Broitman, S.A. Regulation of cholesterol synthesis in 4 colonic adenocarcinoma cell lines. Lipids (in press).
31. Repko, E.M., Maltese, W.A. Post-translational isoprenylation of cellular proteins is altered in response to mevalonate availability. J. Biol. Chem., 264: 9945-9952, (1989).
32. Brown, M.S., Faust, J.R., Goldstein, J.L., Kaneko, I., and Endo, A. Induction of 3-hydroxy-3-methylglutaryl coenzyme A reductase activity in human fibroblasts incubated wuth compactin (ML-236B), a competitive inhibitor of the reductase. J. Biol. Chem. 253, 1121-1128. (1978)
33. Laemmli, U.K. Cleavage of structural proteins during the assembly of head of bacteriophage T4. Nature, 227: 680-685, (1970).
34. Cannizzo, F., and Broitman, S.A. Postpromotional effects of dietary marine or safflower oils on large bowel or pulmonary implants of CT-26 in mice. Cancer Res., 49:4289-4294, (1989).
35. G.C. Ness, C.E. Sample, M. Smith, L.C. Pendleton, and D.C. Eichler. Characterisitics of Rat Liver Microsomal 3-Hydroxy-3-methylglutaryl-coenzyme A Reductase. Biochem. J. *233*:167-172, (1986).

36. D.J. Shapiro, J.L. Nordstrom, J.J. Mitschelen, V.W. Rodwell, and R.T. Schimke. Micro Assay for 3-Hydroxy-3-methylglutaryl-CoA Reductase in Rat Liver and in L-cell Fibroblasts. Biochim. Biophys. Acta *370*:369 (1974)..
37. G.C. Ness, Z. Zhao, and L. Wiggins. Insulin and Glucagon Modulate Hepatic 3-Hydroxy-3-methylglutaryl-coenzyme A Reductase Activity by Affecting Immunoreactive Protein Levels. J. Bio. Chem. *269*:29168-29172 (1994)..
38. Crowel, P.L., Ren, Zhibin, Lin, Shouzhong, Vedejs, E., Gould, M.N. Structure-activity relationships among monoterpene inhibitors of protein isoprenylation and cell proliferation. Biochem. Pharmacol., 47: 1405-1415, (1994).
39. Crowell, P.L., Chang, R.R., Ren, Z., Elson, C.E., Gould, M.N. Selective inhibition of isoprenylation of 21-26-kDa proteins by the anticarcinogen d-limonene and its metabolites. J. Biol. Chem., 266:17679-17685, (1991).
40. Schultz, S., Buhling, F., Ansorge, S. Prenylated proteins and lymphocyte proliferation: inhibition by d-limonene and related monoterpenes. Eur. J. Immunol. 24:301-307, (1994).
41. Hohl, R.J., and Lewis, K. Differential effects of monoterpenes and lovatstatin on ras processing. J Biol. Chem. 270:17508-12, (1995).
42. Ren, Z., Gould, M.N. Inhibition of protein isoprenylation by the monoterpene perillyl alcohol in intact cells and in cell lysates. Proc. Am. Assoc. Cancer Res., 36:585, (1995).

10

ANTITUMORIGENIC EFFECTS OF LIMONENE AND PERILLYL ALCOHOL AGAINST PANCREATIC AND BREAST CANCER

Pamela L. Crowell, A. Siar Ayoubi, and Yvette D. Burke

Department of Biology
Indiana University-Purdue University at Indianapolis
Indianapolis, Indiana 46202

ABSTRACT

Perillyl alcohol is a natural product from cherries and other edible plants. Perillyl alcohol and *d*-limonene, a closely related dietary monoterpene, have chemotherapeutic activity against pancreatic, mammary, and prostatic tumors. In addition, perillyl alcohol, limonene, and other dietary monoterpenes have chemopreventive activity. Several mechanisms may account for the antitumorigenic effects of monoterpenes. For example, many monoterpenes inhibit the post-translational isoprenylation of cell growth-regulatory proteins such as Ras. Perillyl alcohol induces apoptosis without affecting the rate of DNA synthesis in both liver and pancreatic tumor cells. In addition, monoterpene-treated, regressing rat mammary tumors exhibit increased expression of transforming growth factor β concomitant with tumor remodeling/redifferentiation to a more benign phenotype. Monoterpenes are effective, nontoxic dietary antitumor agents which act through a variety of mechanisms of action and hold promise as a novel class of antitumor drugs for human cancer.

DIETARY SOURCES OF MONOTERPENES

Monoterpenes are non-nutritive dietary components found in the essential oils of citrus fruits, cherry, spearmint, dill, caraway, and other edible plants. Their functions may be as chemoattractants or chemorepellents, as they are largely responsible for the plant's pleasant fragrance. These simple 10 carbon isoprenoids are derived from the mevalonate pathway in plants but are not produced in mammals. For example, in spearmint and other plants, *d*-limonene is formed by the cyclization of geranylpyrophosphate by the enzyme limonene synthase.[1] Limonene then serves as a precursor to a host of other plant monocyclic monoterpenes such as carvone, carveol, and perillyl alcohol (Fig. 1). The most abundant

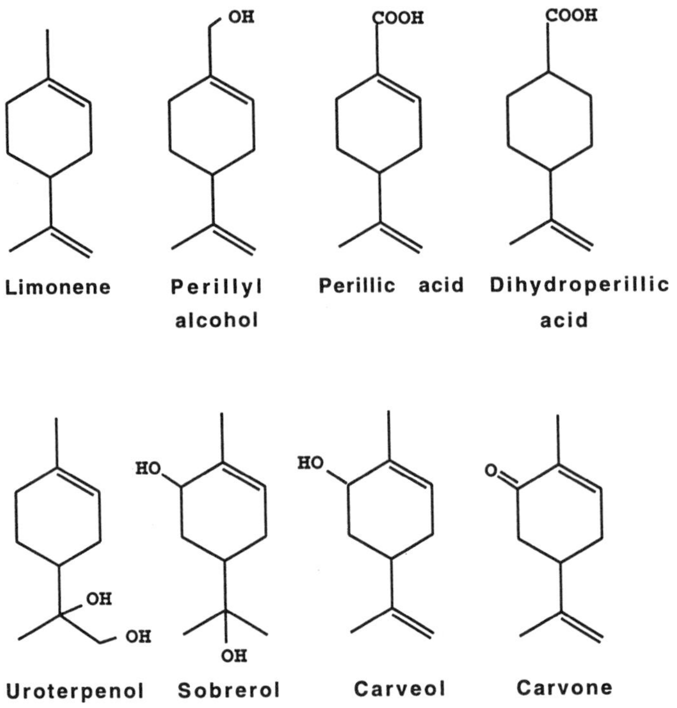

OH COOH COOH

Limonene Perillyl alcohol Perillic acid Dihydroperillic acid

HO HO O OH OH OH

Uroterpenol Sobrerol Carveol Carvone

Figure 1. Monoterpene structures.

food source of limonene is orange peel oil, which is 90-95% *d*-limonene by weight.[2] Other citrus oils are rich in limonene as well.

ANTITUMOR ACTIVITY OF LIMONENE

d-Limonene has a broad range of antitumor activities.[2,3] Limonene has chemopreventive activity against spontaneous and chemically-induced rodent mammary, skin, liver, lung, and forestomach cancers, as well as *ras* oncogene-induced rat mammary cancer.[4] In rat mammary carcinogenesis models, the chemopreventive effects of limonene are evident at both initiation and promotion. Limonene also has chemotherapeutic activity against rat mammary tumors, causing complete regression of >80% of established DMBA- or NMU-induced mammary carcinomas.[5] A Phase I clinical trial testing limonene's cancer chemotherapeutic activity is in progress.[6]

LIMONENE METABOLISM

Limonene is rapidly metabolized *in vivo*. The plasma metabolites of limonene in rodents and humans are perillic acid, dihydroperillic acid, the methylesters of these acids, and limonene-1,2-diol.[7,8] Limonene metabolites are far more abundant than limonene itself in the circulation of limonene-fed rats and humans, even within 20 minutes of limonene ingestion. Thus, limonene is a prodrug, and perillic acid and dihydroperillic acid are the predominant circulating monoterpene species. The plasma half life of limonene (and its metabolites) is 12 hours. Limonene is completely absorbed from the digestive tract, and is eliminated mainly through the urine in the form of uroterpenol and glycine- and glucuronide-monoterpene conjugates.[9]

STRUCTURE-ACTIVITY RELATIONSHIPS AMONG MONOTERPENES

The effects of limonene and those of related monoterpenes, including metabolites, have been compared in several structure-activity analyses. Hydroxylated analogs of limonene, including limonene's urinary metabolite uroterpenol and the dietary phytochemical carveol (Fig. 1), are more potent than limonene in the chemoprevention of DMBA-induced rat mammary cancer during initiation.[10] In addition, the activities of limonene vs. its metabolites as inhibitors of protein isoprenylation have been compared in several systems. Protein isoprenylation entails the post-translational modification of a protein by either a 15 carbon farnesyl or a 20 carbon geranylgeranyl group.[11] Farnesylation of the Ras oncoprotein, for example, is critical to its function in normal and neoplastic cells.[11] Thus, inhibition of protein isoprenylation may be one mechanism by which monoterpenes exert their antitumor activities. All limonene metabolites are more potent inhibitors of protein isoprenylation than limonene itself.[12,13] Further structure-activity studies indicate that monohydroxylated monoterpenes are more effective protein isoprenylation inhibitors than limonene or its major metabolites, perillic acid and dihydroperillic acid.[13] One of the most potent monohydroxylated monoterpenes is perillyl alcohol,[13] the alcohol analogue of limonene and perillic acid. Perillyl alcohol effectively inhibits protein isoprenylation in intact cells,[13,14] cell extracts,[14] and, importantly, directly inhibits farnesyl-protein transferase and geranylgeranyl-protein transferase activities *in vitro*.[15] Thus, among monoterpenes, the same structure-activity relationship (perillyl alcohol>perillic acid>limonene)is observed for the inhibition of human colon carcinoma[13] and rat liver cell proliferation,[16] inhibition of protein isoprenylation in cultured cells,[13,14] and inhibition of prenyl-protein transferases.[15]

CHEMOTHERAPEUTIC EFFECTS OF PERILLYL ALCOHOL

Based on the finding that perillyl alcohol is more potent than limonene in inhibiting protein isoprenylation, Haag and Gould[17] tested the hypothesis that perillyl alcohol would be a more potent chemotherapeutic drug than limonene. Indeed, 2.5% dietary perillyl alcohol caused >75% of established rat mammary tumors to regress,[17] while about 7.5-10% limonene would be required to attain the same rate of tumor regression.[5] Perillyl alcohol also dramatically reduced the appearance of secondary tumors, suggesting that it may also have chemopreventive activity against mammary cancer. Interestingly, the circulating metabolites of perillyl alcohol are identical to those of limonene, but are present in higher concentrations in animals fed 2.5% perillyl alcohol than in rats fed 10% limonene.[17]

We also tested the chemotherapeutic effects of perillyl alcohol on transplantable hamster pancreatic ductal adenocarcinomas. Treatment of tumor-bearing hamsters with orally-administered perillyl alcohol (3% of the diet) resulted in a significant reduction in pancreatic tumor growth.[18] After three weeks of chemotherapy, the average control tumor volume was 3121 ± 1120 mm,^{3} whereas the average volume of perillyl alcohol-treated tumors was only 494 ± 257 mm^3 (mean ± SEM, n=11, $p<0.05$). Perillyl alcohol chemotherapy reduced the average tumor growth rate, measured as the number of tumor volume doublings per week, to less than half that of controls. Moreover, some tumors completely regressed when treated with perillyl alcohol. Sixteen percent (5/31) of the perillyl alcohol-treated pancreatic tumors completely regressed, whereas no control tumors (0/25) regressed ($p<0.05$).

Perillyl alcohol chemotherapy also reduces the growth rate of transplanted prostatic carcinomas in nude mice.[19] Thus, monoterpenes have chemotherapeutic activity against a number of solid tumor types, including pancreatic cancer, one of the most refractory of all human cancers to available cancer therapies. The efficacy of perillyl alcohol chemotherapy against human cancer will be tested in forthcoming Phase I clinical trials (M. N. Gould, personal communication).

MECHANISMS OF ACTION OF PERILLYL ALCOHOL AND LIMONENE

Several mechanisms of action may account for the chemotherapeutic activities of monoterpenes. The post-initiation phase, tumor suppressive chemopreventive activity of monoterpenes may be due in part to the inhibition of isoprenylation of cell-growth associated small G proteins such as p21ras by limonene, perillyl alcohol, and their metabolites.[12,13] This inhibition occurs at the level of the prenyl-protein transferases.[15] In addition, perillyl alcohol affects the mevalonate pathway by inhibiting ubiquinone biosynthesis as well as the conversion of lathosterol to cholesterol.[20] Chemotherapy of chemically-induced mammary tumors with monoterpenes results in tumor redifferentiation.[5] In limonene-treated mammary tumors, expression of the mannose-6-phosphate-insulin-like growth factor II receptor and transforming growth factor β1 are increased in the regressing, differentiating tumors, but not in the small number of tumors which are unresponsive to limonene.[21]

In liver carcinogenesis model systems, the suppressive chemopreventive activity of perillyl alcohol during the promotion phase is associated with a marked increase in tumor cell apoptosis. Under the same conditions, perillyl alcohol has no effect on tumor cell proliferation.[22] To begin addressing possible mechanisms of action of perillyl alcohol against pancreatic tumors, we asked whether perillyl alcohol's antitumor activity was due to modulation of cell proliferation, cell death, or cell differentiation. No overt morphological differences were evident between the control and the perillyl alcohol-treated tumors, indicating that perillyl alcohol did not affect the degree of differentiation of the transplantable tumors. Cell growth was assessed by determing bromodeoxyuridine (BrDU) incorporation into tumor DNA. The percentage of BrDU-labeled tumor cells was not statistically different between the control and the perillyl alcohol-treated group, even though the perillyl alcohol-treated tumors were smaller on average (Table 1). However, treatment of cultured human MIA PaCa2 pancreatic carcinoma cells with 800 μM perillyl alcohol significantly increased the percentage of cells undergoing apoptosis (Figure 2). There is no evidence of increased

Table 1. Effects of perillyl alcohol on tumor cell proliferation and tumor volume

Chemotherapy treatment	% BrDU positive tumor cells	Tumor volume (cm^3)
Control (no drug)	29.7 ± 6.0	3.02 ± 0.62
5% perillyl alcohol	35.1 ± 11.7	1.43 ± 0.63*

Pancreatic tumor-bearing hamsters were treated for 7 days with 5% dietary perillyl alcohol as described previously.[18] Hamsters were injected i.p. with 100 mg/kg BrDU 2 hours prior to sacrifice. BrDU was detected in tumor sections by immunohistochemistry. The data represent the mean ± SEM, n=5. * $p < 0.05$ vs. control by Student's t-test.

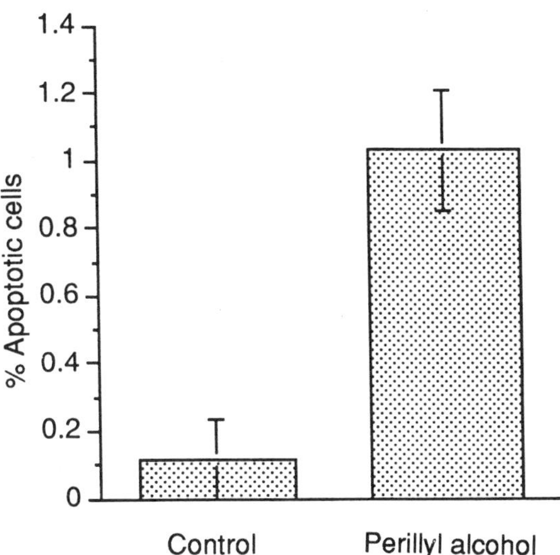

Figure 2. The effect of perillyl alcohol on pancreatic tumor cell apoptosis. Cultured human MIA PaCa2 pancreatic adenocarcinoma cells were treated for 2 days with 0 or 800 μM perillyl alcohol as described previously[18]. Apoptosis was detected by labeling 3' ends of DNA with a deoxynucleotide-digoxygenin conjugate, and digoxygenin was detected by a fluorescein-labeled antibody from Oncor. All cells were counterstained with propidium iodide. The percentage of apoptotic cells is calculated as (number of fluorescein-labeled cells ÷ propidium iodide-stained cells) x 100. The data represent the mean ± SEM, n=4. The control and perillyl alcohol values were significantly different in a two-tailed Student's t-test, $p < 0.01$.

apoptosis in monoterpene-treated mammary tumors.[15,17] Thus, monoterpenes act through multiple mechanisms in the chemoprevention and chemotherapy of cancer, and these mechanisms differ among tumor types.

SUMMARY

The antitumor effects of dietary monoterpenes are attained with little or no host toxicity.[2,3,8,18,23] For example, body weights of tumor bearing, perillyl alcohol-treated hamsters and their pair fed controls are not different.[18] Thus, monoterpenes have considerable antitumor activity, oral bioavailability, low toxicity, and novel mechanisms of action different from those of conventional cancer chemotherapeutic drugs.

REFERENCES

1. Croteau, R. Biosynthesis and catabolism of monoterpenoids. Chem. Rev., 87:929-954, 1987.
2. Elson, C.E. and Yu, S.G. The chemoprevention of cancer by mevalonate-derived constituents of fruits and vegetables. J. Nutr., 124:607-614, 1994.
3. Crowell, P.L., and Gould, M.N., Chemoprevention and therapy of cancer by *d*-limonene. CRC Crit. Rev. Oncogenesis, 5:1-22, 1994.
4. Gould, M.N., Moore, C.J., Zhang, R., Wang, B., Kennan, W.S., and Haag, J.D. Limonene chemoprevention of mammary carcinoma induction following direct in situ transfer of v-Ha-ras. Cancer Res., 54:3540-3543, 1994.
5. Haag, J.D., Lindstrom, M.J., and Gould, M.N. Limonene-induced regression of mammary carcinomas. Cancer Res., 52: 4021-4026, 1992.
6. McNamee, D. Limonene trial in cancer. Lancet, 342:801, 1993.
7. Crowell, P.L., Lin, S., Vedejs, E., and Gould, M.N. Identification of metabolites of the antitumor agent *d*-limonene capable of inhibiting protein isoprenylation and cell growth. Cancer Chemother. Pharmacol., 31:205-212, 1992.
8. Crowell, P. L., Elson, Bailey, H. H., C. E., Elegbede, A., Haag, J. H., and Gould, M. N. Human metabolism of the experimental cancer therapeutic agent *d*-limonene. Cancer Chemother. Pharmacol., 35:31-37, 1994.

9. Kodama, R., Yano, T., Furukawa, K., Noda, K., and Ide, H. Studies on the metabolism of *d*-limonene (p-mentha-1,8-diene): IV. Isolation and characterization of new metabolites and species differences in metabolism. Xenobiotica, 6:377-383, 1976.
10. Crowell, P.L., Kennan, W.S., Haag, J.D., Ahmad, S., Vedejs, E., and Gould, M.N. Chemoprevention of mammary carcinogenesis by hydroxylated derivatives of *d*-limonene. Carcinogenesis, 13:1261-1264, 1992.
11. Clarke S, Protein isoprenylation and methylation at carboxyl-terminal cysteine residues. Ann Rev Biochem, 61:355-386, 1992.
12. Crowell, P.L., Chang, R.R., Ren, Z., Elson, C.E., and Gould, M.N. Selective inhibition of isoprenylation of 21-26 kDa proteins by the anticarcinogen *d*-limonene and its metabolites. J. Biol. Chem., 266:17679-17685, 1991.
13. Crowell, P. L., Ren, Z., Lin, S., Vedejs, E. and Gould, M. N. (1993) Structure-activity relationships among monoterpene inhibitors of protein isoprenylation and cell proliferation. Biochemical Pharmacol., 47:1405-1415, 1994.
14. Ren, Z. and Gould, M.N. Inhibition of protein isoprenylation by the monoterpene perillyl alcohol in intact cells and in cell lysates. Proc. Am. Assoc. Cancer Res. 36:585, 1995.
15. Gelb, M., Tamanoi, F., Yokoyama, K., Ohomashchi, F., Esson, K., and Gould, M.N. The inhibition of protein prenyltransferases by oxygenated metabolites of limonene and perillyl alcohol. Cancer Letters, in press, 1995.
16. Ruch, R.J. and Sigler, K. Growth inhibition of rat liver epithelial tumor cells by monoterpenes does not involve Ras plasma membrane association. Carcinogenesis 15:787-789, 1994.
17. Haag, J.D. and Gould, M.N. (1994) Mammary carcinoma regression induced by perillyl alcohol, a hydroxylated analog of limonene. Cancer Chemother. Pharmacol., 34, 477-483, 1994.
18. Stark, M. J., Burke, Y. D., McKinzie, J. H., Ayoubi, A. S., and Crowell, P. L. Chemotherapy of pancreatic cancer with the monoterpene perillyl alcohol. Cancer Letters, in press, 1995.
19. Jeffers, L., Church, D., Gould, M. and Wilding, G. The effect of perillyl alcohol on the proliferation of human prostatic cell lines. Proc. Am. Assoc. Cancer Res. 36:303, 1995.
20. Ren, Z. and Gould, M.N. Inhibition of ubiquinone and cholesterol synthesis by the monoterpene perillyl alcohol. Cancer Lett. 76:185-190, 1994.
21. Jirtle, R.L., Haag, J.D., Ariazi, E., Gould, M.N. Increased mannose 6-phosphate/ insulin-like growth factor II receptor and transforming growth factor b1 levels during monoterpene-induced regression of mammary tumors. Cancer Res., 53: 3849-3853, 1993.
22. Mills, J.J., Chari, R.S., Boyer, I.J., Gould, M.N., and Jirtle, R. L. Induction of apoptosis in liver tumors by the monoterpene perillyl alcohol. Cancer Res., 55:979-983, 1995.
23. Evans, E., Arneson, D., Kovatch, R., Supko, J., Morton, T., Siemann, L., Cannon, R., Tomaszewski, J., and Smith, A. Toxicology and pharmacology of perillyl alcohol (NSC-641066) in rats and dogs. Proc. Am. Assoc. Cancer Res., 36:366, 1995.

11

MONOTERPENES AS REGULATORS OF MALIGNANT CELL PROLIFERATION

Raymond J. Hohl

Departments of Internal Medicine and Pharmacology
University of Iowa
Iowa City, Iowa 52242

ABSTRACT

Limonene and related monoterpenes display compelling anticarcinogenic activity. The mechanism(s) that underlie this activity is/are as yet unknown. One attractive possibility is that the monoterpenes interact with the RAS signal transduction pathway. The monoterpenes have been shown to impair incorporation of mevalonic acid-derived isoprene compounds, that is farnesyl pyrophosphate, into RAS and RAS-related proteins. As farnesylation is critical for RAS's membrane localization and function, the isoprenylation pathways have received attention as potential targets of anti-RAS pharmacological maneuvers. We have expanded on prior studies and demonstrate that one of limonene's metabolic derivatives, perillyl alcohol, decreases the levels of antigenic RAS in the human-derived myeloid THP-1 and lymphoid RPMI-8402 leukemia cell lines. Both limonene and perillyl alcohol decrease levels of ^{35}[S]-methionine labeled RAS proteins in cells that have been pulsed with radiolabeled methionine for four hours. In contrast, lovastatin, which inhibits hydroxymethylglutaryl coenzyme A reductase and thus depletes cells of farnesyl pyrophosphate, does not diminish levels of total antigenic RAS but rather results in a shift in the RAS protein; levels of farnesylated RAS decrease whereas levels of unmodified/unfarnesylated RAS increase. As limonene and perillyl alcohol do not induce such a shift we conclude that these monoterpenes decrease farnesylated RAS protein levels by a mechanism that is clearly distinct from that of either depleting cells of farnesyl pyrophosphate or inhibiting the enzyme farnesyl protein transferase that catalyzes the posttranslational farnesylation of RAS. These findings are discussed with respect to implications for the monoterpenes to alter RAS protein synthesis and degradation. The results of these studies will likely impact the inclusion of the monoterpenes in clinical anticancer trials.

MONOTERPENES

Limonene (Figure 1) is a monoterpene that is found in a variety of fruits. As a major constituent of many citrus-derived oils, especially orange peel oil, limonene is therefore

Dietary Phytochemicals in Cancer Prevention and Treatment
Edited under the auspices of the American Institute for Cancer Research, Plenum Press, New York, 1996

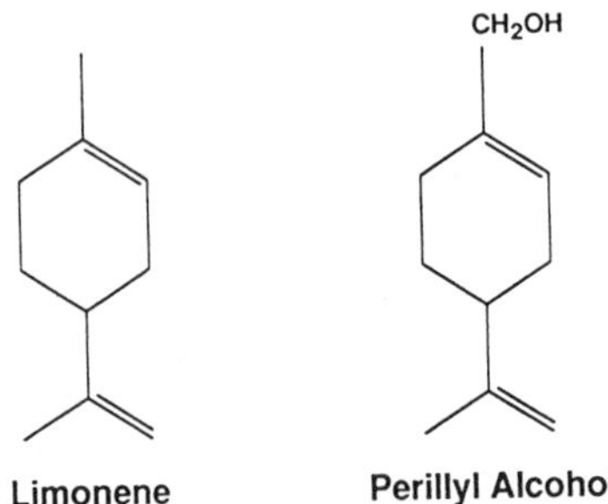

Figure 1. The chemical structures of limonene and perillyl alcohol.

found in most human diets. Limonene has been clinically utilized as a solvent to dissolve gallstones[1] and has been studied in animal models as an anticarcinogen.[2] One particular experimental model that has been extensively studied is the induction of mammary tumors in rats by the gastric administration of 7,12-dimethylbenz[a]anthracene (DMBA).[2] DMBA will reproducibly induce mammary tumors in nearly all animals tested with a dose dependent latency period. In this experimental system, daily dietary supplementation of d-limonene for the one week prior and twenty seven weeks subsequent to DMBA administration significantly prolonged the latency period to tumor development. One-third of the limonene treated animals never developed tumors whereas nearly all control animals developed tumors by twenty seven weeks. Subsequent experiments using a similar system have also shown that limonene caused regression of DMBA-induced mammary tumors.[3] These effects were not due to changes in mammary-relevant endocrine functions, such as estradiol and prolactin levels.[4] Furthermore, this anticarcinogenic effect could be reproduced by feeding nitrosomethylurea(NMU)-treated rats orange oil rather than purified limonene.[5] Limonene is also capable of inhibiting the development of upper digestive tract carcinomas in N-nitrosodiethylamine treated mice.[6] Additional *in vitro* models that illustrate limonene's anticarcinogenic activity that have also been developed including rat mammary organ cultures[7-9] and tracheal epithelial cell cultures.[10]

Limonene is metabolized *in vivo* to a number of oxidized derivatives so that the parent compound is detected at comparatively low levels.[11,12] In humans, the three major metabolic derivatives detected in plasma after single oral doses of 100 mg/kg limonene are perillic acid, dihydroperillic acid, and limonene-1,2-diol.[13] The metabolic precursors to perillic acid and dihydroperillic acid are likely perillyl alcohol and perillyl aldehyde, both of which have potent antiproliferative activities in cell culture systems.[14-18] Perillyl alcohol (Figure 1) has more potent antiproliferative activities in tissue culture than does limonene.[14-18] As a consequence of this increased potency, and increased aqueous solubility as compared to limonene, perillyl alcohol is a useful agent to dissect the mechanism(s) that underlie the antiproliferative properties of the monoterpenes.

RAS AND ITS RELATIONSHIP TO THE CHOLESTEROL BIOSYNTHETIC PATHWAY

RAS is a growth promoting protein[19] that normally binds to guanosine triphosphate (GTP) and has intrinsic GTPase activity that is modulated by its association to a GTP-ase activating protein (GAP).[20,21] Cells transfected with wild-type *RAS* genes grow normally whereas those transfected with RAS genes having point mutations, which either alter the

RAS' protein's GTPase activity or nucleotide binding (i.e. activated *RAS* genes), proliferate in an unrestrained manner consistent with the malignant phenotype.[20,22] An exception occurs in transfected cells in which expression of the wild-type nonmutated RAS protein is greatly enhanced; these cells also display growth patterns of the malignant phenotype.[22] Of interest is that the multidrug resistance (MDR1) gene, the protein product of which is responsible for the resistance of tumor cells to various cytotoxic drugs, can be activated by RAS.[23]

Membrane localization is critical for the normal function of the RAS proteins. This membrane localization is promoted by posttranslational farnesylation of the RAS protein.[22,24] Protein isoprenylation is dependent upon a consensus C-terminal amino acid (AA) sequence that is cysteine-aliphatic AA-aliphatic AA-any AA.[22,24] In transfection experiments using altered *RAS* genes, mutant RAS proteins that lack this consensus sequence are not farnesylated and are therefore not membrane localized and thus accumulate in the cell's cytoplasmic compartment.[19,24,25]

The polyisoprene units that are covalently linked to the RAS proteins are derived from farnesyl pyrophosphate that is an intermediate of the mevalonic acid to cholesterol biosynthetic pathway.[24,25] In this pathway, shown in Figure 2, acetyl coenzyme A is a precursor for hydroxymethylglutaryl coenzyme A (HMG CoA) that is reduced by HMG CoA reductase to mevalonic acid. The six carbon mevalonic acid is subsequently phosphorylated twice and decarboxylated to yield the five carbon isomeric isoprenyl pyrophosphates, dimethylallyl and isopentenyl pyrophosphates. These five carbon isoprene molecules undergo condensation to yield the ten carbon geranyl pyrophosphate. Condensation of another isopentenyl pyrophosphate with geranyl pyrophosphate yields the fifteen carbon atom farnesyl pyrophosphate. Farnesyl pyrophosphate can condense with itself to form the thirty carbon compound presqualene that is reduced to squalene which is a more proximal precursor

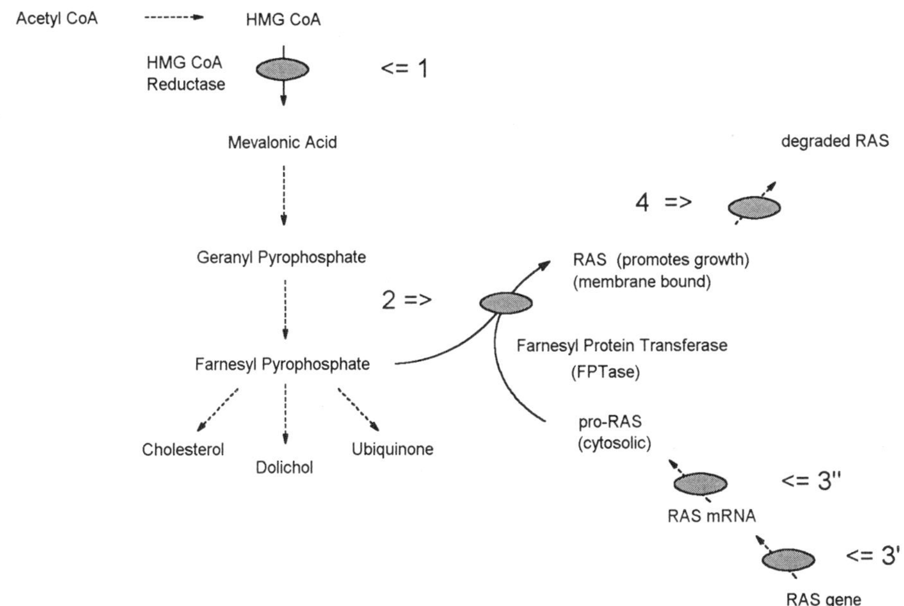

Figure 2. Isoprene biosynthesis and RAS processing. The numbered steps are described in the text and refer to 1-HMG CoA reductase, 2-Farnesyl Protein Transferase, 3'-RAS transcription, 3"-RAS translation, and 4-RAS degradation.

for cholesterol. Farnesyl pyrophosphate also serves as a precursor for other nonsterol products of the mevalonic acid pathway, such as dolichol and ubiquinone.

Relevant to RAS processing is that farnesyl protein transferase catalyzes the covalent linkage of the farnesyl moiety of farnesyl pyrophsophate to a precursor RAS protein (pro-RAS) yielding mature farnesylated RAS protein (Figure 2). Unfarnesylated RAS proteins that are localized to the cytoplasm[24] lack transforming potential in transfection experiments.[19,26] Agents that competitively inhibit HMG CoA reductase, such as lovastatin, decrease mevalonic acid, and therefore farnesyl pyrophosphate biosynthesis.[27] As a consequence these drugs impair RAS farnesylation.[24] Our laboratory has previously shown that lovastatin decreases RAS farnesylation in human derived leukemia cell lines[28] and that the HMG CoA reductase inhibitors alter the *in vitro* growth of freshly obtained malignant cells from patients with leukemia.[29]

As shown in Figure 2, pharmacological strategies to block RAS farnesylation minimally include: 1) depletion of farnesyl pyrophosphate necessary for RAS farnesylation and 2) inhibition of the farnesyl protein transferase that catalyzes RAS modification. As mentioned above, the former strategy is accomplished with high concentrations of the HMG CoA reductase inhibitors[24] and the latter strategy is accomplished through use of a number of novel RAS and farnesyl pyrophosphate analogs.[30-35] Limonene is believed to fit in this scheme by interfering with the incorporation of mevalonic acid-derived products into RAS proteins, possibly at the level of impairing farnesyl protein transferase activity.[36] The monoterpenes have also been reported to downregulate HMG CoA reductase activity[37] and therefore the anti-cancer activity of these agents may be attributable to either depressed levels of growth-promoting isoprenylated proteins, such as those of the RAS family, or depressed endogenous mevalonic acid-derived cholesterol and/or nonsterol products. Finally, apparently unrelated to this pathway, the monoterpenes have been reported to inhibit the activity of transforming growth factor β1 (TGF-β1) in mammary tumor lines.[38]

Application of the monoterpenes to human clinical trials will be greatly facilitated by better understanding of the mechanism(s) that underlie the observed anti-RAS activities that are likely to account for some portion of the monoterpenes' anti-cancer properties.

ANTI-RAS ACTIVITY BY THE MONOTERPENES

Limonene is similar to lovastatin in that it decreases the levels of isoprenylated proteins, including those of the RAS family.[36] The first description of the monoterpenes' activity to decrease levels of isoprenylated RAS were based on experiments that have followed the incorporation of radiolabeled mevalonic acid into protein fractions and immunoprecipitable RAS.[36] This experimental design was based on the studies that originally demonstrated that mevalonic acid, or one or more of its metabolic derivatives, was incorporated into the protein fraction of eukaryotic cells.[39] The initial report demonstrated that limonene decreased levels of radiolabeled RAS proteins in cells that had been pulse labeled with ^{35}S-methionine.[36] The conclusion drawn from this observation was that the monoterpenes likely interfered with farnesyl protein transferase (Figure 2, step 2). While this hypothesis explains the experimental findings there are other possible explanations for the effects of the monoterpenes on RAS levels. These agents could decrease farnesyl pyrophosphate synthesis, by for instance blocking HMG CoA reductase (Figure 2, step 1), or increasing RAS protein synthesis and degradation. The minimal data available described the regulation of *RAS* transcription (Figure 2, step 3'), *RAS* translation (Figure 2, step 3"), or RAS degradation (Figure 2, step 4) in mammalian-derived cells. There is one report that the half-life of the H-RAS protein in NIH 3T3 cells that have been transfected with normal human H-RAS cDNA is approximately 20 hours.[40]

Our laboratory has extended earlier studies of limonene's effects on animal tumor models to human-derived malignancies.[41,42] We have shown that limonene, like lovastatin, impairs DNA synthesis in the human-derived myeloid leukemia cell line THP-1 and in the lymphoid leukemia cell line RPMI-8402 in a concentration dependent manner.[18] Using an alternate experimental system that initially characterized the effects of the monoterpenes on RAS farnesylation, we studied the effects of limonene and perillyl alcohol on *RAS* expression in two human-derived leukemia cell lines.[41] Our studies specifically compared the monoterpenes to lovastatin to better delineate specific alterations of RAS processing. Shown in Figure 3 are results obtained when human-derived THP-1 myeloid and RPMI-8402 lymphoid leukemia cells are incubated with the monoterpenes in tissue culture. Under control conditions there is only one RAS band detectable on Western blot analysis. Incubation of these cells with lovastatin results in the accumulation of a more slowly migrating RAS band that corresponds to unmodified RAS protein. The concentration of lovastatin shown in Figure 3 is sufficiently high to block endogenous cellular mevalonic acid and therefore farnesyl pyrophosphate synthesis; synthesized pro-RAS accumulates and can be detected as a more slowly moving band than the fully modified RAS protein. Apparent in Figure 3 is that increasing concentrations of limonene does not result in the accumulation of an unmodified RAS protein.

Analogous studies using perillyl alcohol are displayed in Figure 4. Again under control conditions there is only one RAS band present. Lovastatin treatment results in the accumulation of a more slowly migrating RAS band again corresponding to the accumulation of unmodified RAS protein. Increasing concentrations of perillyl alcohol do not result in the accumulation of an unmodified RAS band. Of importance is the levels of RAS proteins decrease with increasing concentrations of perillyl alcohol. Thus perillyl alcohol does result

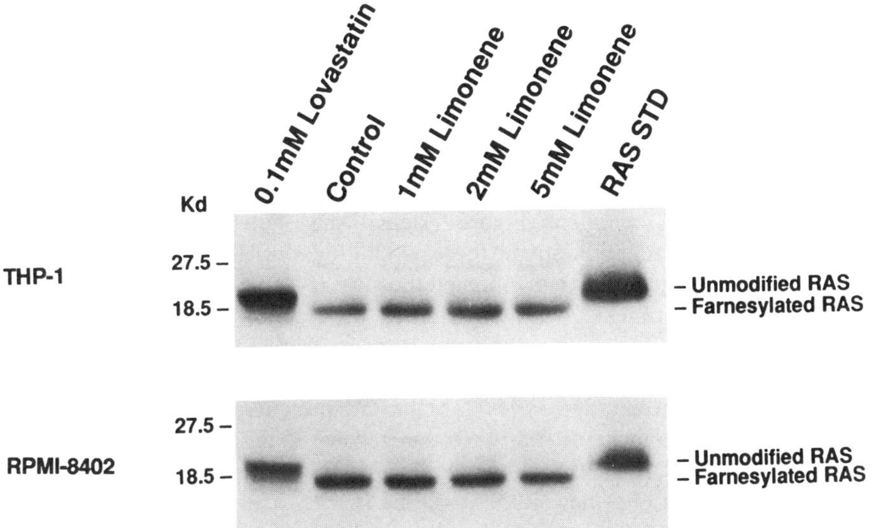

Figure 3. Effects of lovastatin and limonene on RAS isoprenylation in THP-1 and RPMI-8402 cells. These Western blots are developed as described elsewhere.[41] The top panel displays the lysates from the THP-1 cells while the bottom panel displays the lysates from RPMI-8402 cells. Each lane contains 100 μg of protein from cell lysate and is labeled with the concentration of lovastatin or limonene used in the incubations. Control cells were incubated without drug. H-ras standard (*RAS STD*) was also loaded on a separate lane. The standard migrates as a single band with an apparent molecular weight that approximates that of unmodified RAS. Reprinted with permission.[41]

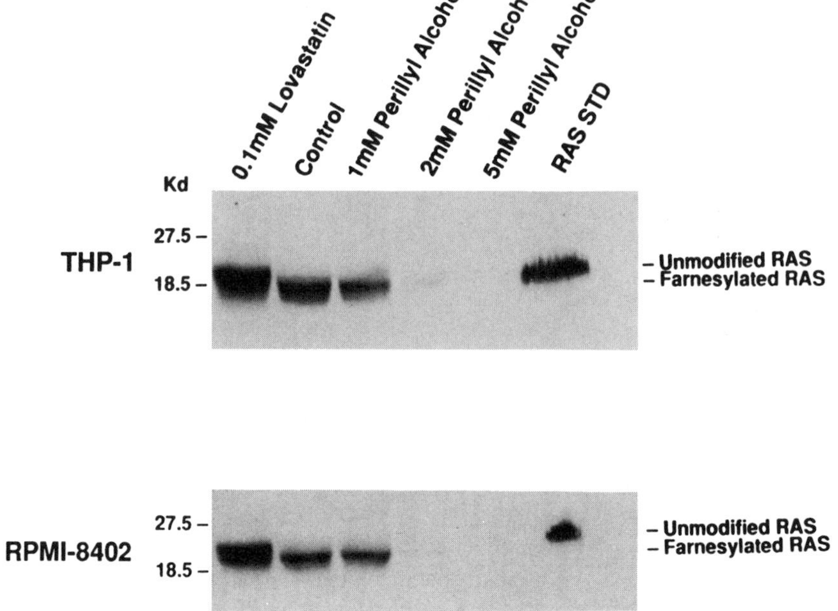

Figure 4. Effects of lovastatin and perillyl alcohol on RAS isoprenylation in THP-1 and RPMI-8402 cells. These Western blots are developed as described elsewhere.[41] The top panel displays the lysates from the THP-1 cells while the bottom panel displays the lysates from RPMI-8402 cells. Each lane contains 100 μg of protein from cell lysate and is labeled with the concentration of lovastatin or perillyl alcohol used in the incubations. Control cells were incubated without drug. H-ras standard (*RAS STD*) was also loaded on a separate lane. Reprinted with permission.[41]

in a lowering of farnesylated RAS levels as previously described[36] but by a mechanism distinct from inhibiting farnesyl protein transferase (Figure 2, step 2).

The mechanisms by which perillyl alcohol decrease RAS levels are either a consequence of decreased RAS synthesis (Figure 2, steps 3' and 3") or increased RAS degradation (Figure 2, step 4). To further investigate these possibilities our laboratory pulse labeled cells with ^{35}S-methionine and immunoprecipitated the RAS proteins. Shown in Figure 5 are the results of such an experiment. *Total RAS* refers to the levels of antigenic RAS present in a constant amount of protein in cellular lysate under control circumstances and with increasing concentrations of perillyl alcohol. RAS levels are relatively stable up to 0.5 mM perillyl alcohol. Above this concentration the levels of the RAS proteins are decreased. *^{35}S-RAS/total protein* refers to the amount of newly synthesized RAS during a four hour pulse with ^{35}S-methionine. New RAS synthesis decreases with relatively low levels of perillyl alcohol. At 0.5 mM perillyl alcohol new RAS synthesis is 40% of that observed under control circumstances. Higher concentrations of perillyl alcohol do not appear to further reduce RAS synthetic rates. These studies show that at low concentrations of perillyl alcohol RAS levels are relatively constant but RAS synthesis is decreased, suggesting that RAS degradation may be impaired (Figure 2, step 4). Higher concentrations of perillyl alcohol result in a decrease in new RAS synthesis without a further reduction in RAS levels, suggesting that RAS degradation is normalized (Figure 2, step 4) but that *RAS* transcription (Figure 2, step 3') or translation (Figure 2, step 3") is impaired. Further studies will better distinguish between these findings.

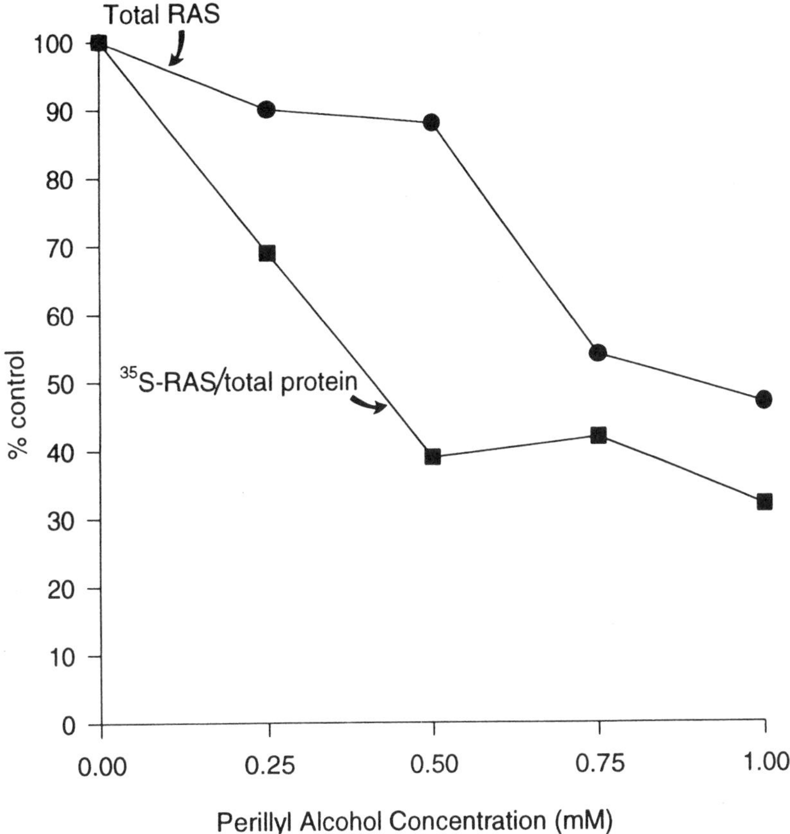

Figure 5. Effects of perillyl alcohol on total antigenic RAS levels and newly synthesized RAS levels. This experiment is described in the text and elsewhere.[41] Total RAS refers to the levels of RAS proteins in the lysates of perillyl alcohol-treated THP-1 cells and is expressed as % as compared to control cells. ^{35}S-RAS/total protein is the amount of radiolabeled methionine incorporated into Ras in 200 μg of cellular protein and is expressed as % as compared to control cells.

Finally, we have also evaluated the effects of the monoterpenes on microsomal HMG CoA reductase using a modification of our prior methods.[29] Shown in Figure 6 are the results of these studies. Lovastatin inhibits HMG CoA reductase activity whereas the monoterpenes have no effect on HMG CoA reductase activity. These findings are not unexpected because we had observed that the monoterpenes did not alter RAS processing in a manner that would be consistent with farnesyl pyrophosphate depletion.

APPLICATIONS OF THE MONOTERPENES TO CLINICAL TRIALS

The importance of these studies to elucidate the mechanisms by which the monoterpenes have laboratory determined anti-cancer properties is that the results obtained will better enable adequate evaluation of their clinical activity. Human trials of the monoterpenes have already begun and there has been a report of the plasma metabolites that follow 100mg/kg single doses of limonene.[13] Of importance to the application of these agents to human studies is that the concentrations of monoterpenes needed to alter protein isopreny-

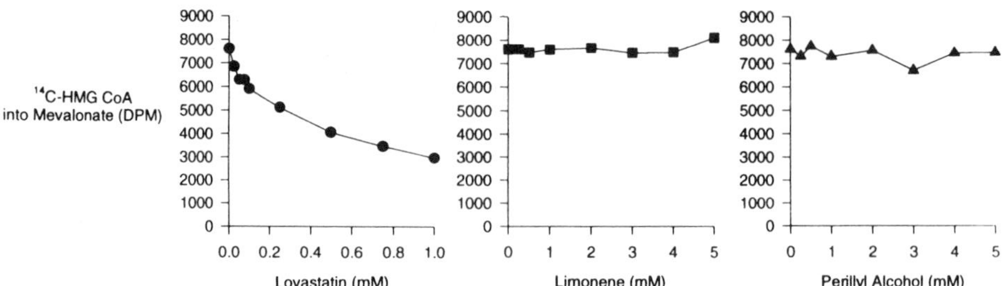

Figure 6. Effect of lovastatin, limonene, and perillyl alcohol on rat hepatic HMG CoA reductase activity. A constant amount of rat hepatic microsomal enzyme was isolated and the HMG CoA reductase activity assayed by modification of our previous methods.[29] Enzyme activity correlates with the amount of ^{14}C-HMG CoA incorporated into mevalonate.

lation in the laboratory are many orders of magnitudes higher than the levels likely achievable after pharmacological dosing, and most certainly after dietary supplementation. Nevertheless, the preclinical animal studies of these compounds demonstrate remarkable, albeit poorly understood, anti-cancer activity. This is perhaps because of the high lipophilicity of these compounds and that relatively low plasma levels may not reflect tissue distribution and concentrations. One strategy that might be of better utility than using the monoterpenes as single agents is to combine the monoterpenes with other compounds that also modulate protein isoprenylation, such as the HMG CoA reductase inhibitors.[42,43] Additionally, the monoterpenes may have better activity as chemopreventive rather than chemotherapeutic agents. There is general appreciation that a diet high in fruits, likely also to be high in the monoterpenes, may have some chemopreventive activity for a variety of human malignancies. However, many other lifestyle factors associated with such a diet may confound actual definition of such activity. Finally, carefully designed human clinical trials are needed that use intermediate endpoints to establish the anti-cancer activity of the monoterpenes. The effects of the monoterpenes on the RAS system suggest that assessment of the RAS pathway might be such an endpoint.

ACKNOWLEDGMENT

This investigation was supported by grants 92SG20 and 93B46 from the American Institute for Cancer Research and from the Roy J. Carver Charitable Trust. R. Hohl is a recipient of an American Society of Clinical Oncology Young Investigator Award, a Pharmaceutical Research and Manufacturers of America Faculty Development Award in Clinical Pharmacology, and a Clinical Associate Physicians Award from the GCRC/NIH.

REFERENCES

1. Igimi, H., Hisatsuga, T., and Nishimura M. The use of d-limonene as a dissolving agent in gallstones. Dig. Dis., 21:926-939, 1976.
2. Elegbede, J.A., Elson, C.E., Qureshi, A., Tanner, M.A., and Gould, M.N. Inhibition of DMBA-induced mammary cancer by the monoterpene d-limonene. Carcinogenesis, 5:661-664, 1984.
3. Elegbede, J.A., Elson, C.E., Tanner, M.A., Qureshi, A. and Gould, M.N. Regression of rat primary mammary tumors following dietary d-limonene. J. Natl. Cancer Inst., 76:323-325, 1986.

4. Elson, C.E., Maltzman, T.H., Boston, J.L., Tanner, M.A., and Gould, M.N. Anti-carcinogenic activity of d-limonene during the initiation and promotion/progression stages of DMBA-induced rat mammary carcinogenesis. Carcinogenesis, 9:331-332, 1988.
5. Maltzman, T.H., Hurt, L.M., Elson, C.E., Tanner, M.A., and Gould M.N. The prevention of nitrosomethylurea-induced mammary tumors by d-limonene and orange oil. Carcinogenesis, 10:781-783, 1989.
6. Wattenberg, L.W., Sparnns, V.L., and Barany, G. Inhibition of N-nitrosodiethylamine carcinogenesis in mice by naturally occuring organosulfu compounds and monoterpees. Cancer Res., 49:2689-2692, 1989.
7. Lin, F.K., Banerjee, M.R., and Cump, L.R. Cell cycle related hormone carcinogen interaction during chemical carcinogen induction of nodule-like mammary lesions in organ culture. Cancer Res., 36:1607-1614, 1976.
8. Mehta, R.G., and Moon, R.C. Characterization of effective chemopreventive agents in mammary gland in vitro using an initiation-promotion protocol. Anticancer Res., 11:593-596, 1991.
9. Russin, W.A., Hoesly, J.D., Elson, C.E., Tanner, M.A., and Gould M.N. Inhibition of rat mammary carcinogenesis by monoterpenoids. Carcinogenesis, 10:2161-2164, 1989.
10. Steele, V.E., Kelloff, G.J., Wilkinson, B.P., and Arnold, J.T. Inhibition of transformation in cultured rat tracheal epithelial cells by potential chemopreventive agents. Cancer Res., 50:2068-2074, 1990.
11. Kodama, R., Yano, T., Furukawa, K., Noda, K. and Ide, H. Studies on the metabolism of d-limonene (p-mentha-1,8-diene). IV Isolation and characterization of new metabolites and species differences in metabolism. Xenobiotica, 6:377-389, 1976.
12. Crowell, P.L., Lin, S.,Vedejs, E., and Gould, M.N. Identification of circulating metabolites of the antitumor agent *d*-limonene capable of inhibiting protein isoprenylation and cell growth. Cancer Chemother Pharmacol 31:205-212, 1992.
13. Crowell, P.L, Elson C.E., Bailey, H.H., Elegbede, A., Haag, J.D., and Gould, M.N. Human metabolism of the experimental cancer therapeutic agent d-limonene. Cancer Chemother Pharmacol 35:31-37, 1994.
14. Crowell, P.J., Ren, Z., Lin, S., Vedejs, E., and Gould, M.N. Structure-activity relationships among monoterpene inhibitors of protein isoprenylation and cell proliferation. Biochem Pharmacol 47:1405-1415, 1994.
15. Ruch, R.J. and Sigler, K. Growth inhibition of rat liver epithelial tumor cells does not involve RAS plasma membrane association. Carcinogenesis 15:787-789, 1994.
16. Schulz, S., Buhling, F., and Ansorge, S. Prenylated proteins and lymphocyte proliferation: inhibition by d-limonene and related monoterpenes. Eur J Immunol 24:301-301, 1994.
17. Bronfen, J.H., Stark, J.M., and Crowell, P.L. Inhibition of human pancreatic carcinoma cell proliferation by perillyl alcohol. Proc Am. Assoc. Cancer Res. 35: 431, 1994. (abstr.)
18. Hohl, R.J. and Lewis, K. *RAS* expression in human leukemia is modulated differently by lovastatin and limonene. Blood 80:299a, 1992. (abstr.)
19. Barbacid, M. RAS Genes. Ann. Rev. Biochem, 56:779-827, 1987.
20. Bos, J.L. *RAS* Oncogenes in Human Cancer: A Review. Cancer Res., 49:4682-4689, 1989.
21. Kitayama, H., Sugimoto, Y., Matsuzak, T., Ikawa, Y., and Noda, M. A *ras*-related gene with transformation supressor activity. Cell, 56:77-84, 1989.
22. Chang, E.H., Furth, M.E., Scolnick, E.M. and Lowy, D.R. Tumorigenic transformation of mammalian cells induced by a normal human gene homologous to the oncogene of Harvey murine sarcoma virus. Nature, 297:478-483, 1982.
23. Chin, K.V., Ueda, K., Pastan, I., and Gottesman, M.M. Modulation of activity of the promoter of the human *MDR1* gene by ras and p53. Science, 255:459-462, 1992.
24. Schafer, W.R., Kim, R., Sterne, R., Thorner, J., Kim, S-H., and Rine, J. Genetic and pharmacologic supression of oncogenic mutations in RAS genes of yeast and humans. Science, 245:379-385, 1989.
25. Hancock, J.F., Magee, A.I., Childs, J.E., and Marshall, C.J. All ras proteins are polyisoprenylated but only some are palmitoylated. Cell, 57:1167-1177, 1989.
26. Willumsen, B.M., Norris, K., Papageorge, A.G., Hubbert, N.L. and Lowy, D.L. Harvey murine sarcoma virus p21ras protein:biological and biochemical significance of the cysteine nearest the carboxy terminus. EMBO J., 3:2581-2585, 1984.
27. Endo, A., Kuroda, M., and Tsujita, Y. ML-236A, ML-236B, and ML-236C, new inhibitors of cholesterolgenesis produced by *Penicillium citrinum*. J. Antibiot., 24:1346-1348, 1976.
28. Hohl, R.J., Mannickarottu, V., and Yachnin, S. The effect of alterations of hydroxymethylglutaryl coenzyme A reductase on the expression of the RAS oncogene. Clinical Research, 38:843A, 1990. (abstr.)
29. Hohl, R.J., Larson, R.A., Mannickarottu, V., and Yachnin, S. Inhibition of hydroxymethylglutaryl coenzyme A reductase activity induces a paradoxical increase in DNA synthesis in myeloid leukemia cells. Blood, 77:1064-1070, 1991.

30. Reiss,Y., Goldstein, J.L., Seabra, M.C., Casey, P.J., and Brown, M.S. Inhibition of purified p21^{ras} farnesyl;protein transferase by Cys-AAX tetrapeptides. Cell 62:81-86, 1990.
31. Gibbs, J., Pompliano, D., Mosser, S., Rands, E., Lingham, R., Singh S., Scolnick, E, Kohl, N., and Oliff, A. Selective inhibition of farensyl-protein transferase blocks ras processing *in vivo*. J Biol Chem 268:7617-7620, 1993.
32. Kohl, N., Mosser, S., deSolms, S., Giuliani, E., Pompliano, D., Graham, S., Smith, R., Scolnick, E., Oliff, A., and Gibbs, J. Selective inhibition of *ras*-dependent transformation by a farnesyltransferase inhibitor. Science 260:1934-1937, 1993.
33. James, G., Goldstein, J., Brown, M., Rawson, T., Somers, T., McDowell, R., Crowley, C., Lucas, B., Levinson, A., and Marsters, J. Benzodiazepine peptidomimetics: potent inhibitors of ras farnesylation in animal cells. Science 260:1937-1941, 1993.
34. Garcia, A.M., Rowell, C., Ackerman, K., Kowalczyk, J.J., and Lewis, M.D. Peptidomimetic inhibitors of ras farnesylation and function in whole cells. J Biol Chem 268:18415-18418, 1993.
35. Nigam, M., Seong, C., Qian, Y., Hamilton, A., and Sebti, S. Potent inhibition of human tumor p21^{ras} farensyltransferase by A_1A_2-lacking p21^{ras} CA_1A_2X peptidomimetics. J Biol Chem 268:20695-20698, 1993.
36. Crowell, P.L., Chang, R.R., Ren, Z., Elson, C.E. and Gould, M.N. Selective inhibition of isoprenylation of 21-26 kDa proteins by the anticarcinogen d-limonene and its metabolites. J. Biol. Chem., 266:17679-17685, 1991.
37. Clegg, R.J., Middleton, B., Bell, C.D., and White, D.A. The mechanism of cyclic monoterpene inhibition of hepatic 3-hydroxy-3-methylglutaryl coenzyme A reductase *in vivo* in the rat. J Biol Chem 257:2294-2299, 1982.
38. Jirtle, R.L., Haag, J.D., Ariazi, E.A., and Gould, M.N. Increased mannose 6-phosphate/insulin-like growth factor II receptor and transforming growth factor β1 levels during monoterpene-induced regression of mammary tumors. Cancer Res 53:3849-3852, 1993.
39. Schmidt, R.A., Schneider, C.J., and Glomset, J.A. Evidence for post-translational incorporation of a product of mevalonic acid into Swiss 3T3 cell proteins. J Biol Chem 259:10175-10180, 1984.
40. Ulsh, L.S., and Shih, T.Y. Metabolic turnover of human c-rasH p21 protein of EJ bladder carcinoma and its normal cellular and viral homologs. Mol Cell Biol 4:1647-1652, 1984.
41. Hohl, R.J., and Lewis K. Differential effects of monoterpenes and lovastatin on RAS processing. J Biol Chem 270:17508-17512, 1995.
42. Kawata, S., Nagase, T., Yamasaki, E., Ishiguro, H., and Matsuzwawa, Y. Modulation of the mevalonate pathway and cell growth by pravastatin and d-limonene in a human hepatoma cell line (Hep G2). Br J Cancer 69:1015-1020, 1994.
43. Raj, M., Kratz, D., Lewis, K., and Hohl, R.J. Effects of combinations of lovastatin and monoterpenes on ras processing. Proc of AACR 36:428, 1995. (abstr.)

12

ORGANOSULFUR COMPOUNDS AND CANCER

Michael A. Lea

Department of Biochemistry and Molecular Biology
University of Medicine and Dentistry of New Jersey
New Jersey Medical School, Newark, New Jersey

INTRODUCTION

Epidemiological and laboratory studies have indicated that consumption of some organosulfides can have cancer preventive action. Compounds of this type occur naturally, particularly in many plant foods. In addition many organosulfides have been considered to be generally recognized as safe food additives. Some of these are presented in Table 1. Garlic and other *Allium* species are known to be rich sources of organosulfides including precursors of diallyl sulfide and diallyl disulfide.[1,2] Subsequent chapters in this book address three important questions about garlic. Firstly, what is in garlic? Secondly, does garlic have any value in the prevention and treatment of cancer? Thirdly, can we improve the chemopreventive value of garlic?

EPIDEMIOLOGICAL STUDIES OF GARLIC AND CANCER

Several reviews have considered the historical evidence that consumption of garlic may be associated with a decreased incidence of cancer.[3-7] The strongest case from epidemiological studies may be made for an inverse relationship between mortality from gastric cancer and consumption of garlic in Italy and China.[8,9] This conclusion is based on case-control studies. In a review Dorant and colleagues concluded that evidence from laboratory experiments and epidemiological studies were not yet conclusive as to the preventive capacity of garlic or garlic constituents. More recent cohort studies in the Netherlands found no association between the consumption of onions, leeks or garlic supplement use and the incidence of lung cancer or breast cancer.[10,11] As outlined in the next section, there is stronger evidence from laboratory studies for the cancer preventive action of garlic constituents.

Table 1. Some organic sulfur-containing compounds that are generally recognized as safe food additives

Allyl isothiocyanate	Dimethyl mercaptan
Allyl mercaptan	Furfuryl mercaptan
Benzyl disulfide	Methyl mercaptan
Benzyl mercaptan	Methyl 2-methylthiopropionate
Benzyl sulfide	Propyl disulfide
Butyl sulfide	2-Thienyl mercaptan
Diallyl disulfide	2-Thienylthiol
Diallyl sulfide	

INHIBITION OF CARCINOGENESIS BY GARLIC AND GARLIC CONSTITUENTS

The development of spontaneous mammary tumors in C3H mice was inhibited by feeding fresh garlic but not with garlic in which allinase had been inactivated.[12] This suggested that allicin rather than alliin was the chemopreventive agent. Topical application of onion or garlic oils inhibited tumor formation on the skin of mice treated with 7,12-dimethylbenz[a]anthracene (DMBA) and later promoted by application of 12-O-tetradecanoyl-phorbol-13-acetate (TPA).[13] Inhibition of TPA-induced tumor promotion has been confirmed by other investigators.[14-16] Dietary supplements of garlic powder reduced the final mammary tumor incidence in rats treated with DMBA and also reduced the formation of DNA adducts.[17,18] Mammary cancer prevention by dietary garlic was confirmed,[19] while oral administration of garlic oil had a chemopreventive action against the induction of cervical tumors in mice by 3-methylcholanthrene (MCA).[20]

Not all the anticarcinogenic effects of garlic need be attributable to sulfur compounds. Antitumor-promoting activity has been reported for allixin, a phenolic compound.[21] However, most studies with single agents derived from garlic have focused on sulfur-containing compounds. Most extensively studied has been diallyl sulfide which, in laboratory animals, inhibited chemical carcinogenesis in colon,[22] liver,[23] esophagus,[24] forestomach,[25,26] lung,[25-27] and several tissues in a multi-organ model.[28] The absence of a protective effect for colon[29] and even a promotional effect on the development of hepatic foci[30] have also been reported. Protective effects have been seen against a variety of carcinogens including 1,2-dimethylhydrazine, benzo(a)pyrene, 4-(methylnitrosamino)-1-(3-pyridyl)-1-butanone (NNK) and other nitrosamines.

The importance of allylic groups as opposed to saturated propyl groups for the inhibitory effects of organosulfides on carcinogenesis was demonstrated.[25] Several organosulfur compounds were examined by Wattenberg and colleagues for their capacity to inhibit carcinogenesis induced by N-nitrosodiethylamine, and the most potent was diallyl disulfide which reduced forestomach tumors by 90%.[26] Dietary diallyl disulfide also decreased the number of azoxymethane-induced colon adenocarcinomas in rats.[31]

EFFECTS OF GARLIC AND GARLIC CONSTITUENTS ON METABOLISM

Carcinogen Metabolism

Activation of microsomal cytochrome P450 by diallyl disulfide was observed using microsomal preparations from 15-day rats.[32] In contrast, others found that diallyl disulfide inhibited microsomal metabolism of N-nitrosomethylbenzylamine.[24] Differential effects on cytochrome P450 isozymes were identified since treatment of rats with diallyl sulfide caused a suppression in the level of cytochrome P450 2E1 but an elevation in the level of cytochrome P450 2B1.[33] Oxidative metabolism of carcinogenic nitrosamines by nasal mucosa was inhibited by treating rats with diallyl sulfide (200 mg/kg body weight)[34] and the metabolism of NNK by mouse liver microsomes was also inhibited by diallyl sulfide.[27] Chronic treatment of rats with diallyl sulfide caused a decrease in N-nitrosodimethylamine demethylase activity and cytochrome P450 2E1 content in liver microsomes.[35] An inhibitory effect of diallyl sulfide on cytochrome P450 2E1 has also been reported with V79 cells.[36]

There was only a slight increase in the levels of cytochrome P450 2B1/2 mRNA when rat hepatocytes were incubated with up to 2 mM diallyl sulfide but 2 mM diallyl sulfone resulted in a greater than ten fold increase and protein levels were also markedly increased.[37] Under these conditions, there was no effect on mRNA or protein levels of cytochrome P450 2E1. Induction of cytochrome P450 2B1/2 in rat liver by diallyl sulfide was shown to be mainly due to transcriptional activation.[38] There were wide differences in tissue response and induction was not seen in the lung.

Other work has reinforced and extended previous studies on the effects of diallylsulfides on drug metabolizing enzymes.[39] The effects of diallyl sulfide and diallyl disulfide on rat intestinal and hepatic drug-metabolizing enzymes were contrasted. Diallyl disulfide had similar effects in the two tissues but diallyl sulfide had lesser effects in the intestine. The tendency was for both allyl sulfides to increase levels of cytochrome P450 1A2 and 2B1/2 and of phase II enzymes but to decrease the level of cytochrome P450 2E1.

A crude garlic extract was shown to inhibit the mutagenic action of aflatoxin B1 and the binding of metabolites to DNA.[40] Of two organosulfur compounds from garlic that were examined under the same conditions, ajoene was more potent than diallyl sulfide. Dietary fortification with S-allylcysteine caused a decrease in the binding of DMBA to DNA.[18]

Polyamine Metabolism

There are data to support positive or negative effects of allyl sulfides on polyamine synthesis in different systems. Garlic oil inhibited the induction of ornithine decarboxylase activity by the tumor promoter, TPA, in mouse epidermal cells.[41] The induction of ornithine decarboxylase in rat stomach by methyl-N'-nitro-nitrosoguanidine was inhibited by administration of diallylsulfide.[42] However, difluoromethylornithine abolished the ability of diallyl sulfide to reduce colonic nuclear damage caused by ionizing radiation.[43] This suggested that diallyl sulfide could act via a polyamine dependent pathway. Further evidence for stimulation of polyamine synthesis by allyl sulfides was provided by other studies on rat liver.[44]

Sufhydryl Groups and Glutathione S-Transferase

The inhibition of enzymes by allicin was generally associated with critical sulfhydryl groups and a similar response was not seen with diallyl sulfide or diallyl disulfide.[45] Some diminution in sulfhydryl groups might have been anticipated but it was reported that garlic oil inhibited the sharp decline in the ratio of reduced/oxidized glutathione caused by TPA in isolated mouse epidermal cells.[41]

Four allylic sulfides were found to increase glutathione S-transferase in the stomach but their saturated analogs produced little or no induction.[25] Dietary supplements of garlic powder increased glutathione S-transferase activity in mammary gland and liver of rats.[17] Moreover, diallyl sulfide increased the activities of phase II drug-metabolizing enzymes including glutathione S-transferase, NAD(P)H-dependent quinone reductase and UDP-glucuronyl transferase in liver and colon.[31]

Other Actions that May Be Related to Cancer

Several actions of garlic or organosulfur compounds have been documented which could conceivably contribute to retarding the growth of cancer. These include effects on the immune system such as stimulation of natural killer cell activity by garlic extracts.[46] Pretreatment of Ehrlich ascites cells with a garlic extract has been reported to enhance immunity in mice against these cells[47] and evidence has been presented that garlic extract may serve as an effective biological response modifier in controlling growth of a transplanted tumor.[5,48]

In some tumor models modulation of eicosanoid metabolism has been found to modify tumor growth. Onion and garlic oils have inhibited the conversion of arachidonic acid to prostaglandins.[49]

Effects of garlic extracts on the vascular system may be mediated by stimulation of nitric oxide as reported recently.[50] It remains to be investigated whether selective effects can be exerted on tumor vasculature.

EFFECTS ON TUMOR GROWTH AND DIFFERENTIATION

Tumor growth in rats and mice was inhibited by compounds structurally related to allicin.[51] The effect was attributed to the blocking of enzyme sulfhydryl groups. Although complete inhibition of ascites tumor cells by allicin *in vitro* was noted, studies with tumor-bearing mice led to the conclusion that alliin and allicin do not appear promising for cancer chemotherapy.[52] An antimitotic effect of garlic extract was reported for ascites sarcoma cells growing in rats but no complete tumor regression was observed.[53] Studies with a line of canine mammary tumor cells gave marked inhibitory effects on proliferation with three oil soluble organosulfur compounds but not with three water soluble compounds at concentrations up to 1 mM.[54] The inhibitory effects of diallyl sulfide, diallyl disulfide and diallyl trisulfide were decreased by glutathione.

S-Allyl cysteine was found to inhibit the growth of LA-N-5 human neuroblastoma cells in culture but it did not induce differentiation.[55] Nevertheless, S-allyl cysteine inhibited proliferation of human melanoma cells and may induce differentiation.[56] We have observed that DS19 mouse erythroleukemia cells are considerably more sensitive to the growth inhibitory effect of diallyl disulfide than other cell lines that we have examined.[57] Incubation of DS19 cells with 0.4 mM or higher concentrations of diallyl disulfide caused a complete inhibition of proliferation and significant inhibitions have been seen with concentrations as low as 0.08 mM diallyl disulfide. We noted that diallyl sulfide was much less effective than the disulfide. Evidence for a differentiating effect of diallyl disulfide was obtained by measuring acetylcholine esterase activity and the percentage of DS19 cells positive for hemoglobin production as judged by benzidine staining. The optimum concentration for induction of differentiated characteristics was approximately 0.5 mM diallyl disulfide. After incubation of DS19 cells with 0.5 mM diallyl disulfide for 96 hours the percentage of benzidine positive cells was 16-33% whereas the value was less than 1% in controls. This differentiating effect is substantial but is less than can be obtained with previously established inducers of differentiation such as butyrate or hexamethylene bisacetamide. K562 cells, a line of human erythroleukemia cells, were less sensitive than DS19 cells to the growth inhibitory action of diallyl disulfide. There was no significant effect of 0.5 mM diallyl disulfide on the proliferation of K562 cells but significant inhibition was seen at the 1 mM concentration. Studies with mouse melanoma cells showed that the incorporation of [^{3}H]thymidine was relatively resistant to the action of diallyl disulfide and suggested that the sensitivity of the DS19 cells was not simply attributable to a species difference in the cell origins. Further studies have shown that a 48 hour incubation with 1 mM diallyl disulfide could cause significant inhibition of the incorporation of [^{3}H]thymidine into DNA with different cell lines including 7800NJ rat hepatoma cells and T47D human breast cancer cells.

Diets containing 5% garlic extracts decreased the growth of Morris hepatoma 3924A transplanted in rats from 10-25%.[58] Injection of garlic extracts slowed tumor growth 30-50% but this was accompanied by toxicity. We have examined the effect of i.p. administration of diallyl disulfide on the incorporation of [^{3}H]thymidine into DNA of rats bearing subcutaneous transplants of three lines of Morris hepatomas (7288CTC, 7777 and 5123C).[57] Incorporation was determined 18 hours after treatment with diallyl disulfide. Marked inhibitory effects were seen in both hepatomas and in the livers of the tumor-bearing rats after treatment at a level of 400 mg/kg body weight. However, this is a toxic dose of diallyl disulfide. At a dose of 200 mg/kg which is better tolerated by the rats, small and statistically insignificant inhibitory effects were obtained in the tumors whereas significant effects were seen in the host livers. The lack of a tumor selective inhibitory response reduces enthusiasm regarding a therapeutic potential.

SUMMARY

There is evidence that organosulfur compounds can inhibit the induction and growth of cancer. Several organosulfur compounds are dietary constituents and *Allium* species are a rich source of such molecules. Some but not all epidemiological studies have suggested that consumption of garlic can decrease cancer incidence. There is substantial evidence that constituents of garlic including diallyl sulfides can inhibit the induction of cancer in experimental animals. Effects on both tumor initiation and promotion have been documented. Some effects may be mediated by modulation of carcinogen metabolism involving altered ratios of phase I and phase II drug metabolizing enzymes.[59,60] Inhibitory actions on the growth of tumor cells have been documented and, for some tumor cells, differentiating effects of diallyl sulfides can occur. A definitive mechanism of action has not been established and evidence exists for effects at several sites in carcinogen metabolism and regulation of tumor growth. It is not always clear that laboratory studies can be extrapolated to reasonable levels of consumption by humans of garlic or other *Allium* species.

REFERENCES

1. Block, E., The chemistry of garlic and onions, *Sci. American*. 252: 114 (1985).
2. Block, E., The organosulfur chemistry of the genus Allium - Implications for the organic chemistry of sulfur, *Angew. Chem. Int. Ed. Engl.* 31: 1135 (1992).
3. Dausch, J.G. and D.W. Nixon, Garlic: a review of its relationship to malignant disease, *Prev. Med.* 19: 346 (1990).
4. Dorant, E., P.A. van den Brandt, R.A. Goldbohm, R.J.J. Hermus, and F. Sturmans, Garlic and its significance for the prevention of cancer in humans: a critical review, *Br. J. Cancer* 67: 424 (1993).
5. Lau, B.H.S., P.P. Tadi and J.M. Tosk, *Allium sativum* (garlic) and cancer prevention, *Nutr. Res.* 10: 937 (1990).
6. Srivastava, K.C., A. Bordia and S.K. Verma, Garlic (*Allium sativum*) for disease prevention, *South African J. Sci.* 91: 68 (1995).
7. Sumiyoshi, H. and M.J. Wargovich, Garlic (*Allium sativum*): A review of its relationship to cancer. *Asia Pacific J. Pharmac.* 4: 133 (1989).
8. Buiatti, E., D. Palli, A. Decarli, D. Amadori, C. Avellini, S. Bianchi, R. Biserni, F. Cipriani, P. Cocco, A. Giacosa, E. Marubini, R. Puntoni, C. Vindigni, J. Fraumeni and W. Blot, A case-control study of gastric cancer and diet in Italy, *Int. J. Cancer* 44: 611 (1989).

9. You, W-C., W.J. Blot, Y-S. Chang, A. Ershow, Z.T. Yang, Q. An, B.E. Henderson, J.F. Fraumeni and T-G. Wang, Allium vegetables and reduced risk of stomach cancer, *J. Natl. Cancer Inst.* 81: 162 (1989).
10. Dorant, E., P.A. van den Brandt, and R.A. Goldbohm, A prospective cohort study on Allium vegetable consumption, garlic supplement use, and the risk of lung carcinoma in the Netherlands, *Cancer Res.* 54: 6148 (1994).
11. Dorant, E., P.A. van den Brandt, and R.A. Goldbohm, Allium vegetable consumption, garlic supplement intake, and female breast carcinoma incidence, *Breast Cancer Res.Treat.* 33: 163 (1995).
12. Kroning, F., Garlic as an inhibitor for spontaneous tumors in mice, *Acta Unio Contra Cancrum* 20: 855 (1964).
13. Belman, S., Onion and garlic oils inhibit tumor promotion, *Carcinogenesis* 4: 106 (1983).
14. Sadhana, A.S., A.R. Rao, K. Kucheria and V. Bijani, Inhibitory action of garlic oil on the initiation of benzo[a]pyrene-induced skin carcinogenesis in mice, *Cancer Lett.* 40: 193 (1988).
15. Nishino, H., A. Iwashima, Y. Itakura, H. Matsuura and T. Fuwa, Antitumor-promoting activity of garlic extracts, *Oncology* 46: 277 (1989).
16. Rao, A.R., A.S. Sadhana and H.C. Goel, Inhibition of skin tumors in DMBA-induced complete carcinogenesis system in mice by garlic (*Allium sativum*), *Ind. J. Exp. Biol.* 28: 405 (1990).
17. Liu, J., R.I. Lin and J.A. Milner, Inhibition of 7,12-dimethylbenz[a]anthracene-induced mammary tumors and DNA adducts by garlic powder, *Carcinogenesis* 13, 1847 (1992).
18. Amagase, H. and J.A. Milner, Impact of various sources of garlic and their constituents on 7,12-dimethylbenz[a]anthracene binding to mammary cell DNA, *Carcinogenesis* 14: 1627 (1993).
19. Ip, C., D.J. Lisk and G.S. Stoewsand, Mammary cancer prevention by regular garlic and selenium-enriched garlic, *Nutr. Cancer* 17: 279 (1992).
20. Hussain, S.P., L.N. Jannu and A.R. Rao, Chemopreventive action of garlic on methylcholanthrene-induced carcinogenesis in the uterine cervix of mice, *Cancer Lett.* 49: 175 (1990).
21. Nishino, H., A. Nishino, J. Takayasu, A. Iwashima, Y. Itakura, Y. Kodera, H. Matsuura and T. Fuwa, Antitumor-promoting activity of allixin, a stress compound produced by garlic, *Cancer J.* 3: 20 (1990).
22. Wargovich, M.J., Diallyl sulfide, a flavor component of garlic (Allium sativum), inhibits dimethylhydrazine-induced colon cancer, *Carcinogenesis* 8: 487 (1987).
23. Hayes, M.A., T.H. Rushmore and M.T. Goldberg, Inhibition of hepatocarcinogenic responses to 1,2-dimethylhydrazine by diallyl sulfide: A component of garlic oil, *Carcinogenesis* 8: 1155 (1987).
24. Wargovich, M.J., C. Woods, V.W.S. Eng, L.C. Stephens and K. Gray, Chemoprevention of N-nitrosomethylbenzylamine-induced esophageal cancer in rats by the naturally occurring thioether, diallyl sulfide, *Cancer Res.* 48: 6872 (1988).
25. Sparnins, V.L., G. Barany and L.W. Wattenberg, Effects of organosulfur compounds from garlic and onions on benzo(a)pyrene-induced neoplasia and glutathione S-transferase activity, *Carcinogenesis* 9: 131 (1988).
26. Wattenberg, L.W., V.L. Sparnins and G. Barany, Inhibition of N-nitrosodiethylamine carcinogenesis in mice by naturally occurring organosulfur compounds and monoterpenes, *Cancer Res.* 49: 2689 (1989).
27. Hong, J-Y., Z.Y. Wang, T.J. Smith, S. Zhou, S. Shi, J. Pan and C.S. Yang, Inhibitory effects of diallyl sulfide on the metabolism and tumorigenicity of the tobacco-specific carcinogen 4-(methylnitrosamino)-1-(3-pyridyl)-1-butanone (NNK) in A/J mouse lung, *Carcinogenesis* 13: 901 (1992).
28. Jang, J.J., K.J. Cho, S.Y. Lee and J.H. Bae, Modifying responses of allyl sulfide, indole-3-carbinol and germanium in a rat multi-organ carcinogenesis model, *Carcinogenesis* 12: 691 (1991).
29. Pereira, M.A. and M.D. Khoury, Prevention by chemopreventive agents of azoxymethane-induced foci of aberrant crypts in rat colon, *Cancer Lett.* 61: 27 (1991).
30. Takahashi, S., K. Hakoi, H. Yada, M. Hirose, N. Ito and S. Fukushima, Enhancing effects of diallyl sulfide on hepatocarcinogenesis and inhibitory actions of related diallyl disulfide on colon and renal carcinogenesis in rats, *Carcinogenesis* 13: 1513 (1992).
31. Reddy, B.S., C.V. Rao, A. Rivenson and G. Kelloff, Chemoprevention of colon carcinogenesis by organosulfur compounds, *Cancer Res.* 53: 3493 (1993).
32. Devasagayam, T.P.A., C.K. Pushpendran and J. Eapen, Diallyl disulphide induced changes in microsomal enzymes of suckling rats, *Ind. J. Exp. Biol.* 20: 430 (1982).
33. Brady, J.F., D. Li, H. Ishizaki and C.S. Yang, Effect of diallyl sulfide on rat liver microsomal nitrosamine metabolism and other monooxygenase activities, *Cancer Res.* 48: 5937 (1988).
34. Hong, J-Y., T. Smith, M.-J. Lee, W. Li, B.L. Ma, S.M. Ning, J.F. Brady, P.E. Thomas and C.S. Yang, Metabolism of carcinogenic nitrosamines by rat nasal mucosa and the effect of diallyl sulfide, *Cancer Res.* 51: 1509 (1991).

35. Chen, L., M. Lee, J.-Y. Hong, W. Huang, E. Wang and C.S. Yang, Relationship between cytochrome P450 2E1 and acetone catabolism in rats as studied with diallyl sulfide as an inhibitor, *Biochem Pharmacol.* 48: 2199 (1994).
36. Fiorio, R. and G. Bronzetti, Diallyl sulfide inhibits the induction of HPRT-deficient mutants in Chinese hamster V79 cells treated with dimethylnitrosoamine in the presence of S-9 of rats induced with acetone, *Environ. Mol. Mutagenesis* 25: 344 (1995).
37. Pan, J., J.-Y. Hong, D. Li, E.G. Schuetz, P.S. Guzelian, W. Huang and C.S. Yang, Regulation of cytochrome P450 2B1/2 genes by diallyl sulfone, disulfiram, and other organosulfur compounds in primary cultures of rat hepatocytes, *Biochem Pharmacol.* 45: 2323 (1993).
38. Pan, J., J.-Y. Hong, B.-L. Ma, S.M. Ning, S.R. Paranawithana and C.S. Yang, Transcriptional activation of cytochrome P450 2B1/2 genes in rat liver by diallyl sulfide, a compound derived from garlic, *Arch. Biochem. Biophys.* 302: 337 (1993).
39. Haber, D., M.-H. Siess, M.-C. Canivenc-Lavier, A.-M. Le Bon and M. Suschetet, Differential effects of dietary diallyl sulfide and diallyl disulfide on rat intestinal and hepatic drug-metabolizing enzymes, *J. Toxicol. Environ. Health* 44: 423 (1995).
40. Tadi, P.P., R.W. Teel and B.H.S. Lau, Organosulfur compounds of garlic modulate mutagenesis, metabolism, and DNA binding of aflatoxin B1, *Nutr. Cancer* 15: 87 (1991).
41. Perchellet, J-P., E.M. Perchellet, N.L. Abney, J.A. Zirnstein and S. Belman, Effects of garlic and onion oils on glutathione peroxidase actvity, the ratio of reduced/oxidized glutathione and ornithine decarboxylase induction in isolated mouse epidermal cells treated with tumor promoters, *Cancer Biochem. Biophys.* 8: 299 (1986).
42. Hu, P.J. and M.J. Wargovich, Effect of diallyl sulfide on MNNG-induced nuclear aberrations and ornithine decarboxylase activity in the glandular stomach mucosa of the Wistar rat, *Cancer Lett.* 47: 153 (1989).
43. Baer, A.R. and M.J. Wargovich, Role of ornithine decarboxylase in diallyl sulfide inhibition of colonic radiation injury in the mouse, *Cancer Res.* 49: 5073 (1989).
44. Takada, N., T. Matsuda, T. Otoshi, Y. Yano, S. Otani, T. Hasegawa, D. Nakae, Y. Konishi and S. Fukushima, Enhancement by organosulfur compounds from garlic and onions of diethylnitrosamine-induced glutathione S-transferase positive foci in the rat liver, *Cancer Res.* 54: 2895 (1994).
45. Wills, E.D., Enzyme inhibition by allicin, the active principle of garlic, *Biochem. J.* 63: 514 (1956).
46. Kandil, O.M., T.H. Abdullah and A. Elkadi, Garlic and the immune system in humans: its effect on natural killer cells, *Fed. Proc.* 46: 441 (1987).
47. Fujiwara, M. and T. Natata, Induction of tumor immunity with tumour cells treated with extract of garlic (*Allium sativum*), *Nature* 216: 83 (1967).
48. Lau, B.H.S., J.L. Woolley, C.L. Marsh, G.R. Barker, D.H. Koobs and R.R. Torrey, Superiority of intralesional immunotherapy with *Corynebacterium parvum* and *Allium sativum* in control of murine transitional cell carcinoma, *J. Urol.* 136: 701 (1986).
49. Vanderhoek, J.Y., A.N. Makheja, and J.M. Bailey, Inhibition of fatty acid oxygenases by onion and garlic oils, *Biochem. Pharmacol.* 29: 3169 (1980).
50. Das, I., N.S. Khan and S.R. Sooranna, Nitric oxide synthase activation is a unique mechanism of garlic action, *Biochem. Soc. Trans.* 23: S136 (1995).
51. Weisberger A.S. and J. Pensky, Tumor inhibition by a sulfhydryl-blocking agent related to an active principle of garlic (*Allium sativum*), *Cancer Res.* 18: 1301 (1958).
52. DiPaolo, J.A. and C. Carruthers, The effect of allicin from garlic on tumor growth, *Cancer Res.* 20: 431 (1960).
53. Kimura, Y. and K. Yamamoto, Cytological effect of chemicals on tumors. XXIII. Influence of crude extracts from garlic and some related species on MTK-sarcoma III, *Gann* 55: 325 (1964).
54. Sundaram S.G. and J.A. Milner, Impact of organosulfur compounds in garlic on canine mammary tumor cells in culture, *Cancer Lett.* 74: 85 (1993).
55. Welch, C., L. Wuarin and N. Sidell, Antiproliferative effect of the garlic compound S-allyl cysteine on human neuroblastoma cells in vitro, *Cancer Lett.* 63: 211 (1992).
56. Takeyama, H., D.S.B. Hoon, D.L. Morton and R.F. Irie, Growth inhibition and modulation of cell markers of melanoma by an aged garlic product, sulfur-allyl cysteine, *Proc. Am. Assoc. Cancer Res.* 32: 426 (1991).
57. Lea, M.A. and U. Ayyala, unpublished observations.
58. Criss, W.E., J. Fakunle, E. Knight, J. Adkins, H.P. Morris and G. Dhillon, Inhibition of tumor growth with low dietary protein and with dietary garlic extracts, *Fed. Proc.* 41: 281 (1982).
59. Kwak, M.K., S.G. Kim, J.Y. Kwak, R.F. Novak and N.D. Kim, Inhibition of cytochrome P4502E1 expression by organosulfur compounds allylsulfide, allylmercaptan and allylmethylsulfide in rats, *Biochem. Pharmacol.* 47: 531 (1994).

60. Kim, N.D., S.G. Kim and M.K. Kwak, Enhanced expression of rat microsomal epoxide hydrolase gene by organosulfur compounds, *Biochem. Pharmacol.* 47: 541 (1994).

13

RECENT RESULTS IN THE ORGANOSULFUR AND ORGANOSELENIUM CHEMISTRY OF GENUS *ALLIUM* AND *BRASSICA* PLANTS

Relevance for Cancer Prevention

Eric Block

Department of Chemistry
State University of New York at Albany
Albany, New York 12222

INTRODUCTION

Since the discovery by Weisberger and Pensky in 1957 of the anti-tumor activity of ethanesulfinothioic acid *S*-ethyl ester (**1**, EtS(O)SEt),[1] there has been considerable interest in the cancer preventative properties of genus *Allium* and *Brassica* plants, known to generate compounds similar to **1** upon cutting or crushing. This interest has been further stimulated by epidemiological studies suggesting that frequent consumption of these plants is associated with a reduction in risk for certain human cancers,[2-4] by research indicating that sulfur compounds from *Allium* spp. can reduce gastric juice nitrite concentrations,[5] that garlic enriched with selenium shows enhanced cancer preventative properties compared to normal garlic,[6-8] and that "allylic" constituents of garlic can both inhibit HMG-CoA reductase[9] and can prevent activation of nitrosamines.[10] Because the cancer preventative properties of genus *Allium* and *Brassica* plants are typically associated with sulfur- and selenium-containing phytochemicals in these plants, the nature of the latter compounds will first be reviewed. This will be followed by a summary of recent epidemiological studies associating garlic consumption with reduced risk of gastrointestinal cancer, and a discussion of the antibacterial, cytotoxic, anti-tumor, and antioxidant activity of garlic relevant to cancer prevention.

ORGANOSULFUR AND ORGANOSELENIUM COMPOUNDS IN *ALLIUM* AND *BRASSICA* SPP.

The characteristic flavors generated when garlic (*Allium sativum*), onion (*A. cepa*) and other genus *Allium* plants are cut or crushed are produced by the action of C-S lyase enzymes (alliinases) on *S*-alkyl cysteine *S*-oxides $RS(O)CH_2CH(NH_2)COOH$ (**2**), where R is variously *methyl*, *n-Propyl*, allyl, or 1-propenyl.[11] The initially released sulfenic acids

Dietary Phytochemicals in Cancer Prevention and Treatment
Edited under the auspices of the American Institute for Cancer Research, Plenum Press, New York, 1996

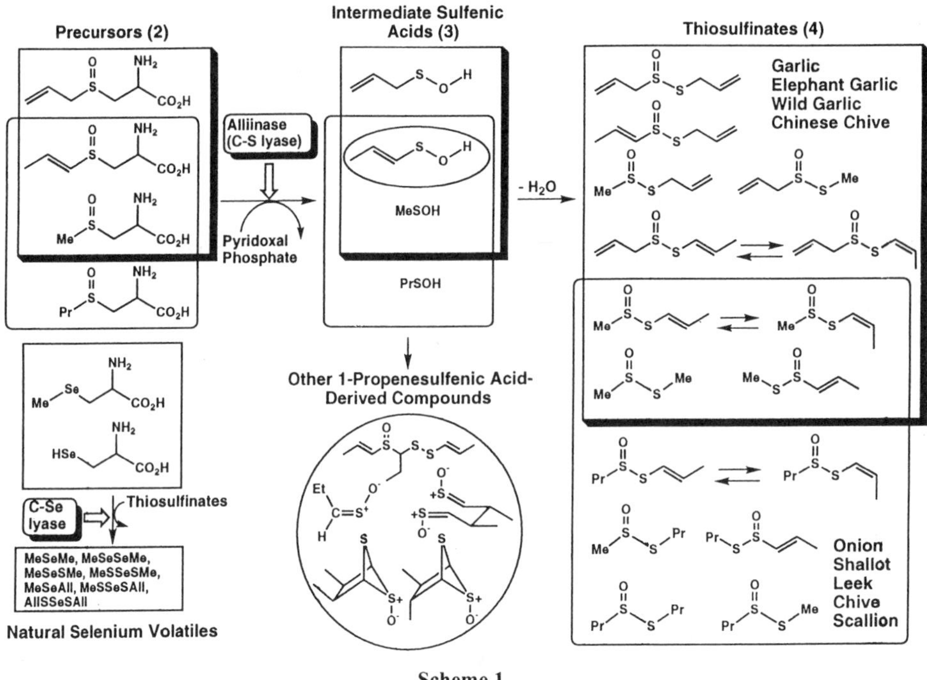

Scheme 1.

RSOH, **3**, condense to form thiosulfinate ester, RS(O)SR' (**4**), analogous to EtS(O)SEt (**1**) studied by Weisberger and Pensky. Scheme 1 illustrates the range of compounds formed on cutting genus *Allium* plants. Compounds of type **2** originate from soil sulfate ion via cysteine through a series of steps culminating in oxidation of γ-glutamyl-*S*-alk(en)yl cysteines by oxidases to the corresponding *S*-oxides followed by cleavage of these to **1** by γ-glutamyl transpeptidases (Scheme 2). Garlic is one of several vegetables containing elevated levels of selenium (Figure 1).[12] The bioavailability of selenium in food products of vegetable origin is high (ca. 60% of total content), although little is known about the form in which selenium is found or about what flavor contributions selenium may make. In 1964, Virtanen reported on the basis of radioisotope studies that onion contained the selenoamino acids selenocystine ($(HOOCCH(NH_2)CH_2Se)_2$) and selenomethionine ($HOOCCH(NH_2)CH_2CH_2SeMe$).[13] Virtanen's discovery suggested that there might be a selenium-based flavor chemistry in *Allium* spp. parallel to that based on sulfur, e.g. originating from soil selenate (SeO_4^{-2}) or selenite (SeO_3^{-2}) (Scheme 3). As noted above, animal studies have shown that selenium-enriched garlic possesses cancer preventative properties.[6-8] We sought to obtain information on the nature and amounts of organoselenium compounds in garlic and related *Allium* and *Brassica* species as well as information on how the body handles the selenium taken in from these plants.

SELENIUM BIOCHEMISTRY

Selenium (Se) is an essential micronutrient whose absence causes skeletal and cardiac muscle dysfunction.[14,15] It is required for the proper function of the immune system and for cellular defense against oxidative damage, and thus may play a role in the prevention of

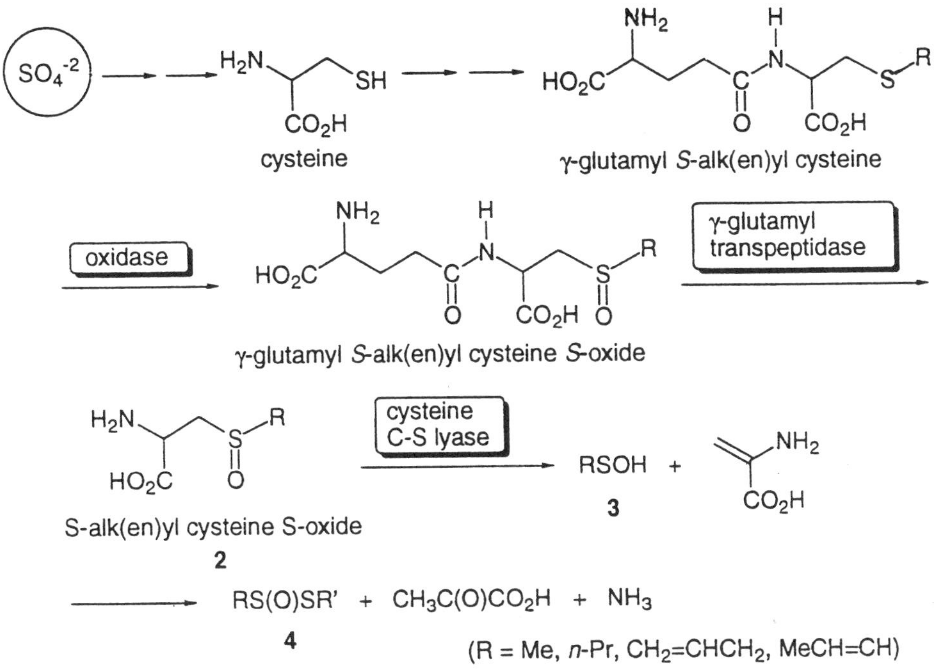

SO_4^{-2} → → H_2N–CH(CO_2H)–CH_2–SH (cysteine) → → HO_2C–CH(NH_2)–CH_2CH_2–C(O)–NH–CH(CO_2H)–CH_2–S–R (γ-glutamyl *S*-alk(en)yl cysteine)

→ [oxidase] HO_2C–CH(NH_2)–CH_2CH_2–C(O)–NH–CH(CO_2H)–CH_2–S(O)–R (γ-glutamyl *S*-alk(en)yl cysteine *S*-oxide) → [γ-glutamyl transpeptidase]

H_2N–CH(HO_2C)–CH_2–S(O)–R (S-alk(en)yl cysteine S-oxide, **2**) → [cysteine C-S lyase] RSOH (**3**) + CH_2=C(NH_2)CO_2H

→ RS(O)SR' (**4**) + $CH_3C(O)CO_2H$ + NH_3

(R = Me, *n*-Pr, CH_2=$CHCH_2$, MeCH=CH)

Scheme 2.

cancer and premature aging.[16] Because the incorporation of selenocysteine ($HSeCH_2$-$CH(NH_2)COOH$; Cys-SeH) into proteins is directed by a UGA codon, it has been called the 21st amino acid essential for ribosome-directed protein synthesis.[17] Selenocysteine is present at the active sites of glutathione peroxidase, 5'-deiodinase and selenoprotein P.[18] In glutathione peroxidase, selenium (as Cys-SeH) removes an O atom from peroxides; in a catalytic

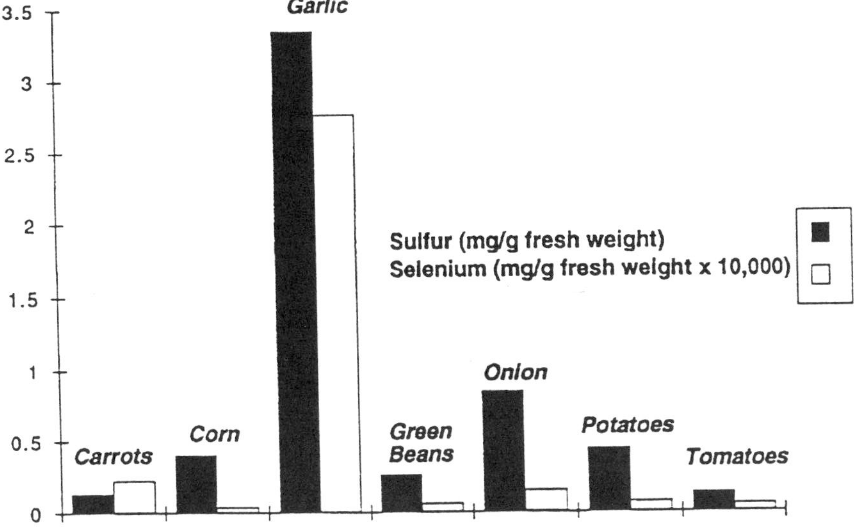

Figure 1. Comparative sulfur and selenium content of common vegetables.

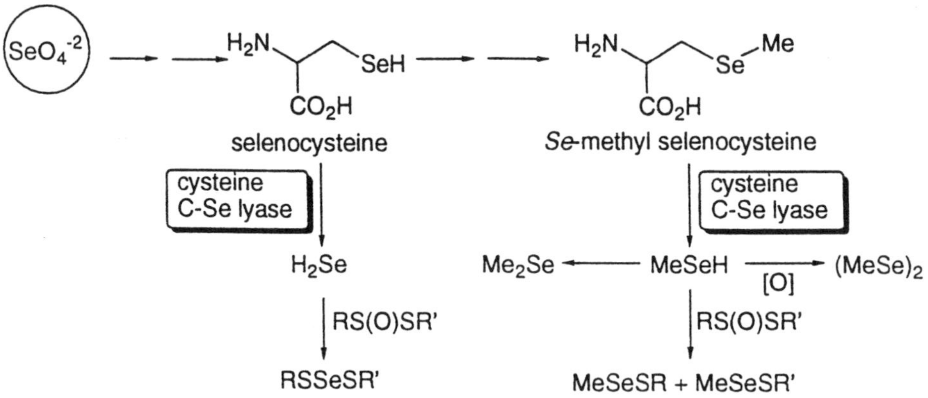

Scheme 3.

cycle the resultant Cys-SeOH is then reduced back to Cys-SeH by glutathione. Because the pKa of free Cys-SeH is 5.2, compared with the >8 value for free cysteine, and because Se^- is more nucleophilic than S^- but at the same time the bonds to Se are weaker than those to S, Cys-SeH can be considered as a "super-active" cysteine when involved in catalytic processes. Selenoproteins[19] and a selenopolysaccharide,[20] are said to be present in garlic. Broccoli is said to accumulate high levels of unknown forms of Se.[21] Cabbage grown with $H_2{}^{75}SeO_3$ is reported to contain various seleno-amino acids, -peptides, and -proteins.[22]

ANALYSIS OF SELENIUM IN *ALLIUM* SPP.

Analysis of flavorants and their precursors from *Allium* spp. plants presents a number of challenges: primary flavorants are formed on cutting or crushing the plants through action of the released enzymes on precursors. The flavorants themselves, which can exist in a variety of sometimes interconverting isomeric forms, are thermally and hydrolytically unstable, decomposing to give mixtures of secondary compounds, some of which are also unstable. Selenium-containing flavorants/flavorant precursors, which may be important both from the standpoints of flavor as well as health benefits, have physical properties quite similar to those of far more abundant homologs containing sulfur. In particular, garlic and onion contain 0.28 and 0.015 μg Se per g fresh weight, respectively,[12] compared to 3.3 and 0.84 mg S, e.g. 12,000 to 56,000 times higher levels of S than Se are present in these *Allium* spp.

We have used gas chromatography (GC) with atomic plasma spectral emission detection (GC-AED) for element specific detection of natural abundance organoselenium compounds in both *Allium* and *Brassica* spp. plants, plant extracts and volatiles, and human exhaled breath following plant consumption. The technique of GC-AED, represented schematically in Figure 2, has the important advantages of high sensitivity, elemental selectivity, and the possibility of simultaneous multi-element analysis.[23] The AED response can flag compounds in the GC effluent which contain specific elements even though these compounds may be present in very small amounts or may coelute with other components. We have used the selenium emission line at 196 nm to detect organoselenium species while concurrently monitoring S and C by lines at 181 and 193 nm, respectively. These assignments were confirmed by GC-MS (mass spectrometry). The above techniques were used to analyze the headspace above homogenized garlic, elephant garlic, onion, Chinese chive and broccoli.[24,25]

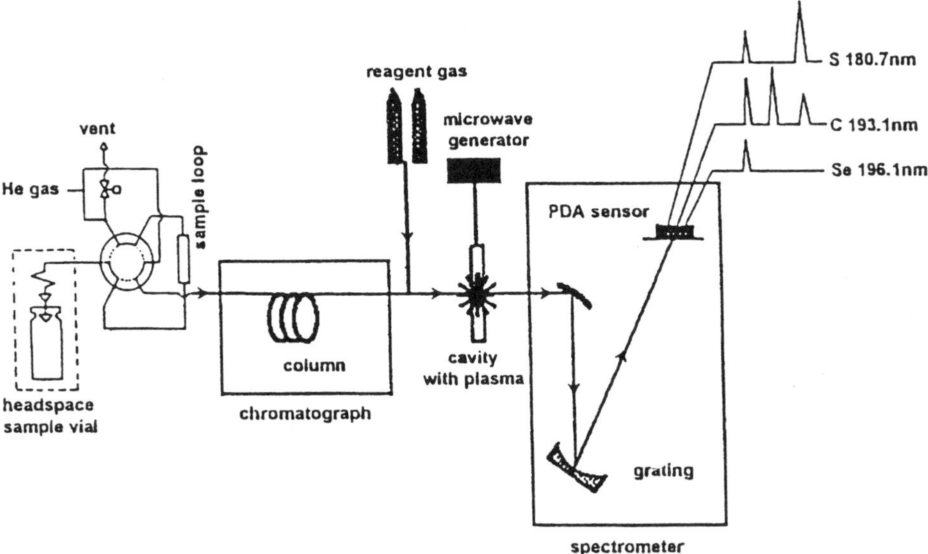

Figure 2. Schematic diagram of headspace-gas chromatograph-atomic emission detector (HS-GC-AED; Hewlett-Packard Co., Palo Alto, CA, USA). Sample is capped in the headspace sample vial which is placed in a heated bath to achieve thermal equilibration. The headspace gas is then driven through the sample loop to the GC injection port. The GC eluent is introduced into a microwave-energized helium plasma that is coupled to a photodiode array (PDA) optical emission spectrometer. The plasma is sufficiently energetic to atomize all of the elements in a sample and to excite their characteristic atomic emission spectra. Up to four elements which have adjacent spectral emission wavelengths can be monitored simultaneously, illustrated here for sulfur (emission at 180.7 nm), carbon (emission at 193.1 nm) and selenium (emission at 196.1 nm).

As the headspace (HS) GC-MS procedure was not sufficiently sensitive and selective to directly detect natural levels of *Allium* and *Brassica* Se compounds, garlic, onion or broccoli grown in a Se-fertilized medium (Se-enriched plants), or garlic homogenates augmented by addition of selenoamino acids (see below) were utilized.

HEADSPACE ANALYSIS

In the case of the HS-GC-AED analysis of volatiles from natural garlic (Figure 3), there are essentially no Se-containing peaks observed in the C or S channels because they are so small as to be lost within the background signals. The S channel shows MeS_nMe, MeS_nAll, and $AllS_nAll$ (n = 1-3, All = allyl), typical of garlic and garlic-like Alliums.[26] The Se channel shows seven peaks: dimethyl selenide (MeSeMe), methanesulfenoselenoic acid methyl ester (MeSeSMe), dimethyl diselenide (MeSeSeMe), bis(methylthio)selenide ($(MeS)_2Se$), allyl methyl selenide (MeSeAll), 2-propenesulfenoselenoic acid methyl ester (MeSeSAll), and (allylthio)(methylthio)selenide (MeSSeSAll). Structures were established by mass spectrometry, in most cases through comparison with spectra of synthetic samples prepared from bis(*N*-benzotriazolyl)selenide and disulfide-diselenide interchange.[27,28] Of the seven compounds, MeSeMe, MeSeSMe, and MeSeSeMe have been previously found in nature.[29-32] The headspace above chopped broccoli, analyzed by GC-AED, showed MeSeMe, MeSeSMe, MeSeSeMe, $MeSeSC_3H_5$, $MeSSeSC_3H_5$, $C_3H_5SSeSC_3H_5$, together with six thus far unidentified peaks, in the Se channel and MeSSMe (major), MeSSSMe, $C_3H_5SSC_3H_5$

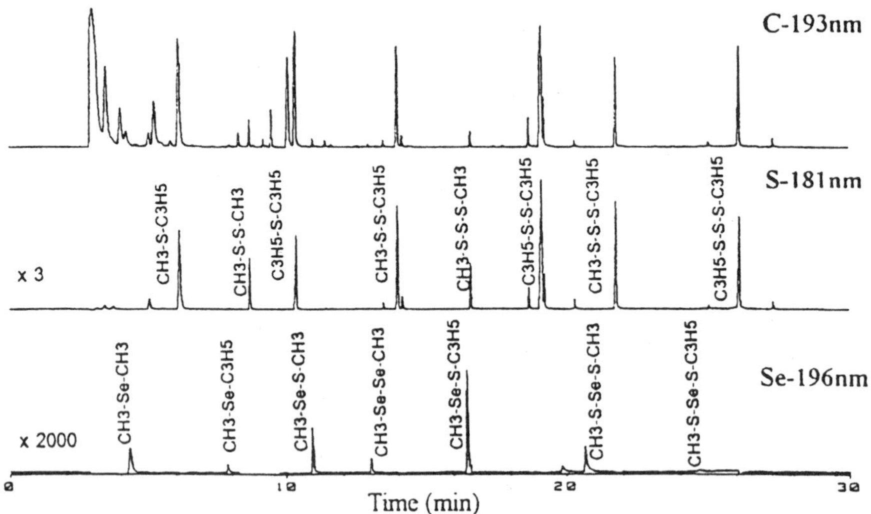

Figure 3. Organoselenium and organosulfur compounds in garlic determined by HS-GC-AED. The sulfur and selenium vertical scales are in units 3 and 2000 times larger, respectively, than the corresponding carbon vertical scale.

and $C_3H_5SSSC_3H_5$ in the S channel. The headspace above chopped onion showed the presence of methyl propyl selenide, MeSePr.

SELENOAMINO ACID ANALYSIS

To determine the origin of the headspace selenium compounds, lyophilized normal garlic (0.02 ppm Se) or moderately *Se*-enriched (68 ppm Se) garlic was derivatized with ethyl chloroformate to volatize the selenoamino acids.[33-35] Subsequent GC-AED analysis showed a single peak in the Se channel, identified as selenocysteine by comparison with the retention time and mass spectral fragmentation of an authentic sample (Figure 4). In garlic more heavily *Se*-enriched (1355 ppm Se), *Se*-methyl selenocysteine was the major selenoamino acid found along with minor amounts of selenocysteine and traces of *Se*-methionine (Figure 5, 6; Scheme 2); the S channel showed 2:1 allyl-cysteine and allyl-cysteine *S*-oxide along with minor amounts of methionine.[36] In contrast to the situation with the selenoamino acids, there were only minor changes in the relative ratios of the sulfur amino acid as the level of selenium was varied from 0.02 to 1355 ppm. Analogous analysis of *Se*-enriched onion (96 ppm Se) revealed the presence of equal amounts of *Se*-methyl selenocysteine and selenocysteine in the Se channel; analysis of *Se*-enriched (150 ppm Se) broccoli showed the presence of ca. 2:1 *Se*-methyl selenocysteine and selenocysteine along with much smaller amounts of *Se*-methionine in the Se channel and *S*-methyl cysteine and lower amounts of methionine in the S channel (the presence of *S*-methyl cysteine *S*-oxide has been previously reported[37]). The presence of *Se*-methyl selenocysteine as the major source of selenium in *Se*-enriched *Allium* and *Brassica* spp. is significant since *Se*-methyl selenocysteine is known to exhibit cancer chemopreventative activity.[38] *Se*-Methyl L-selenocysteine is superior to *S*-methyl L-cysteine as a substrate for L-methionine γ-lyases and *S*-alkylcysteine α,β-lyases in bacteria, due to the superior leaving group ability of $MeSe^-$ compared to MeS^-.[39,40] It is therefore probable that related enzymes in *Allium* and *Brassica*

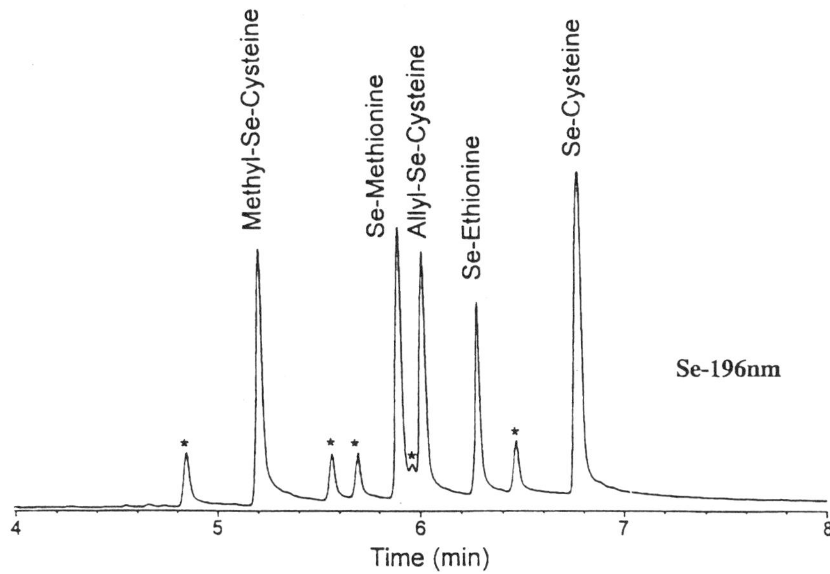

Figure 4. GC-AED analysis (Se channel, monitored at 196.1 nm) of $ClCO_2Et$ derivatized selenocysteine ($HSeCH_2CH(NH_2)COOH$; Cys-SeH), *Se*-methyl selenocysteine ($MeSeCH_2CH(NH_2)COOH$; Cys-Se-Me), seleno-D,L-methionine ($MeSeCH_2CH_2CH(NH_2)COOH$; Se-Met), seleno-D,L-ethionine ($EtSeCH_2CH_2$-$CH(NH_2)COOH$; Se-Eth), and *Se*-allyl selenocysteine (CH_2=$CHCH_2SeCH_2CH(NH_2)COOH$; Cys-Se-All). Peaks labelled with an asterisk are byproducts associated with the derivatization procedure.

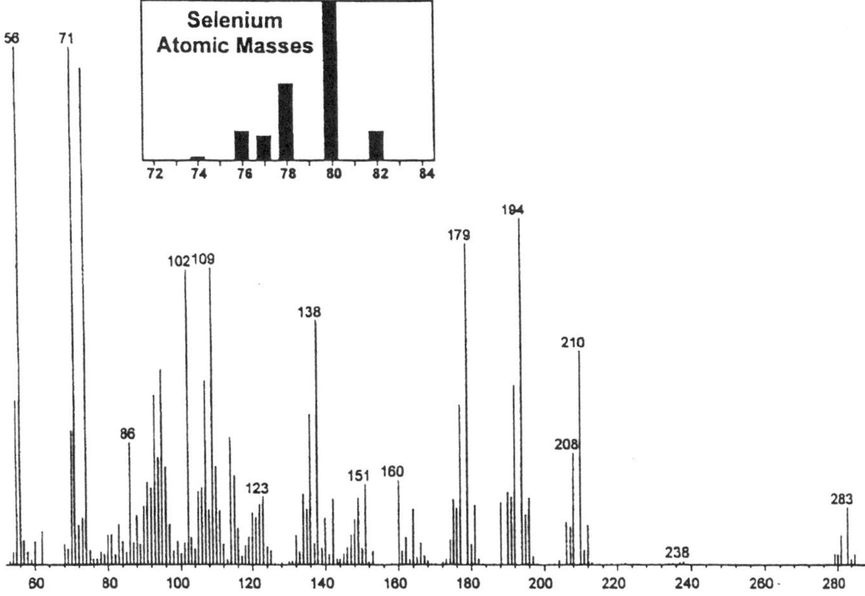

Figure 5. Mass spectrum of $ClCO_2Et$ derivatized *Se*-methyl selenocysteine (Cys-Se-Me), MeSe-$CH_2CH(NHCO_2Et)CO_2Et$ (M^+ 283), showing ions of *m/z* _ 53. Major fragments are seen at *m/z* 210 (M^+ - CO_2Et), 194 (M^+ - $NHCO_2Et$, H^+), 179 (M^+ - $NHCO_2Et$, Me, H^+), 138 ($MeSeCH_2C$_NH^+), 109 ($MeSeCH_2^+$), and 74 ($HCO_2Et^{\cdot+}$). The selenium stable natural abundance isotopic ratios are shown in the inset.

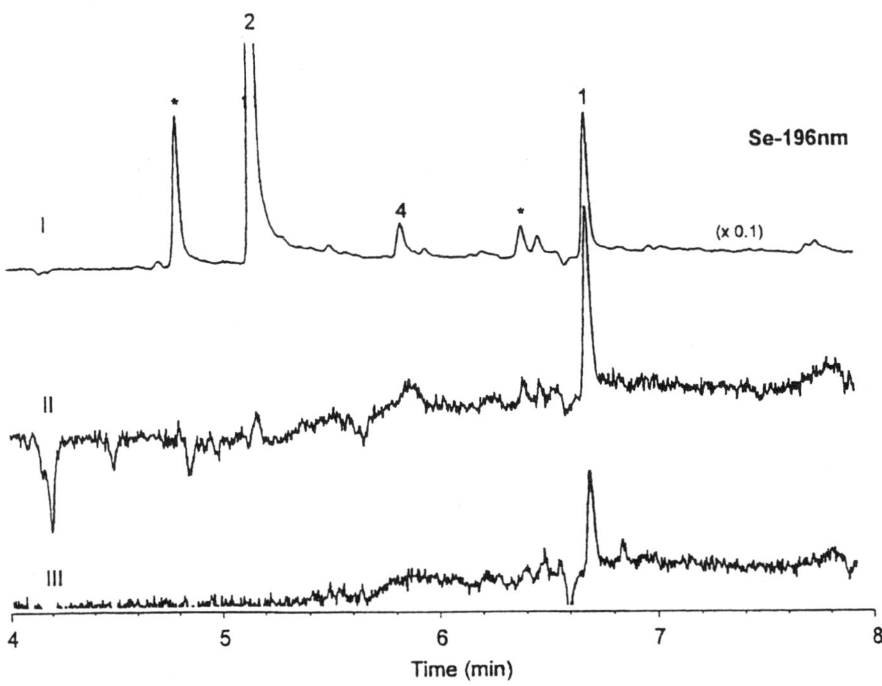

Figure 6. GC-AED analysis (Se channel, monitored at 196.1 nm) of $ClCO_2Et$ derivatized amino acids from lyophilized Se-enriched (I: 1355 ppm Se; II: 68 ppm Se) and unenriched (III: 0.02 ppm Se) garlic. Compound identification: $ClCO_2Et$ derivatized selenocysteine (**1**, $HSeCH_2CH(NH_2)COOH$; Cys-SeH), *Se*-methyl selenocysteine (**2**, $MeSeCH_2CH(NH_2)COOH$; Cys-Se-Me), and seleno-D,L-methionine (**4**, $MeSeCH_2CH_2CH(NH_2)$-COOH; Se-Met). Peaks labelled with an asterisk are byproducts associated with the derivatization procedure.

spp. cleave this selenoamino acid to MeSeH (or $MeSe^-$), which reacts with RSS(O)R' formed when the plant is cut to give MeSeSR and MeSeSR', or to form MeSeSeMe on oxidation. Analogous reaction of H_2Se, released from selenocysteine, could afford RSSeSR', while *Se*-methylation of *Se*-methyl L-selenocysteine followed by enzymatic cleavage would afford MeSeMe (see Scheme 2).

Analysis of the headspace above *Se*-enriched garlic by HS-GC-AED showed a very similar profile of compounds whether or not synthetic *Se*-methyl selenocysteine was added, although all of the peak abundances are enhanced by the addition of the synthetic selenoamino acid. On the other hand, addition of synthetic *Se*-allyl selenocysteine to *Se*-enriched garlic followed by HS-GC-AED analysis showed a profile quite different from that of *Se*-enriched garlic, with the major peaks being All_2Se (major peak; not seen in normal garlic; small peak in *Se*-enriched garlic) and AllSSeSAll (or isomer; small peak in *Se*-enriched garlic). We conclude that *Se*-allyl selenocysteine is not present in our lyophilized garlic samples. Further work is necessary to establish whether or not *Se*-allyl selenocysteine is synthesized to any extent in garlic cloves. We suggest that synthetic *Se*-allyl selenocysteine is cleaved in garlic homogenates probably forming CH_2=$CHCH_2$-SeH, which is oxidized to thermally unstable AllSeSeAll, which in turn loses selenium affording AllSeAll. We further suggest that when garlic is presented with high levels of inorganic selenium fertilizer, the excess selenocysteine formed is *Se*-methylated to give the major selenoamino acid, *Se*-methyl selenocysteine.

HUMAN GARLIC BREATH ANALYSIS

Finally, analysis of human garlic breath by GC-AED (Figure 7) revealed in the Se channel the presence of dimethyl selenide (MeSeMe) as the major component along with one tenth to one fortieth the amount of $MeSeC_3H_5$, CH_3SeSMe and $MeSeSC_3H_5$; the S channel showed the major components to be AllSH, CH_3SAll and AllSSAll with lesser amounts of MeSSMe, $CH_3SSC_3H_5$, an isomer of AllSSAll (presumably MeCH=CHSSAll), $C_3H_5SC_3H_5$ and $C_3H_5SSSC_3H_5$.[41] The composition of the Se and S compounds in garlic breath was examined as a function of time over the course of four hours. After four hours the levels of MeSeMe, AllSSAll, AllSAll and MeSSMe are reduced by 75% from the initial levels of 0.45 ng/L (MeSeMe), 45 ng/L (AllSSAll), 6.5 ng/mL (AllSAll), and 1.8 ng/L (MeSSMe)(Figure 8). The AllSH could only be detected in breath immediately after ingestion of garlic. In view of the reported very low threshhold detection level for low molecular weight organoselenium compounds,[42] it is likely that compounds such as MeSeMe contribute to the overall odor associated with garlic breath. It has been previously reported that dimethyl selenide, which has a garlic-like odor, is found in the breath air of animals fed inorganic selenium compounds and humans who have accidentally ingested selenium compounds.[43]

CANCER PREVENTATIVE ACTIVITY

Epidemiological Evidence

Nine epidemiological studies have thus far appeared relating garlic consumption with diminished risk of gastrointestinal cancer.[3] In a study of 564 stomach cancer patients

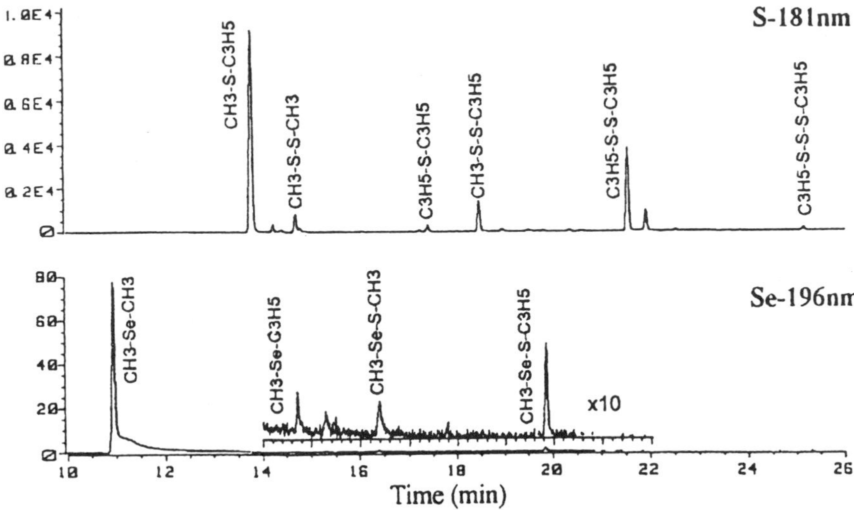

Figure 7. A subject consumed, with brief chewing, 3 g of fresh garlic with small pieces of white bread, followed by 50 mL of cold water. Shown are the organosulfur (upper trace) and organoselenium compounds (lower trace) in the subject's breath one hour after garlic consumption as determined by GC-AED: S, allyl methyl sulfide, dimethyl disulfide, diallyl sulfide, allyl methyl disulfide, diallyl disulfide, diallyl trisulfide; Se, dimethyl selenide, allyl methyl selenide, methanesulfenoselenoic acid methyl ester (MeSeSMe), 2-propenesulfenoselenoic acid methyl ester (MeSeSAll). The sulfur vertical scale is in units 10^4 times larger (shown as "E4") than the corresponding selenium vertical scale.

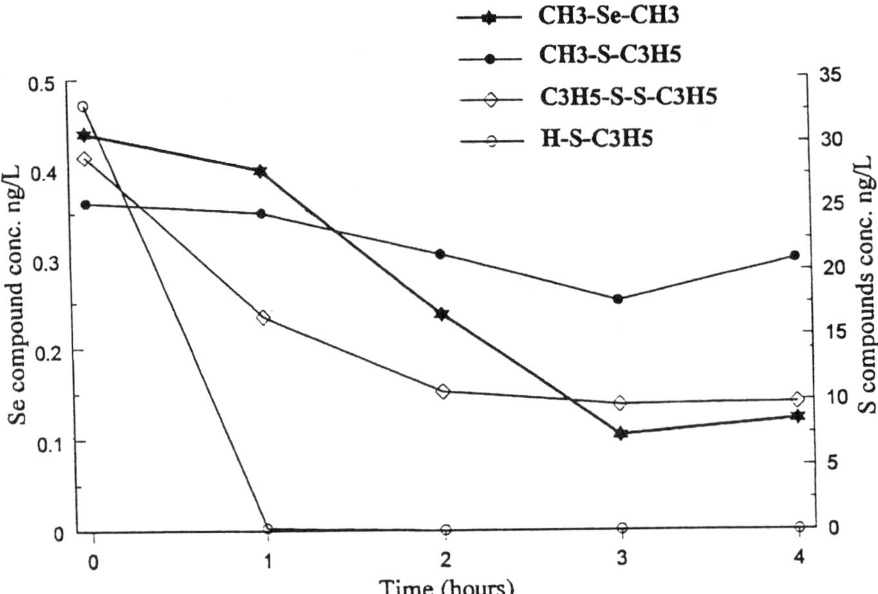

Figure 8. Variation in the concentrations of allyl methyl sulfide, diallyl disulfide, 2-propenethiol and dimethyl selenide in human garlic breath with time.

and 1131 controls in Shandong Province in the People's Republic of China, a significant reduction of stomach cancer risk was found to be associated with increasing consumption of allium vegetables including garlic, scallions and Chinese chives.[2] In a related study comparing the death rate from stomach cancer in two large groups in Shandong Province of China, those living in the province's Cangshan County were found to have a lower death rate (3/100,000) than those living in Qixia County (40/100,000). The residents of the former county eat ca. 20 g raw garlic/day while those in the latter county rarely eat garlic. The most striking finding from an epidemiological study of the incidence of colon cancer among 41,837 Iowa women aged 55-69 years over a period of 5 years was the 35% lower risk observed associated with consumption of one or more servings of garlic per week.[4] Because of the low levels of garlic consumption in this latter study, questions have been raised that "garlic consumption in this group of middle- and older-aged Midwestern women is simply a surrogate of some other variable, perhaps not even diet related".[44] This criticism could, of course, be raised about any epidemiological study. Epidemiological studies involving 120,852 men and women aged 55-69 years over a period of 3.3 years concluded that garlic supplement use, or consumption of onions or leeks, was not associated with lower risks of lung carcinoma or of female breast carcinoma.[45-47] However, in the absence of specific information on the nature or composition of the garlic supplements used by participants in the study, it is difficult to judge the significance of the conclusions *vis-a-vis* garlic usage.

Antibacterial Activity Relevant to Cancer Prevention

Antibacterial properties of garlic may inhibit bacterial conversion of nitrate to nitrite in the stomach, thereby limiting formation of carcinogenic nitrosamines.[5] Alternatively, garlic sulfur compounds may inhibit *N*-nitrosation by rapidly forming *S*-nitrosothiols or thionitrites and nitric oxide; the *S*-nitroso compounds would then decompose giving disulfides.[48] Since high doses of diallyl trisulfide, allyl methyl trisulfide and related sulfur compounds have been found to *promote* diethylnitrosamine-induced neoplasia in rat livers, it is argued that "in evaluating relationships between diet and cancer, it is appropriate to consider not only the possible protective role of garlic and onions but also their enhancing effects".[49] However, one should also consider whether the high doses given the test animals are meaningful in evaluating possible human dosage: the equivalent of 100 mg/kg body weight/day of a trisulfide given to a test animal would be an absurdly high daily human dose of 5.9 L of commercial distilled garlic oil.

Cytotoxicity and Anti-Tumor Activity

Both allicin and ajoene have been found to be cytotoxic to three different fast-growing cell lines, with EC_{50} values in the micromolar range,[50] but less so by two to three orders of magnitude to various liver cells.[9] The low cytotoxicity of both compounds in liver cells is in accord with the observation that garlic consumption, even in high doses, is well tolerated.[9] *S*-Methyl cysteine sulfoxide, and its metabolite MeS(O)SMe, at respective levels of 0.5 and 0.05 mmol/kg body weight, inhibited the formation of benzo[a]pyrene-induced micronucleated polychromatic erythrocytes (MPCEs; an indicator for genotoxicity) by 31-33%, compared with control mice. Higher doses of MeS(O)SMe (0.5 and 1.0 mmol/kg body weight) were toxic to the mice. It is suggested that the inhibition of experimental genotoxicity by these two sulfur compounds may, in part, be responsible for the anticarcinogenic effect of vegetables containing these compounds.[51]

Oil-soluble *S*-allyl compounds are reported to have greater anticancer activity on the *in vitro* growth of human colon tumor cell line (HCT-15) than water-soluble *S*-allyl cysteine, *S*-propyl, or *O*-allyl compounds;[52,53] *Se*-allyl compounds seem to be even more active.[54,55] In connection with the latter work on cancer preventative properties of selenium compounds, Ip and coworkers found *Se*-enriched garlic to show promising activity.[6] The anticancer activity of these compounds may variously be associated with induction of cellular glutathione, a major intracellular antioxidant, stimulation of glutathione-*S*-transferase,[56,57] an enzyme which can detoxify and dispose of harmful chemicals, inhibition of ornithine decarboxylase, an enzyme important in DNA synthesis, where higher activity occurs in neoplasms, or by other activities of these compounds as antioxidants or nucleophiles to provide defense against peroxides and alkylating agents. Glutathione peroxidase, a key enzyme involved in destruction of peroxides, contains selenium in the form of selenol (-SeH) at its active site.

It is reported that "[at 3-12 μg/mL] DATS [diallyl trisulfide] can augment the activation of T cells and enhance the anti-tumor function of macrophage, suggesting that DATS may be potentially useful in tumor therapy". Furthermore, 1-100 μg/mL diallyl trisulfide was found to inhibit production of NO by macrophages.[58] The oil-soluble allylic sulfur compounds diallyl sulfide, disulfide and trisulfide were found to markedly inhibit the growth of canine mammary tumor cells, with diallyl trisulfide being the only compound that was cytotoxic; in contrast the water soluble compounds *S*-allyl cysteine and related compounds had little effect on these cells.[59] These same oil-soluble sulfur compounds have also been shown to inhibit the promotion phase of carcinogenesis in skin tumors induced by 7,12-dimethylbenz[a]anthracene (DMBA)[60,61] and, in the case of diallyl sulfide, carcinogen-induced nuclear damage to colon epithelial cells *in vivo* .[62] Depression in the incidence of skin tumors is accompanied by suppression in the activities of epidermal lipoxygenase and ornithine decarboxylase.[60]

Antioxidant Activity

A number of studies have documented the anti-oxidant and free radical scavenging properties of fresh garlic extracts and commercially available garlic products.[63-67] The apparent radical-scavenging behavior has also been demonstrated by synthetic samples of *S*-allylcysteine and *S*-allylmercaptocysteine.[68] A study of the effect of "aged garlic extract" on human metabolism of acetaminophen, which generates a reactive electrophilic metabolite, led to the conclusion that such a garlic extract "has limited potential as a chemopreventative agent.[69] Pretreatment of human carcinoma cell lines with *S*-allylcysteine followed by *cis*-platin was found to significantly enhance the cytotoxic effect of *cis*-platin; SAC alone had no effect on cell growth.[70]

Tumor Promotion Activity

It has been reported that onion oil (but not garlic oil) is a weak *promoter* of DMBA-initiated papillomas and carcinomas in mouse skin.[71] In addition, oral administration of high doses of fresh garlic homogenate has been found to *enhance* 4-(methylnitrosamino)-1-(3-pyridyl)-1-butanone (NNK)-induced lung tumorigenesis in mice.[72] Furthermore diallyl trisulfide, allyl methyl trisulfide and related sulfur compounds have been found to *promote* diethylnitrosamine-induced neoplasia in rat livers.[49] It is argued that "in evaluating relationships between diet and cancer, it is appropriate to consider not only the possible protective role of garlic and onions but also their enhancing effects".[49] The tumor promoting activity of onion and garlic extracts/oils may be due to the presence of 1-propenyl sulfur compounds, which can act as Michael acceptors.[73] An alternative explanation is suggested by the fact that both 1- and 2-propenyl disulfides have been found to induce hemolytic anemia in animals.[74] It is suggested that by thiol-disulfide exchange with glutathione, the disulfide is reduced to the corresponding thiol. While thiols generally function as antioxidants, they can act as pro-oxidants after one-electron oxidation to the thiyl radical, generating hydrogen peroxide and superoxide anion (see steps 1-3 below). Thus it may be more appropriate to view Allium-derived disulfides and related compounds as redox agents, e.g. antioxidant in some circumstances, pro-oxidant in others.[75]

(1) $RS\bullet + RS^- \rightarrow (RSSR)^{\bullet -}$
(2) $(RSSR)^{\bullet -} + O_2 \rightarrow RSSR + O_2^{\bullet -}$
(3) $RSH + H^+ + O_2^{\bullet -} \rightarrow RS\bullet + H_2O_2$

ACKNOWLEDGMENT

I gratefully acknowledge support from NSF, the NRI Competitive Grants Program/USDA (Award No. 92-37500-8068), and McCormick & Company.

REFERENCES

1. Weisberger, A.S. and J. Pensky, Tumor-inhibiting effects derived from an active principle of garlic (*Allium sativum*), *Science* 126:1112 (1957).
2. You, W.C., W.J. Blot, Y.S. Chang, A. Ershow, Z.T. Yang, Q. An, B.E. Henderson, J.F. Fraumeni Jr., and T.G. Wang, Allium vegetables and reduced risk of stomach cancer. *J. Natl. Cancer Inst.* 81:162 (1989).
3. Reuter, H.D. and H.P. Koch, Therapeutic effects and applications of garlic and its preparations, *in* "Garlic: The Science and Therapeutic Applications of *Allium sativum* L. and Related Species;" H.P. Koch and L.D. Lawson, ed.; Williams and Wilkins, Baltimore, (1995).

4. Steinmetz, K.A., L.H. Kushi, R.M. Bostick, A.R. Folsom, and J.D. Potter, Vegetables, fruit, and colon cancer in the Iowa's Women's Health Study, *Am. J. Epidemiol.* 139:1 (1994).
5. Mei, X., M.C. Wang, H.X. Xu, X.P. Pan, C.Y. Gao, N. Han, and M.Y. Fu, Garlic and gastric cancer: The effect of garlic on nitrite and nitrate in gastric juice, *Acta Nutrimenta Sinica* 4:53 (1982).
6. Ip, C., D.J. Lisk, and G.S. Stoewsand, Mammary cancer prevention by regular garlic and selenium-enriched garlic, *Nutr. Cancer* 17:279 (1992).
7. Ip, C. and D. Lisk, Bioavailability of selenium from selenium-enriched garlic. *Nutr. Cancer* 20:129 (1993).
8. Ip, C. and D. Lisk, Enrichment of selenium in allium vegetables for cancer prevention. *Carcinogenesis* 15:1881 (1994).
9. Gebhardt, R., H. Beck, and K.G. Wagner, Inhibition of cholesterol biosynthesis by allicin and ajoene in rat hepatocytes and DepG2 cells, *Biochim. Biophys. Acta* 1213:57 (1994).
10. Hong, J.Y.; T. Smith, M.J. Lee, W.S. Li, B.L. Ma, S.M. Ning, J.F. Brady, P.E. Thomas, and Yang, C.S., Metabolism of carcinogenic nitrosamines by rat nasal mucosa and the effect of diallyl sulfide, *Cancer Res.* 51:1509 (1991).
11. Block, E., The organosulfur chemistry of the genus *Allium* - Implications for the organic chemistry of sulfur. *Angew. Chem., Int. Ed. Engl.* 31:1135 (1992).
12. Morris, V.C. and O.A. Levander, Selenium content of foods, *J. Nutr.* 100:1383 (1970).
13. C.-G. Spåre and A.I. Virtanen, On the occurrence of free selenium-containing amino acids in onion (*Allium Cepa*), *Acta Chem. Scand.* 18: 280 (1964).
14. Young, V.R., Selenium: A case for its essentiality in man, *N. Engl. J. Med.* 304:1228 (1981).
15. Sathe, S.K., A.C. Mason, R. Rodibaugh, and C.M. Weaver, Chemical form of selenium in soybean (*Glycine max* L.) lectin, *J. Agric. Food Chem.* 40:2084 (1992).
16. Axley, M.J., A. Böck, and T.C. Stadtman, Catalytic properties of an *Escherichia coli* formate dehydrogenase mutant in which sulfur replaces selenium. *Proc. Natl. Acad. Sci. U.S.A.* 88:8450 (1991).
17. Söll, D., Enter a new amino acid, *Nature* 331:662 (1988).
18. Burk, R.F., and K.E. Hill, and Selenoprotein P. A selenium-rich extracellular glycoprotein, *J. Nutr.* 124:1891 (1994).
19. Wang, W., J. Tang, and A. Peng, The isolation, identification and bioactivities of selenoproteins in selenium-rich garlic.*Shengwu Huaxue Zazhi* 5:229 (1989) [*Chem. Abstr.* 111:95847d (1989)].
20. Yang, M., K. Wang, L. Gao, Y. Han, J. Lu, and T. Zou, Exploration for a natural seleni- um supplement – characterization and bioactivities of Se-containing polysaccharide from garlic. *J. Chin. Pharm. Sci.* 1:28 (1992) [*Chem. Abstr.* 118:77092u (1993)].
21. Bañuelos, G.S., D. Dyer, R. Ahmad, S. Ismail, R.N. Raut, and J.C. Dagar, In search of *Brassica* germplasm in saline semiarid and arid regions of India and Pakistan for reclamation of selenium-laden soils in the U.S., *J. Soil Water Cons.* 48:530 (1993).
22. Hamilton, J.W., Chemical examination of seleniferous cabbage *Brassica oleracea capitata*, *J. Agric. Food Chem.* 23:1150 (1975).
23. Uden, P.C., Atomic specific chromatographic detection: An overview, in "Element- Specific Chromatographic Detection by Atomic Emission Spectroscopy," ACS Symp. Ser. 479, American Chemical Society, Washington, DC, 1992.
24. Cai, X.-J., P.C. Uden, J.J. Sullivan, B.D. Quimby, and E. Block, Headspace-gas chromatography with atomic emission and mass selective detection for the determination of organoselenium compounds in elephant garlic., *Anal. Proc.*, 31:325 (1994).
25. Cai, X.-J., P.C. Uden, E. Block, X. Zhang, B.D. Quimby, and J.J. Sullivan, Allium chemistry: Identification of natural abundance organoselenium volatiles from garlic, elephant garlic, onion, and Chinese chive using headspace gas chromatography with atomic emission detection, *J. Agric. Food Chem.* 42:2081 (1994).
26. Deruaz, D., F. Soussan-Marchal, I. Joseph, M. Desage, A. Bannier, and J.L. Brazier, Analytical strategy by coupling headspace gas chromatography, atomic emission spectrometric detection and mass spectrometry. Application to sulfur compounds from garlic. *J. Chromatogr. A.* 677:345 (1994).
27. Ryan, M.D. and D. Harpp, The first selenium transfer reagent: Preparation, and mechanism of formation, *Tetrahedron Lett.* 33:2129 (1992).
28. Potapov, V.A., S.V. Amosova, P.A. Petrov, L.S. Romanenko, and V.V. Keiko, Exchange reactions of dialkyl dichalcogenides, *Sulfur Lett.* 15:121 (1992).
29. Soda, K., H. Tanaka, and N. Esaki, Biochemistry of physiologically active selenium compounds, in "The Chemistry of Organic Selenium and Tellurium Compounds" (ed. S. Patai), Vol. 2, (John Wiley, Chichester, 1987).

30. Evans, C.S., C.J. Asher, and C.M. Johnson, Isolation of dimethyl diselenide and other volatile selenium compounds from *Astragalus racemosus* (Pursh.). *Aust. J. Biol. Sci.* 21:13 (1968).
31. Chasteen, T.G., G.M. Silver, J.W. Birks, and R. Fall, Fluorine-induced chemiluminescence detection of phosphine, alkyl phosphines and monophosphinate esters. *Chromatographia* 30:181 (1990).
32. Chasteen, T.G. Confusion between dimethyl selenenyl sulfide and dimethyl selenone released by bacteria. *Appl. Organomet. Chem.* 7:335 (1993).
33. Husek, P., Rapid derivatization and gas chromatographic determination of amino acids, *J. Chromatogr.* 552:289 (1991).
34. Wang, J., Z.-H. Huang, D.A. Gage, and J.T. Watson, Analysis of amino acids by gas chromatography–flame ionization detection and gas chromatography–MS: simultaneous derivatization of functional groups by an aqueous phase chloroformate–mediated reaction.*J. Chromatogr. A* 663:71 (1994).
35. Janák, J., H.A.H. Billiet, J. Frank, K.C.A.M. Luyben, and P. Husek, Separation of selenium analogues of sulphur-containing amino acids by high performance liquid chromatography and high resolution gas chromatography, *J. Chromatogr. A* 677:192 (1994).
36. Cai, X.-J., E. Block, P.C. Uden, X. Zhang, B.D. Quimby, and J.J. Sullivan, *Allium* chemistry: identification of selenoamino acids in ordinary and selenium-enriched garlic, onion and broccoli using gas chromatography with atomic emission detection. *J. Agric. Food Chem.*, 43:1754 (1995).
37. Marks, H.S., J.A. Hilson, H.C. Leichtweis, and G.S. Stoewsand, *S*-Methylcysteine sulfoxide in *Brassica* vegetables and formation of methyl methanethiosulfinate from Brussels sprouts, *J. Agric. Food Chem.* 40:2098 (1992).
38. Vadhanavikit, S., C. Ip, and H.E. Ganther, Metabolites of sodium selenite and methylated selenium compounds administered at cancer chemoprevention levels in the rat. *Xenobiotica* 23:731 (1993).
39. Takada, H., N. Esaki, H. Tanaka, and K. Soda, The C_3-N bond cleavage of 2-amino-3- (N-substituted-amino)propionic acids catalyzed by L-methionine γ-lyase, *Agric. Biol. Chem.* 52:2897 (1988).
40. Kamitani, H., N. Esaki, H. Tanaka, and K. Soda, Thermostable *S*-alkylcysteine α,β-lyase from a thermophile: purification and properties, *Agric. Biol. Chem.* 54:2069 (1990).
41. Cai, X.-J., E. Block, P.C. Uden, B.D. Quimby, and J.J. Sullivan, *Allium* chemistry: identification of natural abundance organoselenium compounds in human breath after ingestion of garlic using gas chromatography with atomic emission detection, *J. Agric. Food Chem.* 43:1751 (1995).
42. Ruth, J.H. Odor thresholds and irritation levels of several chemical substances: A review, *Am. Ind. Hyg. Assoc. J.* 47:A142 (1986).
43. Buchan, R.F., Garlic breath odor, *JAMA* 227:559 (1974).
44. Ballard-Barbash, R., S. Krebs-Smith, and A.F. Subar, Re: Vegetables, fruit, and colon cancer in the Iowa's Women's Health Study, *Am. J. Epidemiol.* 141:84 (1995).
45. Dorant, E., P.A. van den Brandt, and R.A. Goldbohm, A prospective cohort study on *Allium* vegetable consumption, garlic supplement use, and the risk of lung carcinoma in the Netherlands, *Cancer Res.* 54: 6148 (1994).
46. Dorant, E., P.A. van den Brandt, and R.A. Goldbohm, Allium vegetable consumption, garlic supplement intake and female breast carcinoma incidence, *Breast Cancer Research and Treatment* 33:163 (1995).
47. Dorant, E., P.A. van den Brandt, R.A. Goldbohm, R.J.J. Hermus, and F. Sturmans, Garlic and its significance for the prevention of cancer in humans: A critical view, *British Journal of Cancer* 67:424 (1993).
48. Shenoy, N.R., and A.S.U. Choughuley, Inhibitory effect of diet related sulphydryl compounds on the formation of carcinogenic nitrosamines, *Cancer Lett.* 65:227 (1992).
49. Takada, N., T. Matsuda, T. Otoshi, Y. Yano, S. Otani, T. Hasegawa, D. Nakae, Y. Konishi, and S. Fukushima, Enhancement by organosulfur compounds from garlic and onion of diethylnitrosamine-induced glutathione S-transferase positive foci in the rat liver, *Cancer Res.* 54:2895 (1994).
50. Scharfenberg, K., R. Wagner, and K.G. Wagner, The cytotoxic effect of ajoene, a natural product from garlic, investigated with different cell lines, *Cancer Letters* 53:103 (1990).
51. Marks, H.S.; J.A. Anderson, and G.S. Stoewsand, Effect of *S*-methyl cysteine sulphoxide and its metabolite methyl methanethiosulfinate, both occurring naturally in *Brassica* vegetables, on mouse genotoxicity, *Food Chem. Toxicol.* 31:491 (1993).
52. Sundaram, S.G., and J.A. Milner, Organosulfur compounds in processed garlic alter the *in vitro* growth of human tumor cell lines, *FASEB J.* 8:A426 (1994).
53. Sundaram, S.G., and J.A. Milner, Diallyl disulfide present in garlic oil inhibits both *in vitro* and *in vivo* growth of human colon tumor cells, *FASEB J.* 9:A869 (1995).
54. Ip, C. 1995, personal communication.

55. El Bayoumy, K., C. Ip, Y.H. Chae, P. Upadhyaya, D. Lisk, and B. Prokopczyk, Mammary cancer prevention by diallyl selenide, a novel organoselenium compound, *Proc. Am. Assoc. Cancer Res.* 34:A3322 (1993).
56. Sparnins, V.L., A.W. Mott, and L.W. Wattenberg, Effects of allyl methyl trisulfide on glutathione S-transferase activity and PB-induced neoplasia in the mouse, *Nutr. Cancer* 8:211 (1986).
57. Sparnins, V.L., G. Barany, and L.W. Wattenberg, Effects of organosulfur compounds from garlic and onions on benzo[a]pyrene-induced neoplasia and glutathione *S*-transferase activity in the mouse, *Carcinogenesis* 9:131-4 (1988).
58. Feng, Z.H., G.M. Zhang, T.L. Hao, B. Zhou, H. Zhang, and Z.Y. Jiang, Effect of diallyl trisulfide on the activation of T cell and macrophage-mediated cytotoxicity, *J. Tongji Med. Univ.* 14:142 (1994).
59. Sundaram, S.G., and J.A. Milner, Impact of organosulfur compounds in garlic on canine mammary tumor cells in culture, *Cancer Lett.* 74:85 (1993).
60. Belman, S.; J. Solomon, A. Segal, E. Block, and G. Barany, Inhibition of soybean lipoxygenase and mouse skin tumor promotion by onion and garlic components, *J. Biochem. Toxicol.* 4:151 (1989).
61. Dwivedi, C., S. Rohlfs, D. Jarvis, and F.N. Engineer, Chemoprevention of chemically- induced skin tumor development by diallyl sulfide and diallyl disulfide, *Pharm. Res.* 9:1668 (1992).
62. Wargovich, M.J., and M.T. Goldberg, Diallyl sulfide: a naturally occurring thioether that inhibits carcinogen induced damage to colon epithelial cells *in vivo*, *Mutation Res.* 143:127 (1985).
63. Yang, G.C., P.M. Yasaei, and S.W. Page, Garlic as anti-oxidants and free radical scavengers, *J. Food Drug. Anal.* 1:357 (1993).
64. Popov, I., A. Blumstein, and G. Lewin, Antioxidant effects of aqueous garlic extracts: 1st communication: direct detection using the photochemiluminescence, *Arzneimittel- Forschung* 44:602 (1994).
65. Lewin, G. and I. Popov, Antioxidant effects of aqueous garlic extracts: 2nd communication: inhibition of the Cu^{++}-initiated oxidation of low density lipoproteins, *Arzneimittel-Forschung* 44:604 (1994).
66. Rekka, E.A., and P.N. Kourounakis, Investigation of the molecular mechanism of the antioxidant activity of some *Allium sativum* ingredients, *Pharmazie* 49:539 (1994).
67. Toeroek, B., J. Belagyi, B. Rietz, and R. Jacob, Efectiveness of garlic on the radical activity in radical generating systems, *Arzneimittel-Forschung* 44:608 (1994).
68. Imai, J., N. Ide, S. Nagae, T. Moriguchi, H. Matsuura, and Y. Itakura, Antioxidant and radical scavenging effects of aged garlic extract and its constituents, *Planta Med.* 60:417 (1994).
69. Gwilt, P.R., C.L. Lear, M.A. Tempero, D.D. Birt, A.C. Grandjean, R.W. Ruddon, and D.L. Nagel, The effect of garlic extract on human metabolism of acetaminophen, *Cancer Epidemiology, Biomarkers and Prevention* 3:155 (1994).
70. Yellin, S.A., B.J. Davidson, J.T. Pinto, P.G. Sacks, C. Qiao, and S.P. Schantz, Relationship of glutathione and glutathione-*S*-transferase to cisplatin sensitivity in human head and neck squamous carcinoma cell lines, *Cancer Lett.* 85:223 (1994).
71. Belman, S., A. Sellakumar, M.C. Bosland, K. Savarese, and R.D. Estensen, papilloma and carcinoma production in DMBA-initiated, onion oil-promoted mouse skin, *Nutr. Cancer* 14:141 (1990).
72. Hong, J.-Y., T.J. Smith, W. Huang, Y. Wang, and C.S. Yang, Enhancement of 4- (methylnitrosamino)-1-(3-pyridyl)-1-butanone (NNK)-induced lung tumorigenesis in mice by oral administration of fresh garlic homogenate, Poster Abstract 63, American Institute for Cancer Research 1995 Research Conference.
73. Klopman, G., personal communication.
74. Munday, R., and E. Manns, Comparative toxicity of prop(en)yl disulfides derived from Alliaceae: Possible involvement of 1-propenyl disulfides in onion-induced hemolytic anemia, *J. Agric. Food Chem.* 42: 959 (1994).
75. Herbert, V. Antioxidants, Pro-oxidants and Their Effects, *JAMA* 272: 1659 (1994).

14

ALLIUM VEGETABLES AND THE POTENTIAL FOR CHEMOPREVENTION OF CANCER

Michael J. Wargovich and Naoto Uda

Department of Gastrointestinal Medical Oncology and Digestive Diseases
University of Texas M.D. Anderson Cancer Center
Houston, Texas 77030

INTRODUCTION

A bewildering number of sulfur-containing chemical compounds and mixtures are generated when garlic and onion bulbs are cut and exposed to oxygen. The chemistry of garlic and onion has been the subject of several excellent reviews and the utility of these vegetables as herbal medicines is legendary.[1,2] Nowadays some of these anecdotal reports are finding confirmation in modern research. Because of the highly interesting chemistry, yielding purified agents for screening, and the fact that garlic and onion are consumed world-wide, cancer researchers have been able to investigate the possible use of garlic and onion and their constituent organosulfur compounds as preventives and treatment for the disease.

Since Belman[3] first reported that skin carcinogenesis was strongly inhibited by the painting of garlic and onion oils on the backs of DMBA- initiated SENCAR mice, research into the chemopreventive effects of the odiferous chemicals in *Allium* vegetables has continued to expand to now include proven cancer preventive efficacy for colon, mammary, lung, and other digestive tract organs.[4-7] At the present the most probable mechanism of action in the suppression of tumorigenesis is the ability of *Allium* phytochemicals to modulate carcinogen metabolism and detoxification. Whether these processes, most likely to explain prevention of cancer in animals, also apply to the parallel pathways of prevention in man remains uncertain.

Another unique aspect of the study of the benefits of garlic and onion as preventives of cancer is the fact that, unlike the usual course of events where studies of dietary patterns in specific low-risk populations are subsequently borne out by laboratory studies, the evidence for a benefit for consuming garlic and onions in the diet regularly has *followed* the original basic research reports. Population-based studies now document a reduced risk for stomach cancer in China and Italy when *Allium* vegetables were identified as part of the regular diet; these studies were seconded by two studies that associated a reduced risk for colon cancer, one in Australia, and one in Iowa.[8-11] The first reports of cancer prevention efficacy, however, were from carcinogenicity studies in animals.

Dietary Phytochemicals in Cancer Prevention and Treatment
Edited under the auspices of the American Institute for Cancer Research, Plenum Press, New York, 1996

EARLY STUDIES INDICATING POSSIBLE CANCER PREVENTION EFFECTS FOR ORGANOSULFUR COMPOUNDS

Following the initial report describing the inhibitory effects of garlic and onion oils on tumor promotion in the skin, our laboratory began the first systematic screening of available organosulfur compounds from *Alliums* for possible effects on colon tumorigenesis. The initial assay used during this period focused on the suppression of nucleotoxic effects of the carcinogen dimethylhydrazine in the colons of mice treated with the test agent, usually given orally. Nucleotoxicity is a rapidly expressed phenomenon in intestinal crypts exposed to toxins and/or carcinogens. Within 24 h of exposure to a colon carcinogen such as dimethylhydrazine (DMH), the basal third of colonic crypts is littered with apoptotic bodies and karyorrhectic nuclei.[12] In the early 1980s we first used this assay to rapidly screen agents that might interfere with the DNA damaging effects of colonic carcinogens.[13] Among the first agents to be screened for possible protection were several compounds from garlic (Table 1). It was evident that at least one agent, diallyl sulfide (DAS) was very potent in inhibiting DMH damage to colonic crypt nuclei.[14] It was noted that chemical structure was correlated with protection, for organosulfur volatiles from garlic which contained an allylic side chain were more active in preventing DMH nucleotoxicity than agents that did not have such a grouping.[15] The nucleotoxicity studies provided a strong case for the use of the volatile DAS, as a very potent inhibitor of DMH, while the disulfide (DADS) and volatiles lacking an allylic group had no effect on colonic nuclear toxicity. While this was suggestive that the allylic group was a structural determinant of chemoprevention, the more immediate task was to demonstrate efficacy in the DMH animal colon cancer model. In the experiment we then conducted, DAS was given orally to mice 3 hours prior to each subcutaneous weekly injection with DMH, and DAS treatment was found to be highly effective in inhibiting colon cancer in this model.[4]

To further explore the efficacy of DAS, we reasoned that DAS may have had a profound effect on DMH metabolism since the highly effective three hour pretreatment overlapped with the reported *in vivo* half-life of DMH. DMH is typical of a number of small molecular weight carcinogens activated by the cytochrome P450 oxidation system, and CYP2E1 is generally thought to catalyze DMH oxidation.[16] Through the reports from C.S. Yang's group we were aware that the esophageal carcinogen, nitrosomethylbenzylamine (NMBA) was, as well, metabolized by CYP2E1.[17] In the experiment that followed we found that DAS not only had a prominent inhibitory effect on NMBA metabolism but that pretreatment with the same dose used in the colon study (200 mg/kg) completely suppressed esophageal tumorigenesis in the rat.[18] In later studies we were able to show in the same model that the chemopreventive effect of DAS was dose-related and demonstrable only during the initiation phase of carcinogenesis. Post-initiation studies failed to document any observable effect on tumor incidence or tumor

Table 1. Effects of *Allium* volatiles on colon carcinogen nucleotoxicity

Compound	% Inhibition	Reference
Diallyl sulfide	50-75	14
Diallyl disulfide	No effect	20
Dipropyl sulfide	No effect	20
Allyl methyl sulfide	30%	15
Dimethyl sulfide	No effect	15
Dipropyl sulfone	No effect	15
Dipropyl disulfide	No effect	20

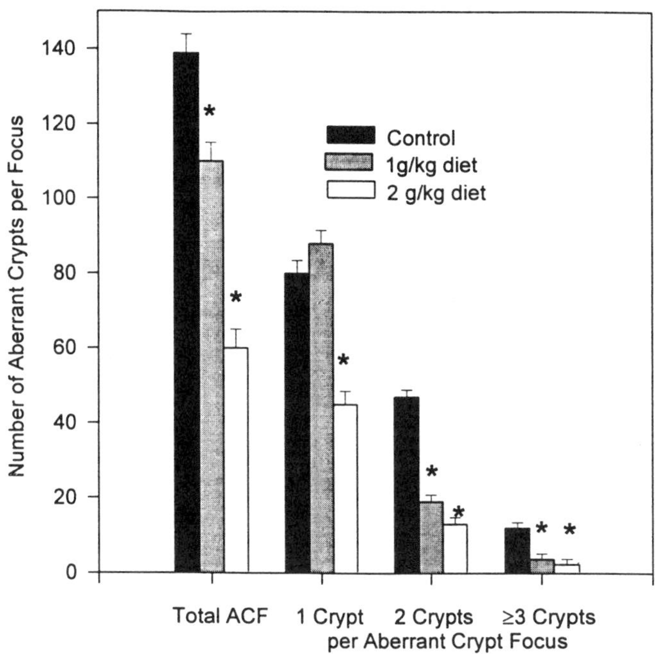

Figure 1. Effect of diallyl sulfide pretreatment on aberrant crypt formation in the colon of F344 rats. Asterisks indicate significant differences from control group (not treated with DAS in the diet).

burden.[18] The experiments again suggested that organosulfur compounds from garlic and onion largely acted as blocking agents for tumorigenesis, a term used by Wattenberg, for agents that predominantly affected tumorigenesis by modulation of metabolism.[19]

Our laboratory has also studied water-soluble organosulfur compounds from garlic as possible chemopreventives. As detailed in the review by Eric Block, the non-volatile fractions of *Allium* vegetables contain a wide array of agents.[2] We have focused our studies on S-allylcysteine (SAC), also known chemically as S-(1-propenyl)-l-cysteine, a non-volatile in garlic that is the precursor of a wide range of flavoring agents upon decomposition. In colonic nucleotoxicity studies SAC was found to be much more effective than cysteine or S-propylcysteine when DMH was used to induce DNA damage in mouse colon; SAC also inhibited DMH induced colon cancer in a dose-related fashion.[20] Short-term studies of SAC also revealed it induced glutathione-S-transferase activity in liver and the colon.

While these studies pointed to the inhibition of carcinogen activation as the mechanism by which the effect of DMH was modulated in the colon, it was soon apparent that garlic and onion volatiles were active in other tumorigenesis systems and that they had notable effects on detoxification pathways.

FURTHER EVIDENCE FOR CHEMOPREVENTION: EFFECTS OF *ALLIUM* VOLATILES IN OTHER ANIMAL MODELS AND CARCINOGEN DETOXIFICATION SYSTEMS

While our work focused on the chemopreventive effects of garlic in gastrointestinal tumorigenesis, other groups contributed new data on the efficacy of garlic and onion compounds in other tumorigenesis assays which indicated that these agents were strong inducers of the glutathione-S-transferase enzyme system. Wattenberg's laboratory found that benzo[a]pyrene-induced forestomach tumors in A/J mice were reduced by 70% when the volatile allyl methyl trisulfide was given.[21] Four agents containing the allylic group generally inhibited B[a]P-induced forestomach tumors whereas little protective activity was found for the non-allylic agents Glutathione-S-transferase (GST) is an important enzyme that catalyzes the detoxification of many electrophilic agents by coupling them to glutathione.[22] In the study in which forestomach tumorigenesis was modulated by certain organosulfur agents,[23] it was additionally noted that allylic-side chain containing agents stimulated GST activity whereas non-allylic agents had little effect on GST. Studies in our laboratory confirmed that DAS efficiently induces GST in the liver and colon of mice, and the induction was apparent 48 h after treatment and was maintained thereafter. Table 2 list the results of available studies on the induction of GST by organosulfur compounds in garlic and onion. Again, the pattern of induction is in favor of agents containing an allylic side chain as part of its chemical structure. Future studies should include an analysis of the long-term administration of garlic and onion derived agents on the maintenance of GST levels in organs susceptible to tumorigenesis.

RECENT STUDIES ON THE EFFECTS OF *ALLIUM*-DERIVED ORGANOSULFUR COMPOUNDS ON CARCINOGEN METABOLISM

One of the promises of chemoprevention by treatment with garlic-derived organosulfur compounds is the apparent ability of these compounds to affect carcinogen metabolism, and their activity toward specific isoforms of P450. Brady *et al.*[24] first reported that oral treatment of Sprague-Dawley rats with DAS resulted in competitive inhibition of N-dimethylnitrosamine demethylase, a CYP2E1 mediated reaction. Suppression of CYP2E1 activity also may explain the selectivity of DAS in inhibiting DMH tumorigenesis since the conversion of DMH to secondary metabolites is governed by CYP2E1.[16] In related work, Yang's laboratory provided data that DAS was an efficient suppresser of the metabolism of a number of important environmental nitrosamines, including nitrosodiethylamine (NDEA) and 4-(methylnitrosoamino)-1-(3-pyridyl)-1-butanone also known as NNK, an important carcinogen in tobacco smoke. This group further confirmed that NNK tumorigenesis in the mouse lung was suppressed by DAS.[25] Extended studies by Kwak *et al.*[26] demonstrated that constitutive and pyrazine inducible CYP2E1 levels in rat liver were reduced by DAS and the garlic volatile allyl methyl sulfide. Further it was found that while CYP2E1 levels were suppressed when measured on Western blots, Northern analysis revealed little effect on CYP2E1 mRNA levels. Several recent studies have further elucidated the role of DAS in

Table 2. Induction of glutathione transferase by organosulfur agents

Compound	Effect on GST	Effect on Tumorigenesis	Reference
Allyl methyl sulfide	Induction	Inhibition	6
Allyl methyl disulfide	Induction	Inhibition	6
Allyl methyl trisulfide	Induction	Inhibition	6
Diallyl sulfide	Induction	Inhibition	4
Diallyl disulfide	Induction	No effect	34
Diallyl trisulfide	Induction	Inhibition	6
Propyl methyl disulfide	No effect	No effect	6
Propyl methyl trisulfide	No effect	No effect	6

influencing ethanol and acetone metabolism (both are metabolized by CYP2E1), and these studies confirm that DAS is a specific inhibitor for this P450 isoform.[27-29]

Taken together the data strongly indicate that organosulfur compounds in garlic and onion have marked effects on carcinogen activation and detoxification. This hypothesis generates a number of important questions:

1. Beyond CYP2E1 what is the effect of DAS and other organosulfur compounds on other P450 isoforms? Are other carcinogen/drug metabolizing enzymes affected?
2. What are the structural determinants of Phase 1 enzyme inhibition/ Phase 2 induction by *Allium* compounds? Is the allylic group the critical determinant? Does sulfur play a role?
3. What are the longer-term consequences of treatment with organosulfur compounds from *Alliums*? Is inhibition of P450s in the short-term reversed by induction in the long term?
4. What are the adverse effects of modulating drug metabolism by naturally occurring chemopreventive agents?

EPIDEMIOLOGICAL AND CLINICAL STUDIES RELATED TO THE CANCER PREVENTIVES EFFECTS OF *ALLIUM* VEGETABLES

Building upon the weight of experimental evidence that *Allium* vegetables abound in cancer preventive compounds, several population-based studies have assessed the possible protective effects diets rich in *Alliums* have on cancer incidence. As reviewed by Steinmetz and Potter[30] there are at least 12 case-control studies in which *Allium* consumption has been analyzed for association with cancer in specific organ sites; of these studies eight reported a negative association with cancer. Some of these studies are listed in Table 3. Of note is the extent of protection for stomach and colon cancer in very different parts of the world and the lack of a protective association for breast and lung cancer.[8-11,31,32] All of these population studies vary considerably in types of *Allium*s consumed, availability of certain vegetables throughout the year, and differences in cooking methods, making direct comparisons difficult. As yet very few leads have been explored to correlate mechanisms of action suggested in animal carcinogenesis experiment with those possibly at work in humans. However, limited studies have been conducted to determine the effect of garlic preparations on substrates commonly metabolized by humans. Gwilt *et al.*[33] found that administration of garlic extract to volunteers did not drastically modify acetaminophen metabolism, but was associated with a slight increase in sulfate conjugation of this commonly used analgesic. Clearly much more work is needed to document whether the profound effects purified organosulfur compounds have on carcinogen and drug metabolism as tested in animals can be verified in humans.

Currently at least one intervention trial is being planned by the National Cancer Institute and will be soon begin in China to test whether a combination supplement of volatile and non-volatile sulfur compounds from garlic reduces the incidence of stomach cancer. From the viewpoint of animal carcinogenesis the phytochemicals in garlic, onion, and related *Allium*s are indeed promising chemopreventive agents and should be investigated for potential to inhibit other neoplastic diseases. Additional research will help to identify the structural moieties that associate with chemoprevention perhaps paving the way for newer trials in humans on these highly interesting phytochemicals.

Table 3. Epidemiological studies on garlic and onion consumption and cancer prevention

Organ Site	Type of Study	Result	Population	Reference
Stomach Cancer	Case-Control	OR of 0.4 for highest quartile of consumers of *Allium* vegetables	China	8
Stomach Cancer	Case-Control	OR of 0.6-0.8 for frequent consumers of garlic and onions	Italy	9
Colon Cancer	Case-Control	OR of 0.72-0.77 for onion consumers	Australia	10
Colon Cancer	Cohort Study	RR of 0.68 for consumers of garlic	United States	11
Lung Cancer	Cohort Study	No effect on risk for breast cancer in users of garlic supplements or consumers of onion and leeks	The Netherlands	31
Breast Cancer	Cohort Study	No effect on risk for lung cancer in consumers of *Allium* vegetables	The Netherlands	32

OR = odds ratio
RR = relative risk

ACKNOWLEDGMENTS

The authors wish to thank the American Institute for Cancer Research and the Wakunaga Pharmaceutical Company for grants in support of research.

REFERENCES

1. Fenwick, G.R. and Hanley, A.B. The Genus *Allium*-Part 2. *Crit Rev Toxicol.* 22:273-377, 1985.
2. Block, E. The organosulfur chemistry of the Genus *Allium* - implications for the organic chemistry of sulfur. *Angew Chem Int Ed Engl* 31:1135-1178, 1992.
3. Belman, S. Onion and garlic oils inhibit tumor formation. *Carcinogenesis.* 4:1063-1065, 1983.
4. Wargovich, M.J. Diallyl sulfide, a flavor component of garlic (*Allium sativum*), inhibits dimethylhydrazine-induced colon cancer. *Carcinogenesis* 8:487-489, 1987.
5. Ip, C., Lisk, D.J. and Stoewsand, G.S. Mammary cancer prevention by regular garlic and selenium-enriched garlic. *Nutr Cancer* 17:279-286, 1992.
6. Sparnins, V.L., Barany, G.L. and Wattenberg, L.W. Effects of organosulfur compounds from garlic and onions on benzo[a]pyrene-induced neoplasia and glutathione S-transferase activity in the mouse. *Carcinogenesis* 9:131-134, 1988.
7. Wargovich, M.J., Woods, C., Eng, V.W., Stephens, L.C. and Gray, K. Chemoprevention of N-nitrosomethylbenzylamine-induced esophageal cancer in rats by the naturally occurring thioether, diallyl sulfide. *Cancer Res* 48:6872-6875, 1988.
8. You, W.C., Blot, W.J., Chang, Y.S., Ershow, A., Yang, Z.T., An, Q., Henderson, B.E., Fraumeni, Jr., J.F., and Wang, T.-G. *Allium* vegetables and reduced risk of stomach cancer. *J Natl Cancer Inst.* 81:162-164, 1989.
9. Buiatti, E., Palli, D., Decarli, A., Amadori, D., Avellini, C., Bianchi, S., Biserni, R., Cipriani, F., Cocco, P., Giacosa, A.. A case-control study of gastric cancer and diet in Italy. *Int J Cancer* 44:611-616, 1989.
10. Steinmetz, K.A. and Potter, J.D. Food-group consumption and colon cancer in the Adelaide Case-Control Study. I. Vegetables and fruit. *Int J Cancer* 53:711-719, 1993.
11. Steinmetz, K.A., Kushi, L.H., Bostick, R.M., Folsom, A.R. and Potter, J.D. Vegetables, fruit, and colon cancer in the Iowa Women's Health Study, *Amer J of Epidemiol* 139:1-15, 1994.
12. Blakey, D.H., Duncan, A.M., Wargovich, M.J., Goldberg, M.T., Bruce, W.R. and Heddle, J.A. Detection of nuclear anomalies in the colonic epithelium of the mouse. *Cancer Res* 45:242-249, 1985.
13. Wargovich, M.J., Goldberg, M.T., Newmark, H.L. and Bruce, W.R. Nuclear aberrations as a short-term test for genotoxicity to the colon: evaluation of nineteen agents in mice. *J Natl Cancer Inst* 71:133-137, 1983.

14. Wargovich, M.J. and Goldberg, M.T. Diallyl sulfide. A naturally occurring thioether that inhibits carcinogen-induced nuclear damage to colon epithelial cells in vivo. *Mutat Res* 143:127-129, 1985.
15. Goldberg, M.T. Inhibition of genotoxicity by diallyl sulfide and structural analogues. In: "*Anticarcinogens and Radiation Protection*," edited by Cerutti, P.A., Nygaard, O.F., and Simic, M.C.New York:Plenum Press, p. 309-312, 1987
16. Sohn, O.S., Ishizaki, H., Yang, C.S. and Fiala, E.S. Metabolism of azoxymethane, methylazoxymethanol and N-nitrosodimethylamine by cytochrome P450IIE1. *Carcinogenesis* 12:127-131, 1991.
17. Brady, J.F., Li, D.C., Ishizaki, H. and Yang, C.S. Effect of diallyl sulfide on rat liver microsomal nitrosamine metabolism and other monooxygenase activities. *Cancer Res* 48:5937-5940, 1988.
18. Wargovich, M.J., Imada, O. and Stephens, L.C. Initiation and postinitiation chemopreventive effects of diallyl sulfide in esophageal carcinogenesis. *Cancer Lett.* 64:39-42, 1992.
19. Wattenberg, L.W. Inhibitors of chemical carcinogenesis. *Adv Cancer Res* 26:197-226, 1978.
20. Sumiyoshi, H. and Wargovich, M.J. Chemoprevention of 1,2-dimethylhydrazine-induced colon cancer in mice by naturally occurring organosulfur compounds. *Cancer Res.* 50:5084-5087, 1990.
21. Sparnins, V.L., Barany, G. and Wattenberg, L.W. Effects of organosulfur compounds from garlic and onions on benzo[a]pyrene-induced neoplasia and glutathione S-transferase activity in the mouse. *Carcinogenesis* 9:131-134, 1988.
22. Habig, W.H., Pabst, M.J. and Jakoby, W.B. Glutathione-S-transferase. *J Biol Chem* 249:7130-7139, 1974.
23. Wattenberg, L.W., Sparnins, V.L. and Barany, G. Inhibition of N-nitrosodiethylamine carcinogenesis in mice by naturally occurring organosulfur compounds and monoterpenes. *Cancer Res.* 49:2689-2692, 1989.
24. Brady, J.F., Wang, M.H., Hong, J.Y., Xiao, F., Li, Y., Yoo, J.-S.H., Ning, S.M., Lee, M.-J., Fukuto, J.M., Gapac, J.M., and Yang, C.S. Modulation of rat hepatic microsomal monooxygenase enzymes and cytotoxicity by diallyl sulfide. *Toxicol Appl Pharmacol.* 108:342-354, 1991.
25. Hong, J.Y., Smith, T., Lee, M.J., et al. Metabolism of carcinogenic nitrosamines by rat nasal mucosa and the effect of diallyl sulfide. *Cancer Res.* 51:1509-1514, 1991.
26. Kwak, M.K., Kim, S.G., Kwak, J.Y., Novak, R.F. and Kim, N.D. Inhibition of cytochrome P4502E1 expression by organosulfur compounds allylsulfide, allylmercaptan and allylmethylsulfide in rats. *Biochem Pharmacol.* 47:531-539, 1994.
27. Morimoto, M., Hagbjork, A.L., Wan, Y.J., Fu, P.C., Clot, P., Albano, E., Ingelman-Sundberg, M., and French, S.W. Modulation of experimental alcohol-induced liver disease by cytochrome P450 2E1 inhibitors. *Hepatology* 21:1610-1617, 1995.
28. Haber, D., Siess, M.H., Canivenc-Lavier, M.C., Le Bon, A.M. and Suschetet, M. Differential effects of dietary diallyl sulfide and diallyl disulfide on rat intestinal and hepatic drug-metabolizing enzymes. *J Toxicol Environ Health* 44:423-434, 1995.
29. Chen, L., Lee, M., Hong, J.Y., Huang, W., Wang, E. and Yang, C.S. Relationship between cytochrome P450 2E1 and acetone catabolism in rats as studied with diallyl sulfide as an inhibitor. *Biochem Pharmacol* 48:2199-2205, 1994.
30. Steinmetz, K.A. and Potter, J.D. Vegetables, fruits, and cancer. I. Epidemiology. *Cancer Causes Control* 2:325-357, 1991.
31. Dorant, E., van den Brandt, P.A. and Goldbohm, R.A. *Allium* vegetable consumption, garlic supplement intake, and female breast carcinoma incidence. *Breast Cancer Res & Treat* 33:163-170, 1995.
32. Dorant, E., van den Brandt, P.A. and Goldbohm, R.A. A prospective cohort study on *Allium* vegetable consumption, garlic supplement use, and the risk of lung carcinoma in The Netherlands. *Cancer Res* 54:6148-6153, 1994.
33. Gwilt, P.R., Lear, C.L., Tempero, M.A., Birt, D.D., Grandjean, A.C., Rudden, R.W., and Nagel, D.L. The effect of garlic extract on human metabolism of acetominophen. *Cancer Epi Bio Prev* 3:155-160, 1994.
34. Reddy, B.S., Rao, C.V., Rivenson, A., and Kelloff, G. Chemoprevention of colon carcinogenesis by organosulfur compounds. *Cancer Res.* 53:3493-3498, 1993.

15

THE ATTRIBUTES OF SELENIUM-ENRICHED GARLIC IN CANCER PREVENTION

Clement Ip[1] and Donald J. Lisk[2]

[1] Department of Surgical Oncology
Roswell Park Cancer Institute
Buffalo, New York 14263
[2] Department of Fruit and Vegetable Science
Cornell University
Ithaca, New York 14853

INTRODUCTION

Selenium supplementation has been reported to suppress carcinogenesis in many different animal models. A summary of these observations can be found in several reviews.[1-6] The effect is not organ specific, since selenium is known to inhibit tumor development in mammary gland, liver, skin, pancreas, esophagus, colon and other sites. In general, there is a dose dependent response, and selenium chemoprevention can be realized in the absence of toxicity. The sensitivity to selenium appears to decrease as cells progress from normal to preneoplastic to neoplastic. In order to achieve maximal prophylaxis, it is necessary to maintain a continuous regimen of selenium administration. This suggests that the active species of selenium with anticarcinogenic potential is generated only when the supply of selenium is sustained at certain levels. Metabolism alters the chemical form of selenium and plays a key role in determining its biological activity. The past collaborative work of Ip and Ganther has produced a substantial body of evidence indicating that the commonly used selenium compounds, such as selenite and selenomethionine, have to be metabolized in order to express their anticancer activity.[6]

Little is known about the form of selenium or the anticarcinogenic efficacy of selenium in foods. A major obstacle is the difficulty in isolating and identifying selenium compounds which are normally present in very low levels (often below 0.1 ppm or 0.1 μg/g) in natural products.[7] In almost all cases where a chemical form was characterized, the starting material (usually a seleniferous plant) contained exceptionally high levels of selenium, and in every instance, selenium was found as analogs of sulfur amino acids such as selenomethionine or selenocystathionine.[8] The fact that plants can convert inorganic selenium in soil to organic selenium compounds following the sulfur assimilatory pathway sows the seed of the high selenium-garlic research project. According to this scenario, vegetables which are a rich source of sulfur might be able to concentrate and utilize selenium biosynthetically if they are so fertilized.

The idea was tested with garlic for the following reasons. Garlic contains an abundance of sulfur derivatives, including a variety of alkyl (e.g. methyl) and alkenyl (e.g. allyl) cysteine sulfoxides.[9,10] These sulfoxides are the precursors to the formation of many volatile sulfides which are responsible for the flavor and pungency of garlic. Not only do these volatile sulfides contribute to the odor of garlic, they are also active anticancer agents.[11] In a series of papers, Wattenberg and colleagues documented the tumor inhibitory activities of diallyl sulfide, diallyl disulfide, diallyl trisulfide, allyl methyl disulfide and allyl methyl trisulfide.[12-14] By substituting the indigenous sulfur containing compounds in garlic with selenium, we had hoped to produce more powerful anticancer agents. This working hypothesis was supported by the finding of Ip and Ganther[15] with structurally related selenium and sulfur analogs that molecule for molecule, selenium is several hundred times more active than sulfur in cancer prevention. In the following sections, data will be presented to show that the high selenium-garlic is capable of delivering selenium both as an anticarcinogen and an essential nutrient.

CULTIVATION OF HIGH SELENIUM GARLIC

The high selenium-garlic used in all our studies was produced at Cornell University, Ithaca, New York. We have been successful in cultivating several crops of high selenium-garlic, both indoor and outdoor, with a selenium concentration ranging from 112 to 1355 ppm dry weight. This was achieved by varying the intensity of selenium fertilization. Natural garlic (i.e. without selenium fertilization) usually contains ~0.03 ppm Se. Cloves of "Valencia" top set garlic were planted in the fall in several polyethylene-lined frames containing a mixture of vermiculite and spaghnum peat moss (2:1 ratio). Starting in early spring, a water solution of sodium selenite and sodium selenate was sprayed onto the plot. The polyethylene lining prevented the leaching of soluble sodium salts into the surrounding soil. Mature bulbs were harvested in the summer. The cloves were separated, freeze-dried and then milled to a fine powder. After thorough mixing in a tumbler, selenium analysis of the entire batch was then determined using the fluorometric method. At the end of the growing season, the vermiculite and peat moss were wrapped up in the lining and disposed of accordingly. Thus every year, we started with a fresh plot. We have found from our experience that customized procedures could be instituted to produce garlic with the desired concentrations of selenium over a fairly wide range.

STUDIES OF MAMMARY CANCER PREVENTION BY HIGH SELENIUM GARLIC

The cancer prevention studies were carried out using the rat dimethylbenz[a]anthracene (DMBA) mammary tumor model in which animals were intubated with a single dose of carcinogen at about 50 days of age. The results from two independent experiments are presented in Table 1. It should be noted that the basic design was identical except for two protocol changes regarding (a) the use of different high selenium-garlic crops: 150 ppm Se-garlic in Experiment A versus 112 ppm Se-garlic in Experiment B, and (b) the dose of DMBA: 10 mg per rat in Experiment A versus 5 mg per rat in Experiment B. In both cases, supplementation of garlic (natural or high selenium) was started 2 weeks before DMBA and continued until the time of sacrifice.

Table 1. Mammary cancer prevention by natural garlic or high selenium-garlic

Expt.	Dietary supplement	Dietary selenium	Dose of DMBA (mg)	Tumor incidence	Total No. of tumors
A	Control	0.1 ppm	10	23/25	90
	Selenite	3.0 ppm	10	12/25[a]	49[a]
	2% Natural garlic	0.1 ppm	10	15/25	60[a]
	2% High Se-garlic	3.0 ppm	10	9/25[a]	34[b]
B	Control	0.1 ppm	5	35/50	72
	0.85% Natural garlic	0.1 ppm	5	27/50	57
	1.7% Natural garlic	0.1 ppm	5	24/50[a]	41[a]
	0.85% High Se-garlic	1 ppm	5	20/50[a]	32[a]
	1.7% High Se-garlic	2 ppm	5	13/50[b]	18[b]

[a]$P<0.05$ compared to the corresponding control.
[b] $P<0.05$ compared to the regular garlic group.

The findings from these two experiments, which have appeared in our previous publications,[16,17] are summarized as follows. (a) The high selenium-garlic was superior to the natural garlic in mammary cancer prevention. This observation was reproducible in both Experiments A and B. (b) Based on the inhbitory effect on the total number of tumors, the high selenium-garlic was just as effective, if not better, than selenite (Experiment A). (c) In Experiment B, supplementation of 1 or 2 ppm Se from the high selenium-garlic produced a 56% or 75% reduction in the total tumor yield, respectively. The demonstration of a dose-dependent relationship is important, because it suggests a specific response to the reagent under investigation. (d) There was no evidence of any adverse effect in rats ingesting the high selenium-garlic, even at a dietary level of 3 ppm Se. The tolerance of the animals to the high selenium-garlic will be discussed in greater detail in Section 6.

THE ANTICARCINOGENIC EFFICACY OF HIGH SELENIUM GARLIC IS PRIMARILY ATTRIBUTABLE TO THE EFFECT OF SELENIUM

In order to answer the question of whether the efficacy of the high selenium-garlic in cancer protection is dependent on the action of selenium, we compared the effects of two batches of garlic with a marked difference in their level of selenium enrichment - 112 ppm Se versus 1355 ppm Se dry weight. Each product was added to the diet to achieve the same final concentration of 2 ppm Se. In this experiment, the supplementation protocol was designed to evaluate the efficacy during the initiation phase (-2 to +1 wks) or post-initiation phase (+1 to +22 wks) of mammary carcinogenesis (the minus and plus signs denote the time before and after DMBA administration, respectively). The results are shown in Table 2. Significant tumor inhibition ($P<0.05$) was evident with either protocol of garlic supplementation. Regardless of the timing of supplementation (Group 2 versus Group 3, or Group 4 versus Group 5), the efficacy of cancer prevention was comparable as measured by the reduction in either tumor incidence or total tumor number. Thus, it can be concluded that both the initiation phase and post-initiation phase of mammary carcinogenesis are equally sensitive to the protective effect of the high selenium-garlic. Furthermore, the magnitude of tumor suppression was very similar with the two batches of garlic, even though the amount of garlic added to the diet varied considerably from each other (1.8% in Groups 2 and 3 versus 0.15% in Groups 4 and 5). This suggests that the anticancer activity of the high

Table 2. Mammary cancer prevention by garlic with different levels of selenium enrichment[a]

Group[b]	Dietary supplement	Amt. of garlic in diet	Dietary selenium	Duration of supplementation[c] (wks)	Tumor incidence	Total No. of tumors
1	Control	——	0.1 ppm	——	27/30	77
2	High Se-garlic (112 ppm Se)	1.8%	2 ppm	-2 to +1	17/30[d]	36[d]
3	High Se-garlic (112 ppm Se)	1.8%	2 ppm	+1 to +22	15/30[d]	30[d]
4	High Se-garlic (1355 ppm Se)	0.15%	2 ppm	-2 to +1	15/30[d]	33[d]
5	High Se-garlic (1355 ppm Se)	0.15%	2 ppm	+1 to +22	14/30[d]	34[d]

[a] The two batches of garlic used in this experiment contained either 112 ppm Se (Groups 2 and 3) or 1355 ppm Se (Groups 4 and 5).
[b] Rats were given 10 mg of DMBA i.g. at 50 days of age and were sacrificed 22 weeks later.
[c] The minus and plus signs denote the time before and after DMBA administration, respectively.
[d] $P<0.05$ compared to the corresponding control value.

selenium-garlic was most likely due to the effect of selenium rather than to the effect of garlic per se.

In view of the above observation that the high selenium-garlic was effective in inhibiting the initiation phase of mammary carcinogenesis, we proceeded to investigate whether changes in DMBA binding to mammary cell DNA was affected. DMBA requires metabolic activation to the ultimate carcinogen, DMBA-3,4-diol-1,2-epoxide, which then reacts with DNA to form adducts. Table 3 shows the results of DMBA binding and DNA adduct formation in mammary cells of rats fed the high selenium-garlic. In this experiment, the garlic supplement was given from 2 weeks before to 2 days after DMBA when the animals were sacrificed. Three specific adducts were characterized: *anti*-dG, *syn*-dA and *anti*-dA. The methodology has been described in a recent publication by Ip et al.[18] It can be seen that both batches of garlic significantly reduced ($P<0.05$) the levels of each of the three DMBA adducts found in mammary cell DNA. There was no difference in the extent of inhibition of adducts between the two supplemented groups, again suggesting that the effect was likely to be accounted for by the intake of selenium present in the garlic.

Quantitatively, the sum of the three adducts totalled 17.2, 9.3 and 9.5 pmol/mg DNA in the control and the two garlic-supplemented groups, respectively. Thus, treatment with the high selenium-garlic resulted in a decrease of these adducts by about 45%. As shown in Table 2, both batches of high selenium-garlic reduced the total number of tumors by slightly over 50% when they were supplemented to the animals for 3 weeks around the time of DMBA dosing (Groups 2 and 4). Thus the magnitude of tumor inhibition correlated reasonably well with the extent of DMBA-DNA adduct suppression. Further investigation will be carried out in the future to delineate the biochemical basis responsible for blocking adduct formation.

NUTRITIONAL BIOAVAILABILITY OF SELENIUM FROM HIGH SELENIUM GARLIC

Nutritional bioavailability refers to the transformation of ingested selenium to a biochemically active form that can be incorporated into a class of mammalian proteins known collectively as selenoproteins.[19] These selenoproteins, which contain one or more seleno-cysteines coded by TGA, include glutathione peroxidase, type I 5'-deiodinase, and plasma

Table 3. DMBA binding and DNA adduct formation in mammary cells of rats fed high selenium-garlic[a]

Dietary supplement	Amt. of garlic in diet	Dietary selenium	Total DMBA binding (pmol/mg DNA)[b]	DMBA adducts (pmol/mg DNA)b,c		
				anti-dG	syn-dA	anti-dA
Control	——	0.1 ppm	41±5.2	10.9±1.8	2.1±0.3	4.2±0.6
High Se-garlic(112 ppm Se)	1.8%	2 ppm	23±2.6[d]	5.7±0.7[d]	1.5±0.2[d]	2.1±0.3[d]
High Se-garlic(1355 ppm Se)	0.15%	2 ppm	21±3.5[d]	6.4±0.9[d]	1.4±0.2[d]	1.7±0.2[d]

[a]High selenium-garlic was given from 2 weeks before to 2 days after DMBA when the animals were sacrificed.
[b]The data are expressed as mean ± SE (n = 4 samples, each sample pooled from 3 rats).
[c]dG, deoxyguanosine; dA, deoxyadenosine.
[d] $P<0.05$ compared to the corresponding control value.

Table 4. Restoration of liver glutathione peroxidase and type I 5'-deiodinase by increasing dietary levels of selenium from either sodium selenite or high selenium-garlic[a]

Selenium supplement (ppm)	Glutathione peroxidaseb		Type I 5'-deiodinasec	
	Selenite	Se-garlic	Selenite	Se-garlic
0	0.11±0.02		6.6±0.7	
0.05	0.75±0.08	0.68±0.08	35.3±5.1	32.8±5.2
0.1	1.6±0.18	1.5±0.17	44.4±5.9	39.5±5.3
0.2	1.7±0.20	1.9±0.21	41.8±4.8	40.1±5.5
0.5	1.9±0.21	1.8±0.20	43.2±6.3	38.0±4.6

[a]Rats were fed a selenium-deficient diet for 4 weeks before repletion with selenite or high Se-garlic. Animals were sacrificed 3 weeks later.
[b]μmol NADPH oxidized/mg protein/min (n=8).
[c]pmol iodide released/mg protein/min (n=8).

selenoprotein P. Selenium deficiency invariably results in marked decreases in their activities or protein levels. Traditionally, the ability of a given selenium compound to restore these selenoproteins in selenium-depleted animals is used as an index of nutritional bioavailability.

In this study, weaning rats were fed a Torula yeast-based selenium-deficient diet containing <0.01 ppm Se.[20] This diet contained a mineral mix without selenium. Torula yeast was used as the protein source instead of casein because there is always a residual amount of selenium in casein. Consequently, a casein-based diet will not produce dietary selenium deficiency of <0.01 ppm. After 4 weeks of depletion, animals were given various levels of selenium from either sodium selenite (positive control) or high selenium-garlic. The repletion of glutathione peroxidase and type I 5'-deiodinase in the liver was determined 3 weeks later.[20] The first enzyme is responsible for the destruction of hydrogen peroxide and lipid peroxides, while the second is responsible for conversion of the prohormone thyroxine (T_4) to the active 3,5,3'-triiodiothyronine (T_3). The results, as shown in Table 4, showed that the high selenium-garlic was just as effective as selenite in restoring the activity of both selenoenzymes. We also have additional information indicating that the progressive increase in enzyme activities as a function of time after the start of selenium repletion was identical with either form of supplementation.[20] Our work therefore conclusively demonstrated that the high selenium-garlic is capable of delivering selenium as an essential nutrient.

TISSUE SELENIUM ACCUMULATION AND TOLERANCE STUDIES IN RATS FED HIGH SELENIUM-GARLIC

As shown in Table 5 (top half of table), rats ingesting the high selenium-garlic at dietary concentrations of 1-3 ppm Se showed only modest increases of selenium accumulation in tissues. These relatively small changes were in contrast to the much larger changes caused by selenite, and particularly by selenomethionine (bottom half of table). Thus the high selenium-garlic is capable of providing effective cancer protection without inviting undue perturbation in tissue selenium levels, an attribute that may alleviate the concern of selenium toxicity at high levels of intake.[21] The low tissue selenium retention characteristics of the selenium-rich garlic might be related to its increased tolerance by the animals. We have found that consumption of the high selenium-garlic at a dietary concentration up to 5 ppm Se did not produce any significant change in weight gain or organ size, including liver, spleen, kidney and uterus. No evidence of histomorphological abnormality was observed upon examination of the above tissues. In contrast, supplementation of either

Table 5. Tissue selenium accumulation in rats fed the high selenium-garlic

		Tissue selenium concentration (μg/g or ml)			
Dietary supplement	Dietary selenium	Liver	Kidney	Mammary	Plasma
Control	0.1 ppm	4.4	5.2	0.14	0.50
High Se-garlic	1 ppm	5.1	6.1	0.12	0.51
High Se-garlic	2 ppm	6.1	7.6	0.15	0.56
High Se-garlic	3 ppm	7.3	8.7	0.20	0.65
High Se-garlic	3 ppm	7.4	8.7	0.29	0.57
Selenite	3 ppm	9.8	12.9	0.37	0.64
Selenomethionine	3 ppm	17.0	29.3	0.57	0.93

selenite or selenomethionine at 5 ppm Se in the diet resulted in growth depression (~15% decrease) and spleen hypertrophy (~35% increase). It appears that the high selenium-garlic is much less likely than selenomethionine to cause undesirable side effects when given at levels that are required for cancer prevention.

DISCUSSION

The National Cancer Institute recommends an increased consumption of fruits, vegetables and grains, and a decreased consumption of fat and animal products, in reducing cancer prevalence in the United States. These dietary guidelines, although laudable in their own rights, are not without some pitfalls. First, it is inherently difficult to change the eating habits of large segments of the population, especially in an economically affluent society. Second, the question of how much change in the intake of certain food groups is necessary in order to reach the desired outcome is still debatable. Third, personal preference, life style, attitude, as well as lack of awareness and resolve, are all tangible barriers to achieving dietary changes. An alternative is to intentionally enrich food with known cancer protective agents through agricultural methods or food processing techniques. In this way, people will be presented with more choices so that it will be convenient for them to use a dietary approach for effective cancer chemoprevention.

One way of fortifying our food supply with selenium is by direct supplementation with an inexpensive chemical such as selenite. This approach is analogous to the supplementation of table salt with iodide. However, the reactivity of selenite may limit its use for such a purpose. Selenomethionine is appropriate in this regard, but the cost of production will be a deciding factor. More importantly, selenomethionine is known to be incorporated nonspecifically into tissue proteins in place of methionine. Total body burden of selenium is consistently higher in rats fed selenomethionine compared to those fed selenite.[22-24] The excessive accumulation of selenium following selenomethionine administration is particularly prominent in skeletal muscle where we have found a 10-fold increase at chemopreventive levels of intake.[24] Such an outcome might potentially lead to certain undesirable effects, especially during periods of accelerated catabolism (e.g. weight loss) when an unbalanced protein turnover rate in favor of degradation will result in a sudden release of selenomethionine and its subsequent conversion to harmful intermediates.

We believe that the high selenium-garlic provides an ideal system for delivering selenium in a food form for cancer prevention. First, it is superior to natural garlic in efficacy and can be easily adapted for wide distribution in a cost effective manner. Second, it is capable of supplying selenium as an essential nutrient in maintaining full activity of functional selenoenzymes such as glutathione peroxidase and type I 5'-deiodinase at nutritional levels of selenium intake. Third, its ingestion at high levels does not result in an exaggerated accumulation of tissue selenium, a concern that is associated with the supplementation of selenomethionine. The low tissue selenium retention characteristic of the high selenium-garlic is likely to increase tolerance and minimize the occurrence of side effects upon continuous feeding. Recently, Se-methylselenocysteine has been identified as a constituent of the high selenium-garlic.[25] By coincidence, this is the same compound that was described by Ip and Ganther a few years ago as having a superior anticancer activity compared to either selenite or selenomethionine.[6] It is reasonable to expect that other selenium-substituted sulfur analogs might also be present in garlic fertilized with selenium. A complete inventory of these selenium compounds would be critical in understanding the potency of the high selenium-garlic in cancer prevention.

ACKNOWLEDGMENT

This work was supported by grant CA 27706 from the National Cancer Institute, NIH, Bethesda, MD.

REFERENCES

1. Milner, J.A. (1985) Effect of selenium on virally induced and transplantable tumor models. Fedn. Proc. *44*: 2568-2572.
2. Ip, C. (1986) The chemopreventive role of selenium in carcinogenesis. J. Am. Coll. Toxicol. *5*: 7-20.
3. Ip, C. and Medina, D. (1987) Current concept of selenium and mammary tumorigenesis. *In*: D. Medina, W. Kidwell, G. Heppner and E.P. Anderson (eds.), Cellular and Molecular Biology of Breast Cancer, pp. 479-494. New York: Plenum Press.
4. Medina, D. and Morrison, D.G. (1988) Current ideas on selenium as a chemopreventive agent. Pathol. Immunopathol. Res. *7*: 187-199.
5. El-Bayoumy, K. (1991) The role of selenium in cancer prevention. *In*: V.T. DeVita, S. Hellman, and S.S. Rosenberg (eds.), Cancer Principles and Practice of Oncology, 4th Edition, pp. 1-15. Philadelphia: J.B. Lippincott, Co.
6. Ip, C. and Ganther, H. (1992) Relationship between the chemical form of selenium and anticarcinogenic activity. *In*: L. Wattenberg, M. Lipkin, C.W. Boone and G.J. Kelloff (eds.), Cancer Chemoprevention, pp. 479-488. Boca Raton, FL: CRC Press.
7. Morris, V.C. and Levander, O.A. (1970) Selenium content of foods. J. Nutr. *100*: 1383-1388.
8. Shrift, A. (1973) Metabolism of selenium by plants and microorganisms. *In*: D.K. Klayman and W.H.H. Günther (eds.), Organic Selenium Compounds: Their Chemistry and Biology, pp. 763-814. New York: Wiley-Interscience.
9. Fenwick, G.R. and Hanley, A.B. (1985) The genus *Allium*, Part 2. CRC Crit. Rev. Food Sci. Nutr. *22*: 273-377.
10. Block, E. (1992) The organosulfur chemistry of the genus *allium* - Implications for the organic chemistry of sulfur. Angew. Chem. *31*: 1135-1178.
11. Wargovitch, M.J. (1992) Inhibition of gastrointestinal cancer by organosulfur compounds in garlic. *In*: L. Wattenberg, M. Lipkin, G. Kelloff, and C. Boone (eds.), Cancer Chemoprevention, pp. 195-203. Boca Raton, FL: CRC Press.
12. Sparnins, V.L., Mott, A.W., Barany, G., and Wattenberg, L.W. (1986) Effects of allyl methyl trisulfide on glutathione S-transferase activity and BP-induced neoplasia in the mouse. Nutr. Cancer *8*: 211-215.
13. Sparnins, V.L., Barany, G., and Wattenberg, L.W. (1988) Effects of organosulfur compounds from garlic and onions on benzo[a]pyrene-induced neoplasia and glutathione S-transferase activity in the mouse. Carcinogenesis *9*: 131-134.
14. Wattenberg, L.W., Sparnins, V.L., and Barany, G. (1989) Inhibition of *N*-nitrosodiethylamine carcinogenesis in mice by naturally occurring organosulfur compounds and monoterpenes. Cancer Res. *49*: 2689-2692.

15. Ip, C. and Ganther, H.E. (1992) Comparison of selenium and sulfur analogs in cancer prevention. Carcinogenesis *13*: 1167-1170.
16. Ip, C., Lisk, D.J., and Stoewsand, G.S. (1992) Mammary cancer prevention by regular garlic and selenium-enriched garlic. Nutr. Cancer *17*: 279-286.
17. Ip, C. and Lisk, D.J. (1994) Enrichment of selenium in allium vegetables for cancer prevention. Carcinogenesis *15*: 1881-1885.
18. Ip, C., Vadhanavikit, S., and Ganther, H. (1995) Cancer chemoprevention by aliphatic selenocyanates: effect of chain length on inhibition of mammary tumors and DMBA adducts. Carcinogenesis *16*: 35-38.
19. Burk, R.F. and Hill, K.E. (1993) Regulation of selenoproteins. Ann. Rev. Nutr. *13*: 65-81.
20. Ip, C. and Lisk, D.J. (1993) Bioavailability of selenium from selenium-enriched garlic. Nutr. Cancer *20*: 129-137.
21. Ip, C. and Lisk, D.J. (1994) Characterization of tissue selenium profiles and anticarcinogenic responses in rats fed natural sources of selenium-rich products. Carcinogenesis *15*: 573-576.
22. McAdam, P.A. and Levander, O.A. (1987) Chronic toxicity and retention of dietary selenium fed to rats as D- or L-selenomethionine, selenite, or selenate. Nutr. Res. *7*: 601-610.
23. Whanger, P.D. and Butler, J.A. (1988) Effects of various dietary levels of selenium as selenite or selenomethionine on tissue selenium levels and glutathione peroxidase activity in rats. J. Nutr. *118*: 846-852.
24. Ip, C. and Hayes, C. (1989) Tissue selenium levels in selenium-supplemented rats and their relevance in mammary cancer protection. Carcinogenesis *10*: 921-925.
25. Cai, X.-J., Block, E., Uden, P.C., Zhang, X., Quimby, B.D. and Sullivan, J.J. (1995) *Allium* chemistry: Identification of selenoamino acids in ordinary and selenium-enriched garlic, onion and broccoli using gas chromatography with atomic emission detection. J. Agric. Food Chem. *43*: 1754-1757.

16

IS THERE A NEED TO CHANGE THE AMERICAN DIET?

Johanna Dwyer

Frances Stern Nutrition Center, New England Medical Center
and Schools of Medicine and Nutrition, Tufts University
Boston, Massachusetts 02111

INTRODUCTION

Phytochemicals are substances found in edible fruits and vegetables, and other foods such as grains and legumes. They have the potential for decreasing chronic degenerative disease risks and thus are of interest. The other parts of this symposium have reviewed the fundamental science, mechanisms of action and effects of dietary fiber, the isothiocyanates, polyphenols, flavonoids, monoterpenes, and organosulfides. Now we turn to more practical questions about phytochemicals in the American diet. First, what should the scientific base be for evaluating the evidence? Second, what are the recommendations of expert groups today on these various phytochemicals? Third, are these recommendations being met? Fourth, is it time for Americans to consider changes in the amount or types of phytochemicals they eat to reduce risks for cancer or other chronic degenerative diseases? Finally, if it is time to implement such changes, how should this be done?

SCIENCE BASE FOR EVALUATING EVIDENCE

In evaluating the evidence for an association between dietary constituents such as phytochemicals and cancer risk reduction, one knowledgeable observer recently remarked, "We are both constrained and empowered by the science. As soon as the science is there, we want to be there as well."[1]

How much evidence is enough to be sure of diet-disease interrelationships and to base proposals for dietary change upon them? The criteria for inferring causality include strength of the association (magnitude of the effect), consistency of the association, temporally correct association, dose-response relationship, specificity of association, and biological plausibility. If surrogate or intermediate markers are available, waiting for the development of disease after making dietary changes may not be necessary. Recommendations for evaluating chemopreventive agents in animal and human studies are available.[2] Also important are the assessment of risks and safety issues as well as benefits. Because most of these issues involve public policy, consultation and discussion within the broad scientific community at for such as this are important, especially in areas where emerging science seems to be particularly promising.

Some groups have made specific statements about the role of phytochemicals in diets. For example, the American Dietetic Association's position is that phytochemicals as naturally occurring functional food components "may have a beneficial role in health as part of a varied diet. The Association supports research regarding the health benefits and risks of these substances." It also urges dietetics professionals to "continue to work with the food industry and government to ensure that the public has accurate scientific information in this emerging field.[3]"

EXPERT RECOMMENDATIONS

What Experts Recommend Today

Some of the earliest recommendations made at the beginning of this century by the nutritionists of President Theodore Roosevelt's administration suggested a plant-based diet and liberal amounts of fruits and vegetables. The rationale was that the diet was inexpensive and healthful, although not all of the substances that made it so were known. What is different today is not that the recommendations are all that different, but that we are on firmer scientific ground for knowing why we make these suggestions. It is not *only* vitamins and minerals or fiber that make plant foods healthful but other substances such as the phytochemicals discussed in this symposium.

Dr. Messina's presentation provides an overview of commonly employed dietary recommendations and what they say about foods containing phytochemicals. The emphasis on a plant-based dietary pervades most of the more recent recommendations. Several make explicit recommendations about specific amounts of fruits, vegetables and grain-based foods. No set of expert recommendations focuses on quantitative or qualitative recommendations for the specific phytochemicals discussed in this symposium.

Are We Meeting Expert Recommendations?

The sources for estimating population intakes of selected nutrients and non-nutrients have recently been summarized. They include levels of foods and nutrients reported in the USDA's Food and Nutrition Supply Series; the Food Additives Survey that summarizes use of food additives and substances generally recognized as safe (GRAS); the USDA's Continuing Survey of Food Intake; the US Department of Health and Human Services' National Health and Nutrition Examination Survey; and special surveys of vitamin and mineral supplement use, among others.[4]

These sources are helpful for assessing consumption of groups of foods. However, food composition tables are not yet refined enough to permit making firm estimates of individual consumption of the different phytochemicals discussed in this symposium. Table 1 presents some of the food sources of these specific substances. We have heard some estimates of consumption. However, much of it is not consumption data but rather food availability and disappearance data based on amounts of commodities available for various purposes. National food availability and consumption data have recently been used to estimate intakes of certain antioxidant nutrients (vitamins A, C, beta carotene and vitamin E) and to describe groups that are likely to be at risk of particularly low or high intakes.[5] It is difficult to perform a similar exercise for many of the specific phytochemical constituents discussed here, because little reliable composition data are currently available. It may be difficult to obtain, especially since the phytochemical content of food varies depending on the basic commodity, subsequent processing, and in some cases metabolic processes in the host.[6] More complete and more definite food tables on specific species of food plants need to be developed to permit such studies in the future.[7]

Today in making estimates of population intakes, many assumptions must be made in deriving the data. Usually food disappearance data are used and estimates for the various commodity groups are made for each, using the often scanty analytical data that exist. Individual dietary intakes may vary considerably from these. For example, tea drinkers and vegetarians have much higher intakes of flavonoids than those eating more conventional diets.

In spite of the limitations of the data, it is apparent that Americans today are failing to meet even the rather broad recommendations now being popularized for consumption of fruits, vegetables and grains, and by extension, it is unlikely that they are receiving large amounts of many of these phytochemicals. The 1977-78 USDA Nationwide Food Consumption Survey data indicated that intakes of both fruits and vegetables failed to meet current recommendations.[8] According to NHANES II data, only 83% of adults reported consuming vegetables and 59% reported consuming fruits the day before they were interviewed. Much smaller numbers of persons ate any specific type of fruit or vegetable.[9] Other analyses reveal that older persons living alone, the poor, and the elderly in general are less likely to consume fruits and vegetables.[10]

Are New Recommendations Needed?

Every few years it is customary to review newly emerging science to make sure that existing recommendations are appropriate. The Dietary Guidelines Committee has just completed its work and has moved fruits, vegetables and grains up a little on the list of do's and don'ts. The Food and Nutrition Board of the Institute of Medicine, National Academy of Sciences, is beginning its review of the Recommended Dietary Allowances. In their next revisions, should expert groups also consider specific recommendations with respect to certain phytochemicals or foods containing them?

Many experts are becoming convinced that more specific and explicit advice on certain foods such as fruits, vegetables and grains, is warranted, although quantitative statements for individual substances other than fiber are rarer. Some experts argue that certain phytochemicals, such as the isoflavones, tocotrienols, and carotenoids, are "candidate nutrients" that may be of health benefit to humans by inhibiting the development of both cancer and possibly atherosclerosis.[11]

Before altering dietary recommendations, fortifying foods, or supplementing with phytochemicals can be considered, it is important to determine what types of data are needed. The unique functions of the phytochemicals need to be better elucidated, the amounts in foods (particularly commonly-eaten foods) determined, animal testing results evaluated, bioavailability in humans assessed, and both short and long term feeding trials completed to determine their potency for health enhancement. Quantitative recommendations, and how much should be used for food fortification proposals and supplements, await the development of such information.

In the next few years, as more specific information becomes available on the biological effects of phytochemicals, and as food composition data improves, it should be possible to develop better estimates of the amounts in commonly available fruits and vegetables, so that we can give people more specific advice. Preliminary information is already available on dose-response relationships of diets high in legumes and allium or low or high in vegetables and fruits and excretion of lignans and isoflavonoids.[12]

Table 1. Some phytochemicals and their sources

Chemical Name Constituent	Primary Food Sources	Actions	Comments
Isothio-cyanates	Found in species of Superorder Violiflorae, Capparales, Brassicaceae, family. Brassicaceae or Cruciferae, cruciferous vegetables (bok choy, broccoli, Brussels sprouts, cabbage, cauliflower, collards, kale, kohlrabi, mustard greens, rutabaga, turnips, watercress) and mustard oils contain both indoles and isothiocyanates which are breakdown products of complex plant compounds called glucosinolates formed during processing, cooking and chewing. Other active substances may also be present such as sulphoraphane, anethol, dithiolthione, etc.[18]	Various compounds vary in their inhibitory pathway.[19] Indoles block carcinogens before they reach their targets in cells; isothiocyanates may suppress tumor growth by blocking phase 2 enzymes.[20,21] Mechanisms of action are now being elucidated.[22,23]	Urine markers are available for allylisothiocynates.[24] Iberin, glucoseminolate hydrolysis product 1-isothiocyanato-3 (methylsulfinyl) propane (IMSP) is consumed at about 1μmol/kg/day, and good effects of cruciferous vegetables may be due to it.[25] Others believe allyl-isothiocyanate or some transformation product is the active compound.[26,27]
Polyphenols	Phenolic acids are present in garlic, green tea, soybeans, cereal grains, Cruciferous, Umbelliferous, Solaraceous, and Cucurbitaceous plants, and also licorice root and flax seed. ECGC, epigallocatechin gallate, is an active polyphenol in green tea.[28,29]	Antioxidant, may reduce lipid peroxidation. [30,31]	
Flavonoids or bioflavonoids	Fruits, vegetables, wine, green tea. Citrus fruits contain bioflavonoids such as hesperidin and naringin (both glycosides), nobiletin and tageretin (methoxylated molecules), and narirutin. The most biologically active compounds out of the 4000 bioflavonoids are in citrus fruits. Quercitin and rutin are two other flavonoids. Onions, apples, kale and beans are also good sources of flavonoids. Distribution includes considerable amounts in green tea, soybeans, cereal grains, Cruciferous, Umbelliferous, Citrus, Solanaceous, and Cucurbitaceous plants, and licorice root and flax seed.	Reduce cancer risk by action as antioxidants, blocking access of carcinogens to cells, suppressing malignant changes in cells, interfering with binding of hormones to cells, chelating metals, inducing enzymes modifying carcinogenicity, stimulating immune response or combinations of these actions.[32]	Intake estimates are crude, probably 1 gm per day.[33,34] The major sources appear to be approximately a third from fruits and juices, a third from wine, beer, coffee, tea, etc., and the rest from herbs, vegetables and other plant foods. Tea is high in flavonoids,[35] with over 100 mg per cup.

Mono-terpenes (such as D limonene and D carvone)[36]	Garlic, citrus fruits (d-limonene), caraway seeds and oils (d-carvone); also Umbelliferous, Solanaceous, and Cucurbitaceous plants; sage, camphor and dill. POH (perillyl alcohol) also appears active. There are many components.	Block action of carcinogens by inducing phase I and II enzymes[37] or during initial inhibition of posttranslateral isoprenylation of growth catalyzing small G protein, slowing promotion and progression, and tissue redifferentiation.[36]	High therapeutic ratio. No toxicity noted at 100 mg/kg limonene in custard in acute toxicity study.[38]
Organo-sulfides (dallyl di-sulfide is one com-pound that appears to be especially potent)	Superorder is Lilliflorae, within Onion family (Alliaceae) containing plants in genus Allium such as garlic, onions, leeks, shallots. Thus most of the sulfides are in garlic and the Cruciferous plants.	Block or suppress carcinogenesis.[39] May also alter serum lipids and platelet aggregation.	In some studies of leek, garlic and onions or garlic supplements, no effects on human breast or lung cancer were observed.[40,41] In others it is suggested that the allium group vegetables may induce pemphigus.[42]
Isoflavones	Phytoestrogens are found in soy beans (large amounts) and many other foods like legumes (Fabiflorae superorder Fabacreae, Leguminosae family) in lower amounts. Phytoestrogens include genistein, biochanin A, daidzein, formononetin, and the gut product equol, among others.	Various effects blocking and suppressing carcinogens, with isoflavones blocking estrogens from entering cells and other actions.[43]	Colonic bacteria convert precursor molecules into active forms.
Lignans	Flaxseed (linseed); whole grain products, vegetables, fruits and some berries. Lignans are one of two types of phytoestrogens (the other being isoflavones).	Some effects appear to be antioxidant in nature.[44] Bind to estrogen receptors and act as weak antiestrogens, reduce estrogen synthesis, increase synthesis of sex hormone binding globulin, and lower circulating levels of free estradiol; in this way or by other actions they may block or suppress cancerous changes.	Colonic bacteria convert precursor molecules into active forms.
Saponins	Most vegetables and herbs, such as soybeans.	Unclear anticancer activity mechanism, although other effects.	
Carotenoids	Dark yellow and orange and deep green vegetables and fruits.		
Grains	Poaceae (Gramineae family) include corn, wheat, oats, barley, rice.		

Dietary Fiber

Note: Superorders of plants are suffixed *-iflorae*, orders with *-ales*, and families with *-aceae*.

Table 2. Pros and cons of various approaches to altering phytochemicals in American diets

Assessment	Status Quo	More Foods High in Naturally Occurring Sources	Foods Fortified with Phytochemicals	Phyto-chemical Supplements
PROS				
Many different phytochemicals provided in a naturally-occuring "cocktail"	Amounts may be too low for desired effects	Amounts may be too low for desired effects	Often only one phytochemical is added	Often only one phytochemical is added
Naturally occurring mixtures may be more effective than single phytochemical sources, or	x	x		
Naturally occurring mixtures coupled with supplements (such as high fruits and vegetables and vitamin E supplements)[45] may have beneficial synergistic effects on decreasing risks of some forms of cancers				x
Intakes within the population are thought to vary widely and if optimal amounts are higher than intake levels, fortification and/or supplementation is necessary			x	x
Possible beneficial effects on many illnesses, not only cancer risk	x	x	x	x
CONS				
Modes of action and mechanisms such as sites of action, balance of agonistic and antagonistic properties, natural potency, short and long term effects are not yet clear; the more that is known about them the easier it is to rule out untoward effects and capitalize on good ones[46]	x	x	x	x
For many products, food composition data are not yet available and analytical methods are only now being developed[47]	x	x	x	x
May be safety problems we are not yet aware of, such as different effects at various times of life and/or in different organs.[48] Some phytochemicals poorly characterized or tested for safety	x	x	x	x
Naturally occurring sources of some (such as folic acid) are bound in foods, while added constituents are more highly bioavailable	x	x		
Requires knowledge of what constituents are effective			x	x

Table 2. (*Continued*)

Assessment	Status Quo	More Foods High in Naturally Occurring Sources	Foods Fortified with Phytochemicals	Phyto-chemical Supplements
Benefits may be present for some groups and harm other specific subgroups (as with folic acid supplementation and risk of masking vitamin B-12 deficiency in the elderly	x	x		

NEXT STEPS

Four Paths

There are several paths we can take with respect to intakes of phytochemicals in the American diet. We can maintain the status quo, wait for more evidence, and let intakes "float" at current levels, or we can act positively to alter intakes. Changes in intakes of phytochemicals might be accomplished by increasing our intakes of existing foods containing them, eating foods fortified with phytochemicals, or taking supplements of these substances. Some of the many pros and cons are presented in Table 2. There is no single solution for all substances.

The most appropriate path to take with respect to altering dietary intake depends on the amount and quality of evidence available. A considerable amount of evidence already exists on some of the phytochemicals with respect to their roles in promoting health, whereas for others only anecdotal evidence is available. Also the appropriate target groups, and narrowness of the therapeutic vis a vis the toxic threshold need to be considered. Finally, what the target groups are, and what food vehicle is most appropriate, are also major issues needing consideration.

Wait for More Evidence to Accumulate and Let Dietary Intakes "Float"

For some of the phytochemicals, enough evidence is not yet available to warrant dietary recommendations. For them, watchful waiting and monitoring of the new science as it evolves is warranted, and in the meantime dietary intakes should be allowed to float.

Safety considerations are such that when we are not sure, caution is warranted. One recent example is the issue about possible deleterious effects of soy phytoestrogens in infant feeding, which was raised recently in New Zealand.[13] Other newspaper clips from New Zealand suggested that the putative declines in the sperm counts of men living in New Zealand might be due to various chemicals, including phytoestrogens, pesticides, and others, although the evidence to support such an assertion is tenuous at best. While neither of these alarms proved to be true, evidence for risk as well as benefit does need to be examined, and it is true that some of the chemicals found in plants enhance carcinogenesis.

Increase Intakes of Foods Naturally Containing Phytochemicals

We can change our dietary intakes of phytochemicals by altering our choices of existing foods. The USDA's current food guide pyramid suggests 2-4 fruit group servings and 3-5 vegetable group servings a day. Profiles of the nutrient composition of food have been developed for fruits, for several categories of vegetables (dark green, deep yellow, dry beans and peas, other starchy and other), and for bread, cereal, rice and pasta (whole grain products and enriched grain products), but we do not yet have such profiles for the phytochemicals of interest in this symposium.[14] The food groups we have today are defined primarily on the basis of the nutrient content of the food, but perhaps in the future, it will be possible to develop food subgroups that provide more guidance with respect to foods particularly rich in specific phytochemicals. In addition, new techniques such as biotechnology or cultivation of vegetables with fertilizers high in nutrients thought to reduce cancer risk may be possible. For example, the use of experimental crops cultivated with selenium fertilization seems to have advantages for the naturally occurring forms in which selenium is delivered versus some other compounds that might be fed.[15]

Vegetarian food grouping systems are also available with greater specificity in the plant food groups.[16]

Dr. Heimindinger's discussion on the National 5-A-Day for Better Health Program addresses the need for increased consumption of fruits and vegetables and points to the considerable successes her group and others have had to date in such efforts. Dr. Richard Black of the Kellogg Company focuses on food components in a healthy diet; what we have and what we need, and also provides useful perspectives in this regard.

Make Phytochemical Fortified Foods More Available and Eat Them

A second choice is to fortify foods with specific phytochemicals. The notion is that the foods that result will be "functional foods" or "hypernutritional foods" (foods modified to provide health benefits beyond the traditional nutrients they contain). If we are to do this we need to know which phytochemicals are the most important in terms of providing desirable health effects and assuring safety, and how much should be used. For example, some of the organosulfur compounds in garlic and onions have protective effects and others may have cancer-enhancing effects.[17] We need to ask if the scientific evidence is strong enough to warrant fortification on the basis of existing evidence, and if so, which food vehicle should be chosen for fortification? Dr. Mark Messina examines some of these alternatives and provides some practical applications of the issue of functional foods in his paper. He gives us a glance into what some experts view as the foods of the future. Dr. John Finley of Nabisco addresses the broader issue of phytochemicals from the food industry's perspective, and what issues fortification poses for new product development.

Promote Phytochemical Supplements

This option must also be considered. The legislation on dietary supplements that was passed last year is now being implemented in regulations. Among the many issues that must be considered in recommending this approach are safety of the compounds and standardization issues, especially for botanical preparations.

CONCLUSIONS

My eating patterns have changed over the past few years to include more grains, fruits and vegetables than ever before. I've begun to eat more main meals, a few a week, that are largely plant-based. I try to snack on fruits and fruit pieces, and to make a meal a day from salads. I still find that I have to concentrate to be sure I eat "Five a Day", especially when eating out.

What should we do next in terms of dietary guidance? I suggest focusing first on Five a Day and then, gradually, on six or even seven! And perhaps by the year 2000 we will know enough to be more specific in our guidance — singling out more specific families of vegetables for attention.

ACKNOWLEDGMENT

This project has been funded at least in part with Federal funds from the U.S. Department of Agriculture, Agricultural Research Service under contract number 53-3K06-01. The contents of this publication do not necessarily reflect the views or policies of the U.S. Department of Agriculture, nor does mention of trade names, commercial products, or organizations imply endorsement by the U.S. Government. I thank Begabati Lennihan for her editorial assistance.

REFERENCES

1. Taylor MR. FDA's public health goals in evaluating health claims. Crit Revs Food Science and Nutrition 35:1-5, 1995
2. El Bayoumy K. Evaluation of chemopreventive agents against breast cancer proposed.Strategies for human clinical trials. Carcinogenesis 15(11): 2395-420, 1994
3. American Dietetic Association. Position of the American Dietetic Association: Phytochemicals and functional foods. J Am Diet Assoc 95:493 1995
4. Interagency Board for Nutrition Monitoring and Related Research. Wright J, ed. Nutrition Monitoring in the United States: The Directory of Federal and State Nutrition Monitoring Activities. Hyattsville, MD: Public Health Service 1992, pg. 77
5. Woteki CE. Consumption, intake patterns, and exposure. Critical Revs in Food Science and Nutrition 35:143-147,1995
6. Dwyer JT. Future directions in food composition studies. *J Nutrition* Suppl 1994;124(9S):1783S-1788S.
7. Ferguson LR, Yee RL, Scragg R, Metcalf PA, Harris PJ. Differences in intake of sppecfic food plants by Polynesians may explain their lower incidence of colorectal cancer compared with Europeans in New Zealand. Nutr Cancer 23: 33-42, 1995
8. Cronin FJ, Krebs-Smith SM, Wyse BW, Light L. Characterizing food usage by demographic variables J Am Dietet Assoc 81: 661, 1982
9. Patterson BH, Block G. Food choices and the cancer guidelines. Am J Public Health 78(3):282, 1988
10. Ryan AS, Martinez GA, Wysong JL, Davis MA. Dietary patterns of older adults in the United States, NHANES II, 1976-80. Am J Hum Biol 1:321, 1989
11. Hendrich S, Lee K, Xu X, Wang HJ, Murphy PA. Defining food components as new nutrients. J Nutrition 124(9 Suppl):1789S-1792S,1994
12. Hutchins AM, Lampe JW, Martini MC, Campbell DR, Slavin JL. Vegetables, fruits, and legumes: effect on urinary isoflavonoid phytoestrogen and lignan excretion. J Am Diet Assoc 95:769-774, 1995
13. Irvine C, Fitzpatrick M, Robertson I, Woodhams D. The potential adverse effects of soybean phytoestrogens in infant feeding. NZ Med J 208-209, 24 May 1995
14. Cronin FJ, Shaw A, Krebs Smith SM, Marsland P, Light L.. Developing a food guidance system to implement the Dietary Guidelines. J Nutr Educ 19:281, 1987
15. Ip C, Lisk DJ. Enrichment of selenium in allium vegetables for cancer prevention. Carcinogenesis 15(9):1881-5, 1994
16. Haddad E. Development of a vegetarian food guide. Am. J. Clin. Nutr. 1994;59:1248S-54S.
17. Takada N, Kitaro M, Yono Y, Otari S, Fukoshama S. Enhancing effects of organosulfur compounds from garlic and onions on hepatocarcinogenesis in rats: association with cell proliferation and elevated ornithine decarboxylase activity. Jap J Cancer Res 85:1267-72, 1994
18. Chen MF, Chen LT, Boyce HW. Cruciferous vegetables and glutathione: their effects on colon mucosal glutathione level and colon tumor development in rats induced by DMH. Nutr Cancer 23:77-83, 1995
19. Zhong Y, Talalay P. Anticarcinogenic activities of organic isothiocyanates: chemistry and mechanisms. Cancer Res 54 (7Supp):1976S-1981S, 1994
20. Zhang Y, Kensler TW, Cho CG. Posner GH, Talalay P. Anticarcinogenic activities of sulforaphane and structurally-related synthetic norbornyl isothiocynates. Proc Natl. Acad. Sci. USA 91(8):3147-50, 1994
21. Zhang Y, Kolm RH, Mannewik B, Talalay P. Reversible conjugation of isothiocyanates with glutathione catalyzed by human glutathione transferases. Biochem Biophys Res Comm 206(2):748-55, 1995
22. Lee MS. Oxidative conversion of isothiocyanates to isocyanates by rat liver. Environ Health Perspectives 102 (S6):45-8, 1994

23. Meger, D.J., Crease, D.J. and Ketterer, B. Forward and reverse catalysis and product sequestration by human glutathione S transferases in the reaction of GSH with dietary aralkyl isothiocyanates. Biochem J 306(2):565-9, 1995
24. Jiao D, Ho CT, Fooiles P, Chang FL. Identification and quantification of the N-acetylcysteine conjugate of allylisothiocynate in human urine after ingestion of mustard. Cancer Epi Biomarkers and Prevention 3(6):487-92, 1994
25. Kore AM, Jeffrey DH, Wallig MA. Effects of 1-isothiocyanato-3-(methylsulfinyl) propane in xenobiotic metabolizing enzymes in rats. Food and Chemical Tox 31(10):723-9, 1993
26. Musk SR, Johnson IT. Allylisothiocyanate is selectively toxic to transformed cells of the human colorectal tumor HT 29 Carcinogenesis 14(10):2079-83, 1993
27. Graham HN. Green tea composition, consumption and polyphenol chemistry. Preventive Med 21:334-50, 1992
28. Komori A, Yatsumoni J, Okake S, Abe S, Hara K, Suganoma M, Kim SJ, Fujiki H. Anticarcinogenic activity of green tea polyphenols. Jap J Clin Oncology 23(3):186-90, 1993
29. Wong ZY, Huang MT, Ho CT, Chang R, Ma W, Ferraro T, Reghl KR, Yang CS, Conney AH. Inhibiting effect of green tea on the growth of established skin papillomas in mice. Cancer Res 52(23):6657-65, 1992
30. Fuhrman B, Lary A, Arion M. Consumption of red wine with meals reduces susceptibility of human plasma and LDL to lipid peroxidation. AJCN 61(3):549-54, 1995
31. Ho CT, Chan Q, Shi H, Zhang KQ,and Rosen R.T., Antioxidativeeffect of polyphenol extract prepared from various Chinese teas, *Prev. Med.* 21:520 (1992)
32. Attaway JA. Medical benefits of juice flavonoids, pp 207-217 in Report of Congress: XI International Congress of Fruit Juice, Sao Paulo, 1991
33. Kuhnau J.The flavonoids. A class of semi-essential food components: their role in human nutrition. World Review of Nutrition and Dietetics 1976;24:117-91
34. Pierpoint WS. Flavonoids in the human diet. In: Cody V, Middleton EJ, Harbone JB, eds. Flavonoids in Biology and Medicine: Biochemical, Pharmacological, and Structure-Activity Relationships. Vol. 213. Alan R.Liss,Inc., New York; 1986; 125-40
35. Herrmann K. Flavonols and flavones in food plants: a review. J Food Technol 1976;11:433-48
36. Crowell PL, Gould MN. Chemoprevention and therapy of cancer by d-limonene, Crit Reviews Oncogenesis 5(1):1-22, 1994
37. Elson CE, YU SG. Chemoprevention of cancer by merulonate derived constituents of fruits and vegetables. Nutr 124:607-14, 1994
38. Crowell PP, Elson CE, Bailey HH, Elegbade A, Haag JD, Gould MN. Human metabolism of the experimental cancer therapeutic agent d-limonene. Cancer Chemotherapy and Pharmacology 35(1):31-7, 1994
39. Redely BS, Rug CV, Revenson A, Kellogg G. Chemoprevention of colon carcinogenesis by organosulfide compounds. Cancer Res 53(15):3493-8, 1993
40. Dorant E, van den Brundt PA, Goldbohm RA. Allium vegetable consumption, garlic supplement intake and female breast carcinoma incidence. Breast Cancer Research and Treatment 33(2):163-70, 1995
41. Dorant E, van den Brandt PA, Goldbohm RA. A prospective cohort study on allium vegetable consumption, garlic supplement use and risk of lung carcinoma in the Netherlands. Cancer Res 54:6148-53, 1994
42. Brenner S, Wolf R. Possible nutritional factors in induced pemphigus. Dermatology 189(4):337-9, 1994
43. Barnes S, Peterson G, Grubbs C, Setchell K. Chapter 10:Potential role of dietary isoflavones in the prevention of cancer pg 135-146 in Jacobs MM, ed. Diet and Cancer: Markers, Prevention, and Treatment. Plenum Press New York, 1994
44. Thompson LU. Antioxidant and hormone mediated health benefits of whole grains. Critical Reviews in Food Science and Nutrition 34(5 &6): 473-497,1994
45. Gridley G, McLaughlin JK, Block G, Blot WJ, Fraumeni JF Vitamin supplement use and reduced risk of oral and pharyngeal cancer . Am J Epidemiol 135:1083, 1992
46. Fotsis T, Pepper M, Adlercreutz H, Hase T, Montesano R, Schweigerer L. Genistein, a dietary ingested isoflavonoid, inhibits cell proliferation and in vitro angiogenesis. J Nutrition 125(3 Suppl):790S-797S, 1995
47. Franke AA, Custer LJ,Cerna CM, Narala K. Rapid HPLC analysis of dietary phytoestrogens from legumes and from human urine. Proc Soc Exper Biol Med 208:18-26, 1995
48. Whitten PL, Lewis C, Russell E, Naftolin F. Potential adverse effects of phytoestrogens. J Nutrition 125(3 Suppl):771S-776S, 199

17

THE NATIONAL 5 A DAY FOR BETTER HEALTH PROGRAM

Jerianne Heimendinger and Daria Chapelsky

National Cancer Institute
Bethesda, Maryland 20892

INTRODUCTION AND RATIONALE

The widespread enthusiasm for phytochemical research is obvious from the participation in this conference. However, the emerging research findings from this field can be confusing to consumers, who report they are frustrated with conflicting advice from the "experts." The media's instant coverage of new discoveries, sometimes without placing the new findings into a broader context, can have a negative impact on consumers' motivations to change unhealthy dietary behaviors. You, as researchers, can help prevent this deleterious effect when you report your research results by referring consumers to the National Cancer Institute's 5 A Day for Better Health Program. We do not yet know which combination of phytochemicals confers protection, but there is adequate evidence to recommend that consumers eat 5 or more daily servings of a variety of fruits and vegetables.

Of all the dietary factors postulated to be related to cancer, the epidemiological evidence is most consistent for an inverse association between the risk of cancer and fruit and vegetable consumption. Both retrospective and prospective studies have demonstrated this association for a variety of cancers, including those of the oral cavity, esophagus, pharynx, stomach, pancreas, colon, rectum, larynx, lung, bladder, endometrium, cervix, and ovary.[1-7] The evidence is strongest for epithelial cancers, especially those of the digestive and respiratory tracts, and more equivocal for hormone-related cancers.[1] In 128 of 156 dietary studies that calculated a relative risk, a statistically significant protective effect of fruits and vegetables was found.[1] Only four of these studies found a positive association between fruit and vegetable association and cancer. For lung cancer, 24 of 25 studies showed a protective effect; for esophageal cancer, 15 of 16; for stomach, 17 of 19; for colorectal, 20 of 27; for oral cancers, 9 of 9; pancreas, 9 of 11; cervix, 7 of 8; and ovary, 3 of 4. The majority of these studies have controlled for possible confounding factors, such as smoking, dietary fat, calories, and alcohol.

It is possible that fruit and vegetable consumption is associated with other demographic or lifestyle factors which are the true causative agents. For example, people who eat more fruits and vegetables may less likely be smokers and are more likely to be better educated. Although these alternative hypotheses cannot be ruled out, many studies have controlled for smoking, socioeconomic status, and other potential dietary confounders such

Dietary Phytochemicals in Cancer Prevention and Treatment
Edited under the auspices of the American Institute for Cancer Research, Plenum Press, New York, 1996

as fat, calories, and alcohol, and the effect of higher consumption of fruits and vegetables remains. It is unlikely that non-dietary factors totally explain the risk. In addition, studies have been conducted in 17 countries with very diverse populations such as those in the Netherlands, China, India, and the United States. It is difficult to imagine what lifestyle correlates would explain similar conclusions in such diverse populations and environments.[1]

The results are not only statistically significant, but also clinically important. In the vast majority of studies, a dose-response relationship was found in which those with lower consumption experience a cancer risk generally at least twice as high as those with higher consumption levels. The consumption levels of persons in the lower thirds, quartiles, or quintiles of the studies reviewed by Block et. al. were less than 1.0 to 1.3 servings per day; the upper ranges were from 3 to 5 servings.[1] However, better estimates of population consumption levels come from national surveys. The NHANES II survey, using a single 24-hour recall, indicated that adults in the bottom quintile of consumption averaged one serving per day; adults in the top quintile averaged 5 servings per day.[8] The 1989 Continuing Survey of Food Intakes of Individuals, using one 24 hour recall and two days of diet records, indicated that the population in the bottom quartile in the nation are consuming 2.1 servings a day or less and those in the highest quartile are consuming 4.8 servings or more.[9]

In addition, the epidemiological studies indicated that a reduced risk of cancer was associated with a variety of fruits and vegetables. These included dark green, yellow and orange fruits and vegetables, cruciferous vegetables, dried fruits, berries, beans, tomatoes, and carrots. These epidemiological studies were not designed to ascertain which foods were the most effective, and therefore, consuming a broad variety is prudent.

Adding to the weight of the evidence is the fact that there are plausible biochemical mechanisms for the effects of fruits and vegetables. Fruits and vegetables are sources of vitamins and minerals (including vitamins A, C, E and folate), carotenoids and other antioxidants, fiber, and various phytochemicals such as dithiolthiones, flavonoids, glucosinolates, indoles, isothiocyanates, phenols, *d*-limonene, and allium compounds. Each of these substances may play a role in reducing risk. More likely, it is a combination of these factors, and others not yet explored, which may confer protection.

Clearly, more research must be done to elucidate the roles of fruits and vegetables in cancer etiology. However, from a public health perspective, there is abundant evidence to suggest that substantial health benefits could be achieved by increasing the population's consumption of fruits and vegetables.

The data indicate that Americans fall far short of the 5 or more servings recommended by the U.S. Dietary Guidelines, the Food Guide Pyramid, and the Diet and Health report.[10-12] Results of the 5 A Day baseline survey of a nationally representative sample indicated that only 23% of adults were consuming the recommended level in 1991. The median level of consumption in this survey was 3.4 servings per day.[13]

Based on the strength of the scientific data discussed above and the low consumption rates in the population, the National Cancer Institute (NCE) initiated the national 5 A Day for Better Health Program in the fall of 1991 to encourage Americans to eat 5 or more servings of fruits and vegetables every day in the context of a low-fat, high-fiber diet. This program addresses one of the nation's health promotion and disease prevention objectives.[14]

PROGRAM DESIGN

The program is a public-private partnership between the NCI of the National Institutes of Health, U.S. Department of Health and Human Services, and the Produce for Better Health Foundation (PBH), a nonprofit consumer education foundation representing the fruit and vegetable industry. Examples of industry participants are supermarket chains, independent grocery stores, growers, shippers, packagers, merchandisers, branded product companies, suppliers, commodity boards, fruit and vegetable marketers, and noncommercial and commercial foodservice operations. Currently, PBH has more than 1200 licensed members. Supermarket chains represent over 35,000 retail stores. Members also include representatives of the frozen, canned, and dried foods sectors of the industry. Prior to the formation of PBH, the fruit and vegetable industry had no structure or funds for such a broad collaboration across industry sectors.

The national program replicated and expanded the design of its progenitor, the California 5 A Day Program. The latter was developed through a capacity-building grant awarded to the California Department of Health Services in 1986 by the (NCI). California established the 5 A Day dietary message, developed and trademarked the 5 A Day program logo, established a public/private partnership between state agencies and the fruit and vegetable industry, and implemented social marketing strategies using the mass media and supermarkets. The enthusiastic response by both industry and state health agencies to the California Program convinced the staff at the NCI to explore expansion of the program to the national level. A concept for the national program was approved by the Board of Scientific Counselors of NCI's Division of Cancer Prevention and Control in the Fall of 1991. The produce industry organized the Produce for Better Health Foundation (PBH) to work with the NCI, and the national 5 A Day public/private partnership was born.

The design of the national 5 A Day Program reflects results of previous research on the effectiveness of media, public/private collaborations and community based health promotion interventions for disease prevention. Various studies have shown that the media plays a vital role in increasing consumer awareness of health issues and, in some instances, changing individual patterns of behavior.[15-17] Public confidence in messages from a credible health agency such as NCI has been shown to be a key factor in affecting consumer buying patterns.[18] The combination of credible health messages promoted through industry via media has been shown to be effective in influencing consumers.[15]

Although use of the media alone can produce behavioral change, its effect is increased when its use is supplemented by other community-based educational efforts.[19-21] The Stanford Three-Community Study, the North Karelia Project in Finland, and the Stanford Five-City Project were successful in reducing cardiovascular risk factors through community-based interventions using mass media, interpersonal and community education strategies.[20-24] Similar mixes of media and social marketing strategies were used in other community-based cardiovascular disease prevention programs, such as the Minnesota and Pawtucket Heart Health Programs and Project LEAN.[25,26]

Constructs or concepts from various behavioral theories aided the design and development of the 5 A Day program. These theories, such as social cognitive theory, consumer information processing, health belief models, and stages of change, can be extremely valuable in building models and establishing measures which identify the diverse factors that affect behavior.[27-29] These models were used to design the different program components, as discussed below.

With national industry participation and the involvement of nearly all state and territorial health agencies, the national program has the potential for reaching all Americans through supermarkets, restaurants, and other specific channels, such as schools, worksites, food assistance programs, churches, and community organizations.

PROGRAM COMPONENTS

The National 5 A Day Program consists of four components: media, retail, community, and research.

Media

The purpose of the media effort is to increase consumer awareness of the program message: "Eat 5 or more servings of fruits and vegetables every day for good health." This component is implemented in a complementary fashion by NCI's Office of Cancer Communications and the Produce for Better Health Foundation. Annual plans consist of events throughout the year to keep the message visible to the public through the media. Press conferences are used to focus on topics or announce new findings, such as the release of the 5 A Day baseline survey in 1992. Special 5 A Day events, such as the World's Largest Fruit Basket, involve the local community and create media interest. (Produce left over from events is supplied to food banks.) Media events may also feature celebrity spokespeople, such as Olympic champions, media tours and public service announcements. One effort included work with the National Football League to appeal to males and to demonstrate that increased consumption of fruits and vegetables is part of the training regimen of football players.

In addition, 5 A Day Week was established, several years ago, as the second week in September and provides the opportunity for all 5 A Day participants to work together to increase the synergy of their efforts. Governors of nearly all 50 states signed Proclamations in support of the 5 A Day effort in 1993, 1994, and 1995.

The theme for the 3rd Annual National 5 A Day Week, September 10-16, 1995, was "Take the 5 A Day Challenge!". The strategy was to mobilize Americans to challenge each other to eat at least 5 servings of fruits and vegetables every day of 5 A Day Week. The intent was that once Americans start eating 5 A Day for one week, they will see how easy it can be to make a simple change for better health. On the national level, PBH and NCI showcased individuals or groups (including media personalities) who took the 5 A Day challenge. A key feature of the challenge concept was to move beyond awareness of the importance of eating 5 A Day, to provide an opportunity for people to practice a new behavior with community support.

Retail

The purpose of the retail component of the program is to reach consumers with informational and motivational messages at the point of purchase. Partnership with industry brought with it the willingness of supermarkets and smaller groceries to participate in the program. Since people in all ethnic, educational, and economic strata shop in grocery stores, this channel provides potential access to the entire U.S. population. More than 35,000 stores across the country are currently licensed to participate in the program, although many more need to be encouraged to participate. Three times a year, the stores receive promotional kits of print-ready materials from the national program (PBH). The kits focus on specific themes, such as "Fast and Easy" and "Fitness" and include new brochures, posters, and copy for use in weekly newspaper ads. Participating stores agree to display materials and attempt to provide interactive events, such as supermarket tours and taste tests. They are encouraged to work with their local health agency to provide the latter.

In addition, industry participants, such as Dole, Sunkist, and the Washington Apple Commission have developed brochures, curricula, kits for conducting supermarket tours, and the CD-ROM "5 A Day Adventures" for children in grammar school. The latter is an outstanding application of the latest technology to nutrition education that makes it fun and memorable for children. These innovations by various industry participants add to the attraction and reach of the program.

The 5 A Day Program is beginning to work with restaurants as another important point of purchase for the consumer. More details of how this component operates will be available as experience is gained in working with this sector of the industry.

Community

The purpose of the community channel is to bring the program to life at the local level, tailoring the nutrition messages and strategies to make them relevant to people in their daily lives and in their own communities. Activation of the community channel provides opportunities to supply the skills, social support, and environmental support components of a program necessary to help create new behaviors.

The community component is structured in the following manner: State Health Agencies are licensed by the NCI to develop state-wide 5 A Day programs, coordinating all 5 A Day activities in the state. States are encouraged to develop state or local coalitions as a forum for collaboration between the public and private sectors. Members might include state departments of education, agriculture, and welfare; cooperative extension, voluntary agencies, hospitals, cancer centers, food banks, and licensed 5 A Day industry participants. The purposes of the collaboration are to increase effectiveness in reaching consumers, use scarce resources for maximum effect, coordinate state media efforts with national efforts, encourage creativity, and to create working relationships between the public and private sectors at the state and local levels.

The NCI has licensed 52 of the 56 State and Territorial Health Agencies to coordinate 5 A Day activities at the state level, using various community channels to effectively reach the public. The flexibility afforded to these licensees with regard to the amount of involvement and types of activities allowed permit them to devise very innovative approaches for implementing their state-based initiatives. They are also encouraged to use their ingenuity in developing supportive materials, within certain guidelines.

NCI, in collaboration with the Center for Disease Control and Prevention, provides training and technical assistance for the State Health Agencies on strategies for implementing the 5 A Day program at the local level. The Centers for Disease Control have also provided funds for implementation of dietary interventions, including 5 A Day activities, by the State Health Agencies.

Research and Evaluation

The evaluation objectives for the 5 A Day Program include both outcome and process evaluation components. The outcome evaluation objective are to 1) measure changes in population awareness, knowledge, stages of change and median consumption of fruits and vegetables between the baseline survey in 1991 and the follow-up survey in 1996; 2) determine the effect of the program on target populations in specific channels, such as worksites, schools, churches, and food assistance programs through the implementation of nine grants with randomized designs; and 3) develop a series of common questions for use by grantees and by other licensees to measure program impact.

The process evaluation objectives are to 1) document program implementation between 1991 and 1996 in each program component (i.e. industry, states, and media); and 2) develop substudies which correlate program outcomes with measures of program implementation (i.e. case studies and comparisons using an index that measures intensity of intervention).

The NCI has included questions on 5 A Day in omnibus telephone surveys which revealed an increased awareness of the need to eat at least 5 servings of fruits and vegetables daily. At baseline in the Fall of 1991, only 8% of respondents thought they should eat 5 or more servings for good health. This percentage increased to 22% in 1992 and 29% in 1993. The proportion of the population who stated they were aware of The 5 A Day Program increased from 10% in 1992 to 15% in 1993. These surveys, however, were conducted immediately following large media efforts that may overestimate the sustained awareness of the general population.

PROGRAM MATERIALS

The 5 A Day Program has developed a number of brochures, posters, point of sale signs, ad copy, and other materials for use by the industry and health licensees. The program encourages use and wide-spread dissemination of such materials to facilitate implementation of various campaigns or simply to raise awareness of the 5 A Day message. For example, in 1993, participants (mostly industry members) bought more than 65,000 5 A Day posters and 2.9 million brochures to get the 5 A Day message to consumers. Approximately 600,000 of the 5 A Day brochures have been made available, free of charge, to State Health Agencies for use in their programs. State Health Agencies are also working closely with the Cancer Information Service (CIS) Outreach Coordinators to obtain copies of these and other materials for the program. The Produce for Better Health Foundation assembles promotional kits that are made available to each licensee which contains sample materials for each promotional theme conducted during the year. Additional copies of promotional materials (including pamphlets, T-shirts, hats, videos, PSA's, and stickers) can be obtained through TryFoods and Remline Corporations. Single copies of 5 A Day brochures produced by NCI are available to consumers through NCI's 800-4-CANCER number.

The State Health Agency licensees develop school curricula, fliers, posters, press kits, videos and other materials for use in their initiatives in various channels. The NCI has established a national clearinghouse of state-developed materials, to allow for material exchange among all 5 A Day participants.

CONCLUSION

The national 5 A Day program is now approaching the final year of its first five years of implementation. With collaborative participation of health and industry licensees, the program message is being enthusiastically delivered to Americans through all channels. Strategic planning for the subsequent five years is underway. The next five years will feature the collection of data from research grant projects; assessment of program implementation; expansion of the national program to commercial foodservice, cross promotions, and to special populations; and continued growth on the community level in various channels.

The program has been viewed as a model for other nutrition efforts, and approximately 15 countries have requested information on how to implement the program in their countries.

As research on phytochemicals continues, the 5 A Day Program can help steady the population's progress in consuming more fruits and vegetables, and thereby a mix of phytochemicals. The program will be a laboratory for improving methods of behavior change operating parallel to the advances being made in phytochemical research. I invite all of you researchers to be involved at the state and local levels with your own state 5 A Day coalition. The name and phone number of your state coordinator can be obtained from the National Cancer Institute at (301) 496-8520.

REFERENCES

1. Block G, Patterson B, and Subar A. Fruit, vegetables, and cancer prevention: a review of the epidemiological evidence. Nutr and Cancer 1992; 18(1):1-29.
2. Steinmetz KA and Potter JD. Vegetables, fruit, and cancer. I. epidemiology. Cancer Causes and Control 1991; 2:325-357.
3. Steinmetz KA and Potter JD. Vegetables, fruit, and cancer. II. mechanisms. Cancer Causes and Control 1991; 2:427-441.
4. Ziegler RG. Vegetables, fruits, and carotenoids and the risk of cancer. Am J Clin Nutr 1991; 53:251S-259S.
5. Ziegler RG, Subar AF, Craft NE, Ursin G, Patterson BH, and Graubard BI. Does beta-carotene explain why reduced cancer risk is associated with vegetable and fruit intake? Cancer Res 1992; 52:2060S-2066S.
6. Willett WC. Vitamin A and lung cancer. Nutr Reviews, 1990; 48:5, 201-211.
7. Negri E, LaVecchia C., Franceschi S, D'Avanzo B and Parazzini F. Vegetable and fruit consumption and cancer risk. Int J Cancer, 1991; 48:350-354.
8. Patterson B, personal communications (1993).
9. Krebs-Smith S., personal communications (1994).
10. U.S. Department of Agriculture and U.S. Department of Health and Human Services. Nutrition and Your Health: Dietary Guidelines for Americans. 1990: Home and Garden Bulletin No. 232. U.S. Government Printing Office: Washington, D.C.
11. U.S. Department of Agriculture. USDA's food guide pyramid. 1992. Home and garden bulletin 249. U.S. Government Printing Office: Washington, D.C.
12. National Academy of Sciences. Diet and health, implications for reducing chronic disease risk. 1989. National Academy Press: Washington, D.C.
13. Subar AF, Heimendinger J, Patterson BH, Krebs-Smith SM, Pivonka E, and Kessler R. 5 A Day for Better Health: A Baseline Study of Americans' Fruit and Vegetable Consumption. Division of Cancer Prevention and Control. National Cancer Institute: Rockville MD.
14. U.S. Department of Health and Human Services. Healthy People 2000: National Health Promotion and Disease Prevention Objectives. DHHS Publication No. (PHS) 91-50212. Washington, DC: Government Printing Office.
15. Levy, A. and Stokes, R. 1987. Effects of a health promotion advertising campaign on sales of ready-to-eat cereals. Public Health Rep 102:398-403.
16. Davis, R. 1988. Health education on the six-o'clock news. JAMA 259:1036-1038.
17. Russo, J., Staelin, R., Nolan, C., Russell, G., and Metcalf, B. 1986. Nutrition information in the supermarket. J of Consumer Res 13:48-69.
18. Hammond, S. 1986. Health advertising: the credibility of organizational sources. Paper presented to International Communication Association's Annual Meeting, Health Communication Division, Chicago, IL.
19. Flay, B. R. 1987. Mass media and smoking cessation: A critical review. Am J Public Health 77:153-60.
20. Farquhar, J., Maccoby, N., and Wood, P. 1977. Community education for cardiovascular disease. Lancet 1:1192-1195.
21. Puska, P., Wiio, J., McAlister, A., Koskela, K., Smolander, A., Pekkola, J., and Maccoby, N. 1985. Planned use of mass media in national health promotion: The Keys to Health TV program in 1982 in Finland. Can J Public Health 76:336-42.
22. Stern, M., Farquhar, J., Maccoby, N. and Russell, S. 1976. Results of a two-year health education campaign on dietary behavior. Circulation 54:826-833.
23. Puska, P., Nissinen, A., Salonen, J., and Tuomilehto, J. 1983. Ten years of the North Karelia project: results with community-based prevention of coronary heart disease. Scand J Soc Med 11:65-68.
24. Farquhar, J., Fortmann, S., Flora, J., Taylor, B., Haskell, W., Williams, P., Maccoby, N., and Wood, P. 1990. Effects of community-wide education on cardiovascular disease risk factors. JAMA 264:359-365.
25. Shea, S., and Basch, C.E., 1990. A review of five major community-based cardiovascular disease prevention programs. Part II: Intervention strategies, evaluation methods, and results, Am J Health Promotion 4:279-287.
26. Samuels S. 1993. Project LEAN-Lessons learned from a national social marketing campaign. Public Health Reports 108:45-53.
27. Achterberg CL, Novak JD, Gillespie AH. Theory-driven research as a means to improve nutrition education. J Nutr Educ 1985;17:179-84.
28. Sims LS. Nutrition education research: reaching toward the leading edge. J Am Diet Assoc 1987;87 (suppl):10-8.

29. Glanz K, Lewis FM, and Rimer BK (eds). Health Behavior and Health Education. Theory, Research, and Practice. 1991. Jossey-Bass: San Francisco, California.

18

NUTRITIONAL IMPLICATIONS OF DIETARY PHYTOCHEMICALS

Mark Messina and Virginia Messina

Nutrition Consultants
1543 Lincoln Street
Port Townsend, Washington 98368

THE GOLDEN AGES OF NUTRITION

It has been only during the past few decades that nutritionists have become concerned about the relationship between diet and chronic disease. For most of this century, the nutrition community focused solely on preventing nutrient deficiency diseases. Between 1910 and 1950, a period often referred to as the *Golden Age of Nutrition*, all of the known vitamins were identified. While these discoveries qualify as some of the most important scientific achievements in the field of nutrition, there were some unfortunate consequences to this very productive period. One, was that the search for new dietary components largely ended. Perhaps nothing better illustrates this point than a decision by Oxford University during the 1940s not to approve the creation of a nutrition department because it was thought that all of the known dietary factors had been discovered and the essential problems related to nutrition had been resolved.[1]

Of course, it is now known that plant foods contain hundreds, perhaps thousands, of biologically active components that may affect chronic disease risk. The recognition of these plant chemicals or phytochemicals, has ushered in a new era in the field of nutrition, which we refer to as the *Second Golden Age of Nutrition*. Phytochemicals are, in a sense, the vitamins and minerals of the 21st century. The Second Golden Age of Nutrition may be thought of as formally beginning in 1982 with the publication of *Diet, Nutrition and Cancer*, a landmark report by the National Academy of Sciences on the relationship between diet and cancer.[2]

Although this report focused on the need to reduce fat intake and increase fiber intake, there was also specific mention of the possible anticancer effects of nonnutritive components. However, most attention was on the phytochemicals in cruciferous vegetables. This, in large part, can be attributed to the pioneering work of Dr. Lee Wattenberg, who was studying the possible anticancer effects of glucosinolates in cabbage, broccoli, and Brussels sprouts [3] Since that time, the wide array of phytochemicals with possible anticancer and other beneficial effects has been recognized.[4]

PHYTOCHEMICALS AND THEIR IMPACT ON DIETARY GUIDELINES

Without question, nutrition professionals need to focus not just on the fiber and nutrients in foods, but also on the phytochemicals. Although our understanding of the biological effects of dietary phytochemicals is still in its infancy, it is clear that these effects have profound implications for the nutrition community. Nutrition training will need to focus more heavily on scientific disciplines that more easily permit understanding of the dietary phytochemicals. Of more immediate concern is whether our knowledge of phytochemicals should change the message we communicate to consumers.

To address this question, we need to consider past and current government dietary guidelines. All of the guidelines issued during the past twenty years have focused on the need for Americans to reduce their intake of high-fat animal products, and to increase their intake of fruits, vegetables, beans and grains. Although they do not suggest that Americans adopt a vegetarian diet, these guidelines do recommend that Americans shift toward a more plant-based diet. Thus, these guidelines are generally consistent with the recognition that plant foods possess potentially beneficial phytochemicals.

Currently, the primary nutrition education instrument for teaching consumers is the USDA's Food Guide Pyramid which continues to emphasize the importance of plant foods and represents an important step forward; it replaces the Four Food Groups, which was better suited for teaching how to prevent nutrient deficiency diseases. The Food Guide Pyramid visually illustrates the relative amounts of foods that should comprise a healthy diet. Grains are the foundation of the diet, followed by vegetables, then fruits, and then dairy and meat (the meat group includes meat, poultry, fish, eggs, dried beans, seeds and nuts) (Table 1). Fat and sweets are at the top of the pyramid and are listed as optional and are to be used sparingly. The Food Guide Pyramid still allows for generous portions of animal foods. One may consume up to 24 oz of milk (or equivalent servings of other dairy products) and 9 oz of meat per day and still conform to the guidelines of this teaching tool.

In our opinion, this pattern is inconsistent with healthy eating, and represents a continued over-reliance on animal foods. We suggest that the food guide pyramid would be more effective in teaching the principles of optimal nutrition with the following revisions (table 1): 1) change the recommended servings from the dairy group from 2-3 to 0-2, thereby making these foods optional. A concomitant emphasis on plant foods that are rich in calcium is necessary to help consumers meet calcium needs 2) develop a separate food group for beans, nuts and seeds since these foods have different nutrient and phytochemical profiles from meat, poultry, fish and eggs 3) reduce the recommended servings for the new meat

Table 1. Current and suggested recommended number of servings to be consumed from each food group

Food Group	USDA Food Guide Pyramid (Servings/day)	Suggested Food Guide Pyramid (Servings/day)
Grains	6-11 servings	6-11 servings
Vegetables	3-5 servings	3-5 servings
Fruits	2-4 servings	2-4 servings
Dairy	2-3 servings	0-1 serving
Meat (meat, poultry, fish, eggs, legumes, nuts, seeds)	2-3 servings	NA*
Meat (meat, fish, poultry, eggs)	NA*	0-1 serving
Legumes, nuts, seeds	NA*	1-3 servings

*not applicable

group from 2-3 to 0-1 and recommend the consumption of 1-3 servings per day from the beans, nuts and seeds groups. These types of changes would automatically result in diets that are lower in saturated fat, and higher in fiber and phytochemicals.

The first suggestion is to emphasize the importance of obtaining adequate amounts of calcium. In fact, there is a concern among some experts that the current RDA is too low for some groups,[5] although it is likely that calcium needs of Westerners are so high because of our high intake of dietary factors that increase calcium requirements, such as protein and sodium.[6] But calcium needs can be met through the consumption of calcium fortified foods and plant foods. Many plant foods are high in calcium and calcium absorption from plants foods overall is quite good. In the case of green leafy vegetables such as broccoli and kale, calcium absorption is actually better than from milk.[7] An advantage to focussing on plant sources of calcium is that these foods also provide antioxidants, fiber and phytochemicals. Dairy products can also potentially increase total and saturated fat, and cholesterol intake, although low-fat and non-fat dairy products are widely available.

Currently, many young and adult women do not meet the calcium RDA,[8,9] and most Americans do not consume the recommended amount of vegetables.[10] Consequently, any attempt to de-emphasize dairy foods as the primary source of dietary calcium must be accompanied by more vigorous nutrition education efforts to encourage the consumption of plant foods high in calcium.

A greater emphasis on plant sources of protein is also advantageous. For example, legumes are protein-rich foods that are low in fat and cholesterol-free. While nuts are high in fat (mostly unsaturated) they are also rich in fiber and possess an interesting array of phytochemicals. There is little reason to expect that reducing meat intake will lead to nutrient deficiencies and it is clear that three servings of meat are excessive even for those concerned about meeting iron and zinc needs.[11]

CHANGING VIEWS OF DIET AND FOODS

There are several other implications of the Second Golden Age of Nutrition that need to be considered and that change the traditional view of diet. First, it is no longer appropriate to evaluate foods solely on the basis of their fiber and nutrient content. Individuals with similar nutrient intakes may be at markedly different risks for chronic disease as a result in differences in phytochemical intakes (Table 2).

Although adding a fiber supplement to skim milk is one way to achieve a low-fat, high-fiber diet, it is not the same as consuming whole grains and beans, which are also low in fat and high in fiber but also contain numerous phytochemicals. More than ever, it is important to focus not on macronutrients, micronutrients or even phytochemicals, but on *foods*. Much of our understanding about the relationship between diet and disease is based on observations of populations. It is difficult to specifically identify the food components in

Table 2. Nutritional implications of dietary phytochemicals

1. Foods can no longer be evaluated solely on the basis of their macronutrient and fiber intakes, therefore diets comprised of similar levels of fiber, macronutrients and micronutrients may be associated with different chronic disease risks.
2. Some foods may warrant special emphasis in the diet because of their unique or relatively unique phytochemical composition
3. Some foods, in addition to contributing to nutrient intake, may now be viewed as important for their medicinal properties.

a given diet that are most associated with disease risk, particularly in the case of chronic diseases such as cancer.

The second implication of this new age of nutrition is that some foods may warrant special emphasis in the diet because of their unique or relatively unique phytochemical composition. In one sense, this conflicts with traditional dietary advice. When considering only nutrients, there are really no instances where one food needs to be specifically emphasized since nutrients can be obtained from a wide variety of foods. The case is different for phytochemicals, however. For example, soybeans and soyfoods are essentially unique sources of isoflavones[12] and although many grain products contain lignans, flax seed is an extremely rich source, containing many times higher levels than other foods.[13]

Finally, some foods may be viewed primarily for their medicinal properties or therapeutic effects. For example, soy protein is hypocholesterolemic in hyper-cholesterolemic individuals [14] and soyfoods, because of their phytoestrogen content, may be able to relieve menopausal symptoms.[15] Historically, certain foods have been used much like medicine or drugs. The phytochemicals have given nutritionists a scientific basis for understanding the possible effectiveness of this approach. Dietitians will undoubtedly now be faced with situations in which they will be asked to advise clients on the effectiveness of using certain foods as medicinal agents.

PHYTOCHEMICAL SUPPLEMENTS AND PHYTOCHEMICAL FOOD FORTIFICATION

There are three issues arising from developments in the food and dietary supplement industries in relation to phytochemicals that will also impact professionals providing nutritional advice (Table 3). First, phytochemical supplements are increasingly available and it is certain that this trend will continue. Nutrition professionals will be called upon to counsel clients using these supplements and will be faced with addressing these issues even while our understanding of phytochemicals is so limited.

Second, there is the likelihood that foods will be fortified with phytochemicals or with food extracts highly concentrated in certain phytochemicals. We will be called upon to answer questions about the advisability of developing and using such foods. The fortification issue has already been debated in relation to vitamins and minerals and products are already on the market. For example, some brands of orange juice are fortified with calcium because calcium intake for many groups is below the RDA. Fortification is in a sense, an acknow-ledgement that people will not eat foods in the appropriate amounts to satisfy nutrient requirements. Is it desirable for orange juice to be fortified with indole-3-carbinol because people may not eat enough cruciferous vegetables? It may in the end be true that people will not adhere to dietary guidelines regardless of the benefits. But thus far, there has not been enough time, money and resources encouraging people to do so, to determine whether or not this is in fact the case.

Finally, through traditional breeding and newer methods such as genetic engineering, it is possible to increase markedly the phytochemical content of foods that naturally contain

Table 3. Issues related to phytochemicals facing nutrition professionals

1. Phytochemical supplements
2. Foods fortified with individual phytochemicals or with food extracts concentrated in particular phytochemicals.
3. Foods with markedly increased levels of phytochemicals naturally present in those foods.

those substances. If foods are marketed on the basis of their higher phytochemical content, should dietitians encourage consumption of those foods?

In our opinion, understanding of the possible benefits and risks of individual phytochemicals is far too limited to encourage the use of phytochemical supplements, or of foods fortified with phytochemicals. It is worthwhile noting that the consumption of individual phytochemicals in isolation may exert different physiological effects, both with respect to risks and benefits, than the consumption of a similar amount of a phytochemical via food that naturally contains that phytochemical. It is also easy to envision that supplements or foods fortified with phytochemicals could lead to over consumption of phytochemicals, defined here as levels that exceed the intake of a phytochemical reasonably possible by consuming foods naturally containing that phytochemical.

SUMMARY

Although increasing the levels of phytochemicals in foods via traditional breeding or genetic engineering would appear to pose less of a risk, there is still the possibility of over consumption given the ease with which phytochemical content can be increased.

According to the recent position paper by the American Dietetic Association on phytochemicals, "the dietetics professional ... is the specialist who should make recommendations concerning appropriate dietary intake to optimize the potential benefits of phytochemical-rich or functional foods in overall health."[16] For dietitians to make these recommendations, it will be necessary for the nutrition community to make a concerted effort to incorporate information on phytochemicals into nutrition manuals and professional resources. As it is, most of the research and discussion of phytochemicals is conducted by professionals not related or only peripherally related to the field of nutrition.

This is truly an exciting time to be involved in the nutrition field. We have moved ahead from thinking that only a relatively small number of dietary factors possess biological activity to recognizing that there are hundreds and perhaps thousands of such factors. It will be many years before our understanding of phytochemicals approaches our knowledge of vitamins and minerals — and it is worth noting that our knowledge of vitamins and minerals is still incomplete. But it does appear that the phytochemicals help explain why plant-based diets, in general, are associated with a reduced risk of many chronic diseases. And in fact, this may be the most important outcome from use of phytochemicals. The nutrition community now has more reasons for encouraging consumers to eat plant-based diets.

REFERENCES

1. Hegsted DM. A look back at lessons learned and not learned. J Nutr 124:1867S-1870S, 1994.
2. The Committee on Diet, Nutrition and Cancer, Assembly of Life Sciences, National Academy of Sciences. Diet, Nutrition, and Cancer. National Academy Press, Washington, DC, 1982.
3. Wattenberg LW. Inhibition of neoplasia by minor dietary constituents. Cancer Res (suppl) 43:2448s-2453s, 1983.
4. Steinmetz KA, Potter JD. Vegetables, fruit, and cancer. I. Epidemiology. Cancer Causes Control 2:325-357, 1991.
5. NIH Consensus Statement. Optimal Calcium Intake. Vol 12, Number 4, June 6-8, 1994, National Institutes of Health.
6. Barger-Lux MJ, Heaney RP. Caffeine and the calcium economy revisited. Osteoporosis Int 5:97-102, 1995.
7. Weaver CM, Plawocki KL. Dietary calcium: adequacy of a vegetarian diet. Am J Clin Nutr 59 (suppl) 1238S-1241S, 1994.

8. USDA (U.S. Department of Agriculture). 1987. Nationwide Food Consumption Survey. Continuing survey of food intakes of individuals. Women 19-50 years and their children 1-5 years, 4 Days, 1985. Report No. 85-4. Nutrition Monitoring Division, Human Nutrition Information Service, Hyattsville, Md, 182 pp.
9. USDA (U.S. Department of Agriculture). 1984. Nationwide Food Consumption Survey. Individuals in 48 States, Year 1977-78. Report No. 1-2. Consumer Nutrition Division, Human Nutrition Information Service, Hyattsville, Md, 439 pp.
10. U.S. Consumption of Fruits, Vegetables, and Grains. J Natl Cancer Inst 87: 477, 1995.
11. Johnson JM, Walker PM. Zinc and iron utilization in young women consuming a beef-based diet. J Am Diet Assoc 92:1474-1478, 1992.
12. Coward L, Barnes NC, Setchell KDR, Barnes S. Genistein, daidzein, and their β-glycoside conjugates: antitumor isoflavones in soybean food from American and Asian diets. J Agric Food Chem 41:1961-1967, 1993.
13. Thompson L. Mammalian lignan production from various foods. Nutr Cancer 16: 43-?, 1991.
14. Anderson JW, Johnstone BM, Cook-Newell ML. Meta-analysis of the effects of soy protein intake on serum lipids. N Engl J Med 333:276-282, 1995.
15. Murkies AL, Lombard C, Strauss BJG, Wilcox G, Burger HG, Morton MS. Dietary flour supplementation decreases post-menopausal hot flushes: effect of soy and wheat. Maturitas 21:189-195, 1995.
16. Position of the American Dietetic Association: phytochemicals and functional foods. J Am Diet Assoc 95:493-496, 1995.

19

DESIGNER FOODS

Is There a Role for Supplementation/Fortification?

John W. Finley

Nabisco, Inc.
East Hanover, New Jersey 07936

DESIGNER FOODS

The concept of designer foods is clearly growing rapidly in the United States. There are now several meetings a year and almost all food industry meetings discuss designer or functional foods. The definition of these foods and the nomenclature is often somewhat ambiguous. What we will refer to as designer foods have been described by a wide variety of terms. Table 1 lists some of the terms that have been suggested to describe foods or fortified foods with disease preventative attributes. The term hyper-nutritious has been used to describe foods fortified with ingredients that take the food beyond what might be considered normal nutrition. For the purpose of this discussion I will use the term hypernutritious to describe foods that are supplemented in some way to improve their disease preventative characteristics. We will define hypernutritious foods as foods that have been augmented to provide preventative or health improving activity, specifically, with emphasis on ingredients or components with cancer preventative activities. The question then becomes should foods be supplemented with these bioactive materials, and if so, what and how should this be accomplished. Interest in these foods is coming from the consumer, the medical community and the research community.

The United States food industry has been somewhat slow in embracing and incorporating materials with pharmacological effects into foods. This reticence, however is changing because of consumer demand. Consumer interest in foods that protect health or enhance performance is driven by the over all goals of a healthier life style. The consumer is now becoming convinced that diet and exercise can significantly impact health and quality of life. The increased cost of health care is causing much greater interest in both preventative and curative components found in foods or related natural products.

Many consumers are "taking charge" of their health and well being. In part, this is being accomplished by controlling the diet in new and different ways. Now diet is not just fuel; the role of food is to make the consumer feel better, perform better longer, and to have improved quality of life for extended lifetimes.

The food industry is the keeper as well as the provider of our nation's food supply. Our diet has evolved as being the healthiest in the history of mankind. We in the food industry,

Table 1. Names for functional foods

Medicinal Foods	Superfoods
Neutraceuticals	Longevity Foods
Functional Foods	Pharmaceutial Foods
Prescriptive Foods	Pharma Foods
Neutraceutics	Designer Foods
Nutritional Foods	Medifoods
Fitness Foods	Foodiceuticals
Therapeutic Foods	

as "keepers" of our nation's food supply, are entrusted with providing adequate nutrients to ensure good health and wellness for everyone. We approach this social responsibility seriously, and with deliberate thought. In fact, food in modern America has been more abundant, varied, and accessible than at any other time in our history, thanks in large part to industry initiatives in agriculture, in food processing, and in distribution techniques.

Nutrition has taught us how to formulate and use foods which prevent deficiency diseases. In most parts of our country we do not see beri beri, pellagra or scurvy. The 20th century has seen significant increases in the average lifespan of Americans, coincident with decreases in both nutrient deficiency diseases and even chronic, degenerative diseases. Flour enrichment, cereal enrichment, salt iodization, vitamin D supplementation in milk, and reductions in fats, cholesterol, and sodium across product lines have all played a major role in these health trends.

Our post-modern society is on the verge of entering into a new relationship with the food it eats. There are reports that our Recommended Dietary Allowances are high enough for us to avoid deficiency disease states, but some may be too low for us to achieve our full health potential.[1] There is growing evidence that other phytochemicals beyond the scope of traditional vitamins may be important in preventing chronic disease; thus, we are revisiting fortification.

We are now learning how various components in foodstuffs, particularly phytochemicals, can have a significant impact on chronic disease. We are at the rather exciting point of beginning to identify which components may be active in chronic disease prevention. This has evolved into an interesting and refreshing change in the way researchers look at food. Ten or fifteen years ago the literature had many papers pointing out all the carcinogens and antimetabolites in foods. The result was a public perception that all food was harmful, particularly processed food. Now we are seeing food as a curative, not a toxicant. This promise can be almost as dangerous as the concerns raised a decade ago. While we are all very excited about the potential of phytochemicals to enhance the quality of life and prevent chronic disease, we must be cautious not to over promise the public with prevention or cures that are not carefully documented and substantiated by carefully controlled clinical research.

It has long been known that fruits and vegetables were healthy foods. Generally nutritionists have considered them as excellent sources of vitamins, minerals and fiber. More recently we have seen enormous interest in the phytochemicals in plants which appear to offer protection from chronic disease. The support for the benefits of various plant sources range from folklore, to anecdotal, to epidemiological, to hard scientific evidence.[2] Table 2 summarizes the results of multiple epidemiological studies reporting the cancer preventative vs cancer causing effects from consumption of fruits and vegetables. The point of the table is that the overwhelming body of evidence suggests that plant tissue can help improve resistance to various types of cancer. Identification of active components from a variety of plant sources has been documented elsewhere in this symposium. It is clear that there is justification for consumption of plant materials which will deliver bioactive phytochemicals along with fiber, vitamins and minerals. We may also wish to consider the likelihood that

Table 2. Epidemiological studies of fruit and vegetable intake and cancer prevention

Cancer Site	Protect	Harmful
All Sites	132	6
All Sites Except Prostate	128	4
Lung	24	0
Larynx	4	0
Oral Cavity, Pharynx	9	0
Esophagus	15	0
Stomach	17	1
Colorectal	20	3
Bladder	3	0
Pancreas	9	0
Cervix	7	0
Ovary	3	0
Breast	8	0
Prostate	4	2

the potential for cancer prevention comes from a combination of the components in plants and it is less likely that there is a magic bullet.

DIETARY GUIDELINES

The USDA has used a food pyramid to suggest dietary guidelines as shown in Figure 1. These guidelines are to reduce fat consumption and increase fruits, vegetables and cereal based products in the diet. From the pyramid it can be seen that modest consumption of meat and full fat dairy products is suggested. In addition it suggests consumption of three

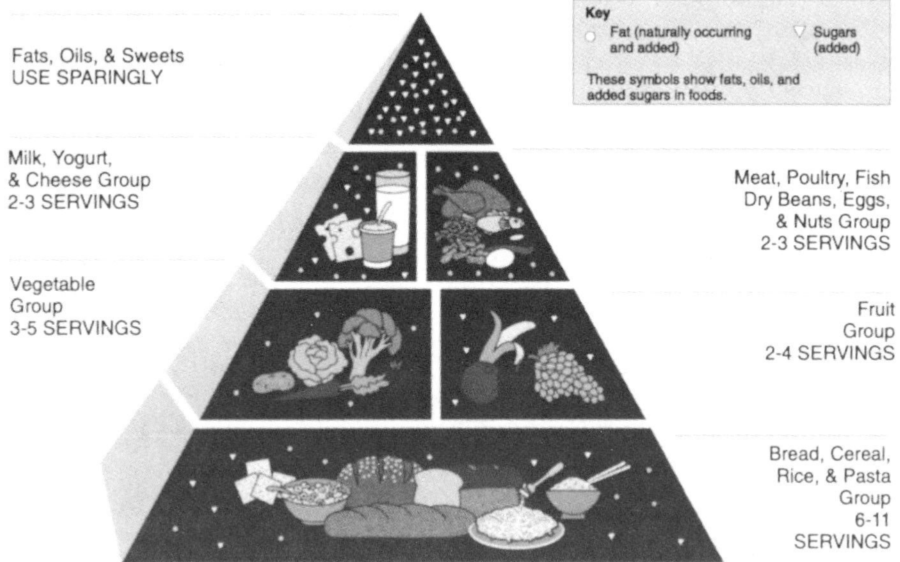

Figure 1. USDA food pyramid: a guide to daily food choices.

servings of vegetables and two servings of fruit per day. How are we doing as a nation in reaching that goal? Using data from the 1989-90 food consumption survey, Patterson *et al.*[3] report the patterns for fruit and vegetable consumption, but it can be seen that consumption is far below the suggested guidelines. Not suprisingly, when the total daily fruit and vegetable consumption is considered the results are even more discouraging. The Table 3 indicates that 25 to 30% of the population meet the guidelines for fruit and vegetable consumption and that 38% of the total population meet one or the other guidelines. Only 9% of the population were meeting both goals in 1990. Hopefully this trend is changing and will be seen in the next food survey. The 53% of the population that do not meet the guidelines represent both concerns and opportunities. If this group can be reached through nutrition education programs sponsored by organizations such as AICR, schools, the popular press, and government agencies and taught to consume more fruits and vegetables, it is very likely our national health can be significantly improved. There is also great opportunity for manufacturers to provide healthier products if the consumers will purchase them. Products made with and derived from plants will contain the phytochemicals that provide chemoprotection from disease. As an example, the National Cancer Institute is currently investigating several classes of phytochemicals for anti-cancer activity. The various classes of compounds and the plants from which they are derived are presented in Table 4. This table illustrates the enormous task of trying to understand the biochemistry of a large number of classes of materials from a variety of sources. Each chemical class can have from dozens to hundreds of chemicals. The entire field is complicated by the fact that none of these materials appear alone. Ultimately the interactions with food, food processing and other food components must be layered on top of this already complex set of investigations. The early results many discussed in this symposium are clearly showing the benefit of the phytochemicals in prevention of cancer cell growth. Caragay[4] proposed a pyramid that develops a hierarchy of potential cancer preventative foods, as illustrated in Figure 2. From the figure it can be seen that garlic, cabbage and licorice appear to have tremendous potential as sources of anti-cancer phytochemicals. Simply studying this chart one can imagine opportunities to develop a variety of new or unique food products which would take advantage of current food materials as useful ingredients in providing the benefits of hypernutritious foods. Soy, for example is rich in genistein which seems to be well documented for chemo-prevention (Barnes, this volume) and soy fractions offer great potential as food ingredients. There is particular advantage in soy protein isolate since the genistein is carried throughout the isolation process with the protein. It is interesting to speculate on the possibility of extending hamburger by

Table 3. Percentage of individuals meeting dietary guidelines

Consumption	Percent of Population
Vegetables (three or more)	
None	22
1 serving	26
2 servings	26
3 or more	27
Fruits (two or more)	
None	45
1 serving	28
2 or more	29
One of the guidelines	38
Both guidelines	9
Neither	53

Table 4. Potential anticancer phytochemicals under NCI investigation

	SULFIDES	PHYTATES	FLAVONOIDS	GLUCARATES	CAROTENOIDS	COUMARINS	MONO TERPENES	TRI TERPENES	LIGNANS	PHENOLIC ACIDS	INDOLES	ISOTHIOCYANATES	PHTHALIDES	POLYACETYLENES
GARLIC	X						X	X		X				
GREEN TEA			X	X		X				X				
SOYBEANS		X	X		X	X		X	X	X				
CEREAL GRAINS		X	X	X	X	X		X		X				
CRUCIFEROUS	X		X	X	X	X	X	X		X	X	X		
UMBELLIFEROUS			X		X	X	X	X		X			X	X
CITRUS			X	X	X	X	X	X		X				
SOLANACEOUS			X	X	X	X	X	X		X				
CUCURBITACEOUS			X		X	X	X	X		X				
LICORICE ROOT			X		X	X		X		X				
FLAX SEED			X			X			X	X				

30% with hydrated soy isolate. Making a quarter pound hamburger with 30% replacement of hamburger by soy would result in delivery of 30 mg of genistein and a fat reduction of 5 g per serving. The advantages would include lower fat, lower saturated fat, a good source of the potential cancer preventative compound genistein, and a burger that is indistinguishable in taste from the full fat burger. This is just one example of how ingredients rich in chemo-preventitive phytochemicals are currently available and could be used in mainstream

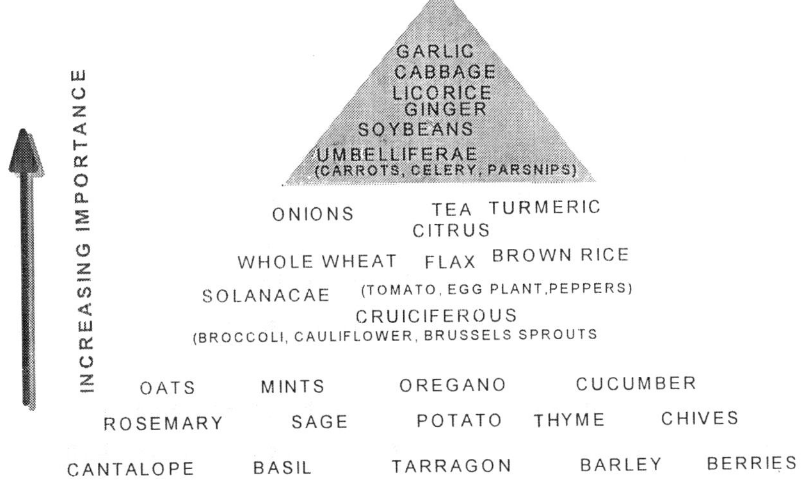

Figure 2. Possible cancer preventing foods and ingredients (Caragay, 1992).

Table 5. Benefits of hypernutritious foods

Better Nutrition
Improved Quality of Life
Disease Prevention
Better Performance
New Markets
New Business
Lower Health Care Cost

food products without going to extravagant dietary changes or introducing new components into the food supply.

Let's consider a dietary change including increased consumption of cereal, fruit and vegetable foods. Some of the benefits of consumption of increased fruits and vegetables or hypernutritious foods are listed in Table 5. First, increased fruits and vegetables would provide better overall nutrition supplying higher levels of dietary vitamins, minerals and fiber. The improved nutritional status of the individual and the population will improve overall quality of life. The fact that such a diet change will inevitably result in increased consumption of chemo-preventative phytochemicals, disease prevention, or delay of onset, would be observed in the population as a whole. From the perspective of the food processor the development of such a opens rings the door of opportunity. First we must acknowledge that the public is not likely to change dietary habits dramatically overnight. Therefore the processed food industry should and is evolving toward healthier foods. This currently is translated as lower fat. The public is responding! Consider the commercial success of Con Agra's Healthy Choice and Nabisco's SnackWell's products.

The next challenge is to provide products rich in phytochemicals in a form that will be eaten. Despite George Bush's renouncing broccoli, the consumption of this vegetable is increasing dramatically, which might be considered a small step for nutritional improvement. In reality one only needs tour the local supermarket to see the increased emphasis on fruits, vegetables and lower fat products. While much of the population has not yet accepted the message, the signs are clearly encouraging.

With the increased consumer interest in eating healthier, what about fortification? Fortification is not new; fortification of milk and vitamin fortification of cereals are well accepted. Calcium is an example of a nutrient where fortification of foods offers new opportunities and significant health impact by prevention or delay of osteoporosis (NIH, 1994).[5]

For example, when diet and health conferences and NIH consensus documents alert us to the need to pay attention to calcium nutrition throughout life—that is, to build a large peak bone mass early on and to minimize bone mass depletion later on—the food industry acknowledges the responsibility to do its part. It is important to emphasize that the need for calcium and the benefits of supplementation are clearly demonstrated. Similar demonstrations of need and efficacy of supplementation are necessary before the fortification or extensive use of dietary supplements should be considered. Examples where supplementation is currently justified based on need and data are tocopherol, β-carotene and omega 3 fatty acids.

It's a fact of commercial life that the food industry is consumer-driven and must be responsive to consumer demands. If consumers say "give us an added health benefit," then someone, somewhere will oblige. In general, this has proven to be a good dynamic, from both the financial and the public health points of view.

But this dynamic can also open up the risk that marketing decisions will be made based on fashion and perceived need rather than on consensus science. Marketers may be motivated to titrate the desired effect—to increase the amount of a compound added to products as long as the effects include increasingly energized sales and maximized profits. There may not always be a cautious regard for safe upper consumption limits. As we see increased interest in bio-active phytochemicals, care must be taken to clearly document the efficacy. Beyond efficacy, however, an appropriate level of safety testing must be conducted. There is always concern about upper safety limits and whether other less desirable components may enter or be increased in the food supply when products are supplemented with extracts or heretofore underconsumed plants.

Here is where an up-dated national fortification policy would be valuable. The current fortification policy of the Food and Drug Administration is geared toward eradicating classical nutrient deficiencies such as beri beri and goiter. The food supplementation act clearly opens the door for supplements and plant extracts. The impact on fortified foods is still not clear. The food and pharmaceutical industries are now left with the question as to where supplements end and fortified foods begin. We need a fortification policy that expands its mandate to include guidelines for the prevention of chronic degenerative disease. Without an up-dated policy, how can we best determine which nutrients are legitimate candidates for fortification? Which fortification levels should companies aim for in the various food categories ranging from cereals, to beverages, to snacks, to condiments or, whether or not certain food categories are inappropriate vehicles for fortification based on their consumption frequency or targeted age groups?

While many of these same questions have been addressed by the FDA over the years, they take on new meaning when applied to issues of 'health optimization' as opposed to 'life maintenance.' It's one thing to modify the food supply on a grand scale to avoid premature deaths and disabilities. It's another thing to do so to keep seventy-year-olds active into their eighties and nineties.

MATERIALS WITH PERCEIVED EFFICACY

The field of designer or hypernutritious foods is complicated by a plethora of anecdotal reports of preventative or curative effects from various foods. One can not discount the placebo effect of materials perceived to be of value, but it is important that real effects be documented with sound scientific data. In Table 5 some of the reported benefits derived from plant sources are listed along with current nature of the evidence. Perhaps the most interesting information from Table 4 is that a large percentage of plants that provide pharmacoactive components are already in the food supply to varying degrees. Although purely speculative at this point, one needs to ask the question, why do plants make these compounds? Many are toxic at various levels (i.e. they kill cancer cells in model systems). One proposal is that some phytochemicals are produced for protection or to discourage the predator. For this reason if any enrichment of phytochemicals is to be considered, the toxicology must be investigated thoroughly. One might consider that a delicate balance has evolved in plants and we do not want to change these chemical balances through genetic manipulation, deliberate enrichment or extraction for fortification. The point is that for each toxin in a plant there may be an chemopreventative agent. Frequently these may be the same compounds.

Where does this morass leave us now? I would like to suggest some alternative approaches short of fortification of foods with pharmacologically active phytochemicals or extensive use of dietary supplements. First, as has been suggested by others (Dwyer, this volume, Messina, this volume) encourage the consumption of plant-based foods including cereals, fruits and vegetables. While consumption of fresh fruits and vegetables is preferable, there also are considerable benefits to be derived from processed products. A second step is for the industry to develop snack foods with improved levels of beneficial components. This can range from fortification with tocopherol to producing healthier snacks. Certainly the acceptance of granola bars demonstrates interest in healthier snacks by the consumer. One might imagine convenient snacks with real fruits and vegetables in the product. Carrying this healthier food notion one step further, one might consider enhanced foods as suggested by Smith.[6] These foods would be processed foods which were restored to the original nutrient value by refortifying with nutrients lost during processing. Thus an enhanced food would deliver the same level vitamins (and perhaps other phytochemicals) as the fresh product. Improving snacks and other foods to make them more like fresh fruits and vegetables in concert with increased use of fresh product will bring the consumer much closer to meeting the five a day goal. It is likely that if this scenario were played out with the general public that the need for fortification or hypernutritious foods would be minimal. Clearly as more scientific evidence accumalates specific fortification opportunities are likely to emerge. These will have to be identified, carefully evaluated and responsibly implemented.

CONCLUSION

At this juncture only minimal fortification of foods seems necessary. Knowledge of phytochemicals is a rapidly emerging science that will undoubtedly provide the basis for much healthier diets in the future. Currently it appears that encouraging consumption of a wide variety of fruits , vegetables and grains seems to be the most rational approach to a healthier overall diet. The health professionals, marketers, manufacturers and consumers should be cautioned against over fortification and excessive use of dietary supplements. The science is emerging rapidly but currently it appears that the beneficial phytochemicals can be derived from a varied diet based on consumption of currently available plant materials.

REFERENCES

1. Food and Nutrition Board "How Should the recommended dietary allowances be revised?" A concept paper from the Food and Nutrition Board. *Nutrition Reviews* 1994, 52 (6), 216.
2. Block, G.; Patterson, B.; and Subur, D. "Fruit, Vegetable and Cancer Prevention: A Review of the Epidemiological Evidence." *Nutr. Cancer* 1992, 18 (1), 1-29.
3. Patterson, H.; Block, G.; Rosenberger, W.F.; Pee, D.; Kahle, L.L. "Friuit and Vegetable in the American Diet: Data from the NHANES II Survey." *Am. J. Public Health* 1990, 80 (12), 1443-1449.
4. Caragay, A. B. "Cancer-Preventive Foods and Ingredients." *Food Technology* 1992, April, 65-68.
5. Optimal Calcium Intake. NIH Consens Statement 1994 Jun 6-8; 12(4): 1-31.
6. Smith, R. E. "Functional Foods: The Food Industry's Views and Concerns." Presented at the 208th National Meeting of the American Chemical Society, Washington, DC; American Chemical Society: Washington, DC, 1994.

20

WHEAT BRAN, COLON CANCER, AND BREAST CANCER

What Do We Have? What Do We Need?

Richard M. Black

Kellogg Canada Inc.
and
Department of Nutritional Sciences
Faculty of Medicine, University of Toronto
Toronto, Ontario Canada.

INTRODUCTION

"Health maintenance is the key not only to human health, but to human culture's economic sustenance, and the human diet is central to health maintenance."[1]

In North America, and in Western Society in general, the food industry plays a pivotal role in translating the benefits of nutritional and food science research into commercially viable foods for the "human diet" and thus human health. In recent years, it has become obvious that the food industry must extend its horizons beyond the North American continent to uncover foods and food sources with unique chemical/ physiologic effects, effects which have the potential to influence the course of chronic disease or improve health through disease prevention. Notwithstanding the search for alternatives, there are many foods which have been in the North American market for many years, foods which contain biologically active "non-nutritive" substances (food components which do not have a classically defined nutrient function, or are not considered *essential* for human health), and these can have a significant impact on the course of disease. Whether or not these foods are termed "Functional Foods" (in fact, all foods have a function, and so the utility of that moniker seems limited), the point remains that certain foods, or food components, can have a significant, beneficial impact on many disease processes.

Food components such as fiber (soluble and insoluble, as well as other less commonly identified fractions) lignans, phenolic compounds, phytoestrogens and phytic acid are present in a large number of foods to varying degrees.[2] These non-essential nutrients, sometimes referred to as anti-nutrients[3] appear to influence our quality of life via their ability to enhance health. This differs dramatically from a more traditional view of the role of foods (nutrients/micronutrients) which are regarded as preventing or correcting a state of deficiency. While some of these substances have been linked to a reduction in cancer risk, it has only been within the last 5 to 10 years that scientists from such fields as epidemiology, nutrition, oncology, biochemistry and microbiology have been able to demonstrate the potent health effects of many of the foods that are a traditional part of the North American diet.

This chapter will focus on wheat (one of the most widely utilized grains in North America) and wheat products, and the potential for these to have a positive impact on health. This is not to suggest that there are no other foods of equal relevance to this discussion. Certainly other grains, different legumes and members of the allium family deserve mention in any discussion of diet and health. Rather, by restricting the review to wheat as an illustrative example, a more detailed analysis is possible. Specifically, the areas of colon cancer and breast cancer will be discussed, and the role that wheat and its constituents (fiber, phytochemicals) can play in reducing the risk of developing these diseases.

Of equal importance to understanding the relationship between a food and disease risk reduction, is the manner in which this information is ultimately conveyed to the public. It is essential that the information be truthful and honest, but more importantly, the information must have real utility. If conveyed in a manner which makes it unlikely that the consumer will utilize or even understand the information, then regardless of the potential health benefits, the information and the opportunity it presents will have been squandered. This highlights the need to work with various groups to develop meaningful messages which can be shared with the lay public, enabling them to work actively towards improving their own health.

COLON CANCER

Wheat and wheat products have an important role to play in the area of protection against colorectal cancer. The second leading cause of cancer death in the United States,[4] 155,000 new cases of colon cancer and 65,000 deaths due to colon cancer or complications arising therefrom are reported each year. Since there has been little change in the efficacy of treatment for colon cancer (as measured by five year survival rates), prevention will likely prove to be the most powerful tool in reducing the mortality of the disease.[5]

Increased dietary fiber has been linked to a reduced risk of colorectal cancer. Greenwald *et al.*[6] reviewed 40 epidemiologic studies which probed for an association between dietary fiber and colon cancer incidence. In 95% of these studies, there was a significant inverse relationship between measures of fiber intake and colon cancer risk. The next step then might be a Public Health focus, designed to increase fiber awareness and fiber intake. Some have suggested that the most efficacious and cost effective manner by which to achieve this is to increase the intake of cereal grains which are already widely (though not heavily) consumed, namely wheat bran and oat bran.[7-9] In the case of wheat bran, this may be particularly apt advice.

Certainly, animal studies support a unique role for wheat bran, as it has been shown to significantly affect biomarkers of colon cancer. Alabaster *et al.*[10] fed rats a high fat (20% w/w), low calcium diet (0.18% w/w) diet, supplemented with 1%, 4% or 8% dietary fiber from wheat bran for two weeks prior, one week during and 22 weeks after injection with azoxymethane (2 injections of 15 mg/kg body weight, separated by 7 days). At the completion of the study, there were significant decreases in both tumor incidence and tumor multiplicity as a function of increased dietary fiber (Figure 1). Alabaster and his research team have gone on to examine the effect of processing the wheat bran into one of two different forms of breakfast cereal: a flake or an extruded loop (personal communication). While the processed food is not as efficacious as the raw wheat bran, colon cancer risk is reduced as cereal intake is increased, when foci of aberrant crypts/cm^2 of colon are used as a biomarker. This type of research, using products which are available in the market place and enjoy wide

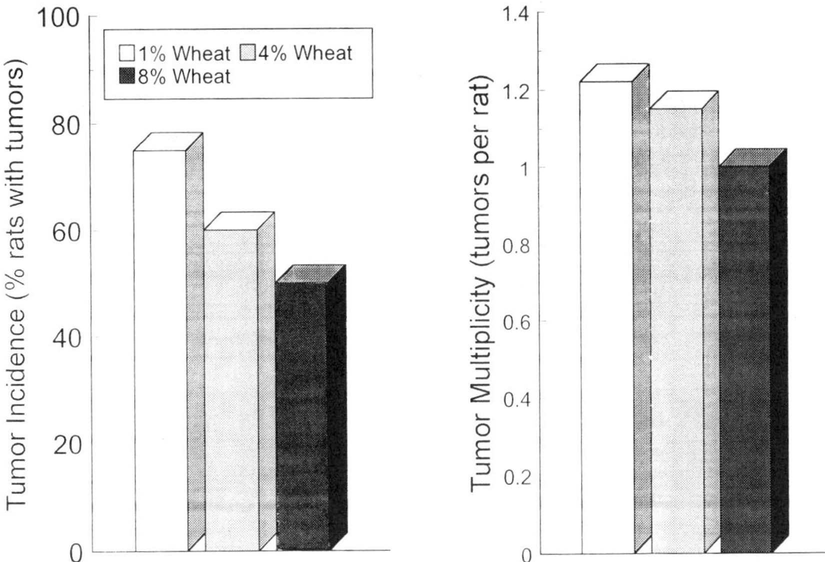

Figure 1. Incidence and multiplicity of colon tumors in rats. Diets were high fat, low calcium, with 1%, 4% or 8% fiber from wheat bran. Cancer was induced following 2 s.c injections of azoxymethane, separated by 7 days. Adapted from Alabaster *et al.*[10]

acceptance, is a key step in determining the possible Public Health benefits of a food regardless of its wheat bran content. If a raw, unprocessed food shows clear anti-tumorigenic effects, but the processed food shows nothing or perhaps even worse, shows an effect but is unacceptable to the consumer, then one is no further ahead. This is the point reached in translating research results into tangible benefits, when the food scientist must work closely with the oncologists, microbiologists, geneticists and nutritionists, to develop a food which can have an impact on health, be cost effective, and be acceptable to the public.

While mechanisms for these anticarcinogenic effects are not yet clearly understood, many hypotheses have been suggested. One prominent hypothesis points to the production of short chain fatty acids, butyrate in particular, during the fermentation of the wheat bran in the colon as the prime anti-tumor agent. For example, Boffa *et al.*[11] examined the effects of short chain fatty acid (SCFA) production on colonic crypts in a rat model of colon cancer by varying the amount of wheat bran in the diet. Animals consumed a diet that contained either 0%, 5%, 10% or 20% wheat bran (by weight). Thus the level of fiber ranged from 0% to 8.5% in these diets (wheat bran contains ~42% fiber by weight). At the end of a two week period, the animals were sacrificed, cell proliferation measured, and luminal contents removed, frozen and analyzed by gas-chromatography. The data showed that wheat bran increased SCFA levels in the lumen, with 5% and 10% wheat bran levels producing the greatest amount of butyrate. Perhaps more importantly, there was a significant inverse correlation between butyrate levels and colonic cell proliferation, as measured through the incorporation of labelled thymidine in the epithelial cells of the crypt. As SCFA levels increased, colonic cell proliferation decreased, an indication of reduced colon cancer risk. Boffa *et al.*[11] concluded that wheat bran modulates colonic butyrate levels, which in turn modulate DNA synthesis in the proliferative compartments of colonic crypts.

McIntyre *et al.*[12] have also examined the role of butyrate in colonic cell proliferation in rats. They compared the effect of guar gum, oat bran and wheat bran (all at ~ 10% w/w) on butyrate production and colon tumor incidence in the rat following administration of 30 μg/g body weight of 1,2-dimethyl hydrazine. The experimental diets were administered three weeks prior to, 10 weeks during and 20 weeks following administration of the carcinogen. Their results were significantly fewer tumors, and lower tumor mass, in rats consuming wheat bran, compared to guar gum, oat bran, or no additional fiber. Moreover, although there was little difference between the diets in terms of SCFA concentrations in total cecal contents, the wheat bran diet resulted in significantly greater SCFA concentrations in fresh feces. This was true for both acetate and butyrate, where levels were roughly twice those in any of the other groups. Since others have reported that butyrate has the potential to directly modulate gene expression,[13] McIntyre *et al.*[12] concluded that the production of butyrate from fermentable fibers at the site of tumor formation may be a key component in wheat bran's protective effects against colon cancer.

Reddy and co-workers[14] have also reported that wheat bran differed from other sources of dietary fiber in its action on the colon, though they examined the effects on bile acid mutagens. In a blinded randomized cross-over design, they found that the concentration of fecal secondary bile acids and of fecal mutagenic activity were significantly lower during wheat bran supplementation (10 g/day of dietary fiber from wheat bran) compared to a control diet, whereas an oat bran diet supplemented at a level to achieve 10 g/day of fiber had no impact on these measures. They concluded that the increased fiber intake in the form of wheat bran could reduce the formation and/or excretion of mutagens, as well as decrease the concentration of secondary bile acids. Because these have been linked to the development of colon cancer, Reddy *et al.*[14] concluded that increasing wheat bran in the diet leads to a reduction in colon cancer risk.

Others have also pointed to the ability of the wheat to increase fecal bulk and decrease transit time, thereby minimizing any contact of potential carcinogens with the lumen.[15-17] There are two fundamental properties of wheat, or more specifically wheat bran, which determine its ability to influence laxation: water holding capacity, and resistance to colonic degradation/digestion. As water holding capacity and resistance to degradation increase, stool weight increases and transit time through the colon decreases. On average consuming about one gram of wheat bran fiber (~ 2.5 g of wheat bran) increases stool bulk by 5.7 g in healthy subjects, roughly double the effect of cellulose, and over four times that of pectin. Milling of the wheat bran can also affect the laxative properties of the bran, with larger particle sizes generally being associated with increased bulking and shortened transit times.

Current work in this area is now focused on the mechanism of risk reduction. Although the notion that the fiber is simply "scraping clean" the colon has long been abandoned, the complete mechanism has yet to be fully understood. None the less, despite the incomplete nature of information on mechanism, the available data provide very strong evidence that colon cancer risk is reduced when dietary fiber derived from wheat bran is increased. This is what we have. What we need to do is develop a consensus on this issue in order that an honest, *simple* and *understandable* message can be communicated to the public.

BREAST CANCER

While the data linking high fiber diets to a reduced risk of colon cancer are quite strong, those linking fiber intake to a reduction in breast cancer risk are just beginning to emerge. This hypothesis, that increased dietary fiber may lead to a reduced risk of breast cancer, has developed from an evaluation of the epidemiologic data.[18,19] Generally, in countries where fiber intake is high, breast cancer rates are relatively low, and *vice versa*. However, these observations are confounded by the fact that high fiber diets are typically low fat diets, and high fiber diets are also typically high in fruit and/or vegetable consumption. Either one of these factors could potentially explain the relation between high fiber intake and low breast cancer rates. However, more rigidly controlled experimental studies have also provided support for the role of dietary fiber in the prevention of breast cancer.

In one highly publicized report, there was little or no relationship detected between fiber intake and breast cancer risk.[20] However, this negative finding may be due in part to a limited range of fiber intakes among study participants. In other words, the entire population under study had what could be termed a "low fiber" diet, making it unlikely that any conceivable protective effect could be detected.

Since elevations in bioavailable estrogen have been linked to an increased risk of developing breast cancer,[21-23] and estrogens are acknowledged as a necessary factor in the genesis of breast cancer (if not the growth),[24] affecting estrogen metabolism could very well alter breast cancer risk. If it is possible to reduce the levels of circulating estrogens, or perhaps more importantly the amount of circulating *bio-available* estrogen, then it may be possible to reduce the risk of developing breast cancer.[21,23]

Goldin and Gorbach[25] studied estrogen metabolism in pre- and post-menopausal omnivores and vegetarians. They reported that estradiol and estriol fecal excretion were higher in vegetarians (who consumed a high fiber diet) compared to omnivores (consuming a low fiber diet). The result was lower plasma estradiol levels in vegetarians compared to omnivores. However, whether these changes in estrogen metabolism were due to the large differences in dietary fiber between omnivores and vegetarians, or some other dietary or lifestyle factor, could not be determined.

In a more direct probe of fiber's effect on plasma estrogens, Rose and colleagues[26] reported that wheat bran (but not corn or oat bran) significantly reduced the level of circulating estrogens in the plasma, specifically serum estrone and estradiol levels in premenopausal women. This was accomplished by supplementing the diet with an average of 15 to 30 g/day of wheat bran, representing 6 to 12 g/day of dietary fiber. Furthermore, these changes were accomplished without any attempt to reduce dietary fat intake. Others have made similar observations.[27-30] The anti-tumorigenic effects of fiber have been documented in animal models of breast cancer with a 30% decrease in tumor incidence in rats fed a diet containing 10% wheat bran compared to rats receiving no supplemental wheat bran.[31] Similarly, Arts *et al.*[32] found that while tumor incidence did not decrease with increased fiber from wheat bran, tumor weights and multiplicity were significantly reduced in the high fiber diet group. Taken together, these studies provide encouraging data and support linking increased fiber with reduced breast cancer risk.

The mechanism for these effects may be a result of an alteration in the enterohepatic cycling of estrogens.[28,33] When conjugated estrogens are secreted in the bile, wheat bran minimizes deconjugation by β-glucuronidase enzymes, an otherwise essential digestive process in the reabsorption of the estrogens from the lumen into the hepatic circulation. It may be that wheat bran renders much of the conjugated estrogens inaccessible to the enzymes by binding the estrogens in some manner.[34] Alternatively, wheat bran may directly inhibit the activity of the enzymes themselves, perhaps through a subtle alteration of the intestinal environment rendering conditions less than optimal for the β-glucoronidase enzymes. It is even possible that the enzymes successfully cleave the conjugated estrogens, but the wheat bran subsequently binds the "free" estrogen and prevents re-uptake. The net result is an increased excretion of estrogens in the feces, and a reduction in circulating levels (Figure 2).

Once again, as with colon cancer, some researchers have ascribed a unique role to the consumption of wheat products as compared to other grains. For example, with regard to breast cancer and lifestyle, Weisburger and Kroes[19] concluded, "The most appropriate preventive measures are a limited fat intake, daily vegetable and fruit intake, and wheat bran fiber, the avoidance of obesity, and regular exercise." Following an assessment of diet and breast cancer risk in Australia which uncovered an inverse relationship between fiber intake

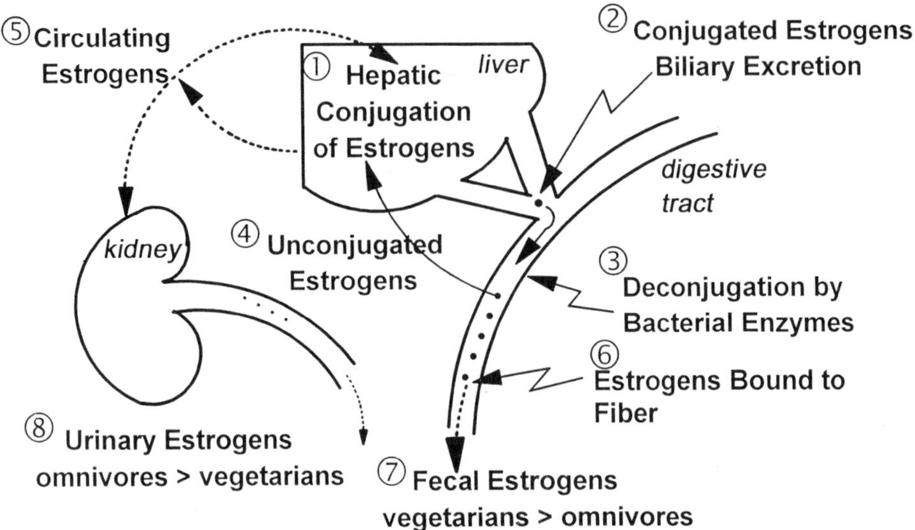

Figure 2. Enterohepatic circulation of estrogens. ① Estrogens are conjugated in the liver. ② The conjugated estrogens are excreted in the bile as part of the digestive process. ③ In the lumen, β-glucoronidase enzymes act to deconjugate the estrogens which permits ④ the re-uptake of free estrogens into the hepatic circulation and ⑤ general circulation. ⑥ However, wheat bran fiber i) may act to inhibit the action of the β-glucoronidase enzymes, ii) may bind, and prevent the enzymes from accessing, the conjugated estrogens; or iii) may bind the free estrogens following deconjugation, all of which prevent or inhibit re-uptake. ⑦ Thus vegetarians with high fiber diets excrete more estrogens in the feces, compared to omnivores (low fiber diets), ⑧ with the result that urinary excretion of estrogens is somewhat reduced in vegetarians (total estrogen excretion, fecal and urinary combined, is greater in vegetarians). Adapted from Rose.[30]

Table 1. Hormone levels

	Case	Control	% Difference
Estrone (pg/ml)	14.0*	11.4	18.5%
Total Estradiol (pg/ml)	33.6**	27.7	17.6%
Free Estradiol (pg/ml)	0.47**	0.36	23.4%
Albumin-Bound Estradiol (pg/ml)	19.4**	14.7	24.2%
SHBG-Bound estradiol (pg/ml)	12.8	11.9	7.0%
Free Estradiol (%)	1.42**	1.34	5.6%
Albumin-Bound Estradiol (%)	58.9**	54.1	8.1%
SHBG-Bound Estradiol (%)	39.7**	44.6	11.0%

Mean levels of estrone, total estradiol, bioavailable (free) estradiol and bound estradiol, in women with breast cancer (case) and breast cancer free (control). For absolute measures, geometric means were calculated, for percentage measures, arithmetic means were calculated.

* $p < 0.01$, paired t test

** $p < 0.001$, paired t test

Adapted from Toniolo *et al*[36]

Table 2. Breast cancer associations

	Quartile				
	I	II	III	IV	*p*
Estrone (pg/ml)	1.0	2.2	3.7	2.5	<0.10
Total Estradiol (pg/ml)	1.0	0.9	1.8	1.8	<0.10
Free Estradiol (pg/ml)	1.0	1.4	3.0	2.9	<0.01
Albumin-Bound Estradiol (pg/ml)	1.0	1.1	2.7	2.2	<0.01
SHBG-Bound estradiol (pg/ml)	1.0	1.0	1.1	1.3	ns
Free Estradiol (%)	1.0	1.7	1.9	2.0	ns
Albumin-Bound Estradiol (%)	1.0	1.3	2.1	3.3	<0.001
SHBG-Bound estradiol (%)	1.0	0.70	0.40	0.32	<0.01

Adjusted (for Quetlet index) odds ratios for breast cancer for quartiles of various measures of estrone and estradiol: total, bioavailable (free) and bound. Adapted from Toniolo *et al.*[36]

and breast cancer risk, Baghurst and Rohan[35] recommended that, "...increased use of breakfast cereals (especially those with a bran base), peas, beans (including soya beans), dried fruits and nuts can significantly elevate an individual's daily fiber intake."

While no one has as yet demonstrated that these changes will lead to a reduction in the risk of developing breast cancer, recent data do show that women with breast cancer have higher levels of circulating estrogens, and higher levels of bioavailable estrogens, than do women who are breast cancer free[36] (Table 1). Furthermore, risk of breast cancer increases as bioavailable estrogen increases (Table 2). Thus the implication is that if one were able to reduce the level of bioavailable estrogens, perhaps through increased consumption of high fiber wheat products, thereby interrupting the enterohepatic circulation of estrogens, the risk of developing breast cancer may also be reduced.

PHYTOCHEMICALS

Phytochemicals present in wheat, such as inositolphosphate (phytic acid), may also have a role to play in health promotion.[2,3,37] Although this is a relatively new area of investigation in the cancer/nutrition field, independent research teams have reported that increased dietary phytic acid reduces colon cancer incidence in the rat. However, as with breast cancer research, much remains to be done to firmly establish this phenomenon.

CONCLUSION

Increasingly, the issue for the food industry becomes one of understanding these benefits and of targeting research towards elucidating the effects of physiologically active components found in well known food sources, in addition to those less readily recognized. Wheat, as well as soya and perhaps garlic, are just three of the many foods with potential health benefits. Research designed to understand and *enhance* these benefits is critically important. Clearly, such a targeted approach could lead to the development of "Designer Foods" or "Functional Foods". However, it is difficult at this stage to precisely determine i) the impact of such foods on the health of the general public (it will depend upon acceptance of the food by the public); or ii) the food industry's desire to become involved in this type of research (to pursue a health claim in the United States could require an investment well in excess of $2 million). Because of the potential health benefits, we are already witnessing a rapid increase of research in this area, and of subsequent translation of these findings from the lab table to the kitchen table.

We *have* many foods currently available in our diet with demonstrated physiologic effects inasmuch as health risk is concerned. We *need* to better understand these foods since they are readily available in North America, are economical, and are widely accepted by consumers. We *have* a good understanding of how to communicate a health/diet message to the public. We *need* partnerships between those in Industry, Government, Non-profit Organizations, and Academia to achieve scientific consensus on these issues in order that truthful, useful, health/diet messages can be communicated and implemented.

REFERENCES

1. Hendrich S, Lee KW, Xu X, Wang HJ, MurphyPA. Defining food components as new nutrients *J Nutr* 1994 124:1789S-1792S.
2. Thompson LU. Antioxidants and hormone-mediated health benefits of whole grains. *Critical Reviews in Food Science and Nutrition* 1994 34: 473-497.
3. Thompson LU. Potential health benefits and problems associated with antinutrients in foods. *Food Research Internl* 1993; 26:131-149.
4. American Cancer Society. Cancer facts and figures. New York, 4-8. (1990).
5. Vargas PA, Alberts DS. Primary prevention of colorectal cancer through dietary modification. *Cancer* 1992; 70:1229-1235.
6. Greenwald P, Lanza E, Eddy GA. Dietary fiber in the reduction of colon cancer risk. *J Am Diet Assoc* 1987 87:1178-1188.
7. Alberts DS, Einspahr J, Rees-McGee S, Ramanujam P, Buller MK, Clark L, Ritenbaugh C, Atwood J, Pethigal P, Earnest D, Villar H, Phelps J, Lipkin M, Wargovich M, Meyskens, FL Jr. Effects of dietary wheat bran fiber on rectal epithelial cell proliferation in patients with resection for colorectal cancers. *J Natl. Cancer Inst.* 1990; 82:1280-1285.
8. Ho EE, Atwood JR, Benedict J, Ritenbaugh C, Sheehan ET, Ambrams C, Alberts DS, Meyskens FL,Jr. A community-based feasibility study using wheat bran fiber supplementation to lower colon cancer risk. *Preventive Medicine* 1991 20, 213-225.
9. Witkin R, Greenwald KM. Familial adenomatous polyposis: a nutritional intervention trial. *J Natl Cancer Inst* 1989 81:1272-1273.
10. Alabaster O, Tang ZC, Frost A, Shivapurkar N. Potential synergism between wheat bran and psyllium; enhanced inhibition of colon cancer. *Cancer Letters* 1993 75 53-58.
11. Boffa LC, Lupton JR, Mariani MR, Ceppi M, Newmark HL, Scalmati A, Lipkin M. Modulation of colonic epithelial cell proliferation, histone acetylation, and luminal short chain fatty acids by variation of dietary fiber (wheat bran) in rats. *Cancer Research,* 1992 52:5906-5912.
12. McIntyre A, Gibson PR, Young GP. Butyrate production from dietary fiber and protection against large bowel cancer in a rat model. *Gut* 1993; 34:386-391.
13. Candido EP, Reeves R, Davies JR. Sodium butyrate inhibits histone deacetylation in cultured cells. *Cell* 1978 14: 105-13.
14. Reddy B, Engle A, Katsifis S, Simi B, Bartram HP, Perrino P, Mahan C. Biochemical epidemiology of colon cancer: effect of types of dietary fiber on fecal mutagens, acid, and neutral sterols in healthy subjects. *Cancer Res* 1989; 49:4629-35.
15. Kritchevsky D. Fiber steroids and cancer. *Cancer Res* 1983; 43:2491S-2495S.
16. Kritchevsky D. Fiber and cancer. *In*: Vahoun GV, Kritchevsky D. Eds. *Dietary Fiber: Basic and Clinical Aspects,* Plenum, New York, 1986: 427-432.
17. Klurfeld DM. Dietary fiber-mediated mechanisms in carcinogenesis. *Cancer Research,* 1992 52.2055s-2059s.
18. Rose DP. Dietary factors and breast cancer. *Cancer Surv* 1986 5:671-87.
19. Weisburger JH, Kroes R. Mechanisms in nutrition and cancer. *Euro J Cancer Prevent,* 1994 3:293-298.
20. Willett WC, Hunter DJ, Stampfer MJ, Colditz G, Manson JA, Spiegelman D, Rosner B, Hennekens CH, Speizer FE. Dietary fat and fiber in relation to risk of breast cancer: An 8 year follow-up. *JAMA* 1992 268:2037-2044.
21. Jones LA, Ota DM, Jackson GA, Jackson PM, Kemp K, Anderson DE, McCamant SK, Bauman DH. Bioavailability of estradiol as a marker for breast cancer risk assessment. *Cancer Res* 1987 47: 5224-5229.

22. Langley MS, Hammon GL, Bardsley A, Sellwood RA, Anderson DC Serum steriod binding proteins and the bioavailability of estradiol in relation to breast diseases. *J Natl Cancer Inst* 1985 75:823-829.
23. Ota DM, Jones LA, Jackson GL, Jackson PM, Kemp K, Bauman D. Obesity, non-protein-bound estradiol levels, and distribution of estradiol in the sera of breast cancer patients. *Cancer 1986 57:558-62.*
24. Nandi S, Guzman RC, Yang J. Hormones and mammary carcinogenesis in mice, rats, and humans: a unifying hypothesis. *Proc. Natl. Acad. Sci. USA* 1995 92:3650-3657.
25. Goldin BR, Gorbach SL. Effect of diet on the plasma levels, metabolism, and excretion of estrogens. *Am J Clin Nutr* 1988 48:787-790.
26. Rose DP, Goldman M, Connolly JM, Strong LE. High-fiber diet reduces serum estrogen concentrations in premenopausal women. *Am J Clin Nutr* 1991 54: 520-525.
27. Goldin BR, Woods MN, Spiegelman DL, Longcope C, Morrill-LaBrode A, Dwyer JT, Gualtieri LJ, Hertzmark E, Gorbach SL. The effect of dietary fat and fiber on serum estrogen concentrations in premenopausal women under controlled dietary conditions. *Cancer,* 1994 74:1125-1131.
28. Gorbach SL, Golden BR. Diet and the excretion and enterohepatic cycling of estrogens. *Preventive Medicine* 1987 16:525-531.
29. Heber D, Ashley JM, Leaf DA, Barnard RJ. Reduction of serum estradiol in postmenopausal women given free access to low-fat high-carbohydrate diet. *Nutrition* 1991 7:137-140.
30. Rose DP. Dietary fiber, phytoestrogens, and breast cancer. *Nutrition* 1992 8: 47-51.
31. Cohen, LA, Kendall ME, Zang E, Meschter C, Rose DP. Modulation of N-nitrosomethylurea-induced mammary tumor promotion by dietary fiber and fat. *J Natl Cancer Inst* 1991 83:496-501.
32. Arts CJM, Albert TH, De Bie HJ, Van Den Berg H, Van't Veer P, Bunnik GSJ, Thijssen JSH. Influence of wheat bran on NMU-induced mammary tumor development, plasma estrogen levels and estrogen excretion in female rats. *J Steroid Biochem. Molec. Biol.* 1991 39:193-202.
33. Gorbach SL, Estrogens, breast cancer, and intestinal flora. *Rev Infect Dis.* 1984 6:S85-90.
34. Arts CJM, Govers CARL, Van Den Berg H, Wolters MGE, Van Leeuwen P, Thussen JHH. *In vitro* binding of estrogens by dietary fiber and the in vivo apparent digestibility tested in pigs. *J Steroid Biochem. Molec. Biol.* 1991 38:621-628.
35. Baghurst PA, Rohan TE. High-fiber diets and reduced risk of breast cancer. *Int. J Cancer* 1994; 56:173-176.
36. Toniolo PG, Levitz M, Zeleniuch-Jacquotte A, Banerjee S, Koenig KL, Shore RE, Strax P, Pasternack BS. A prospective study of endogenous estrogens and breast cancer in postmenopausal women. *J Natl Cancer Inst* 1995; 87:190-197.
37. Shamsuddin AM. Inositol Phosphates have novel anticancer function. *J Nutr.* 1995 125: 725S-732S.

21

DIET, NUTRITION, AND CANCER PREVENTION*

Research Opportunities, Approaches, and Pitfalls

Chung S. Yang,[1] Barbara C. Pence,[2] Michael J. Wargovich,[3] and Janelle M. Landau[1]

[1] Laboratory for Cancer Research
College of Pharmacy, Rutgers University
Piscataway, New Jersey 08855
[2] Department of Pathology
Texas Tech University Health Science Center
Lubbock, Texas 79430
[3] Digestive Disease and Gastrointestinal Oncology
Anderson Hospital and Tumor Institute
The University of Texas System Cancer Center
Houston, Texas 77030

INTRODUCTION

C.S. Yang opened the workshop by welcoming the participants, and he stated the purposes of the workshop. Research on diet, nutrition and cancer prevention is recognized as an important area of research, yet this area of research is not well funded. The purpose of the workshop is to strengthen this area of research by discussing how to select more relevant problems, develop attractive hypotheses, and use effective approaches to obtain research aims. Other important tasks are to convey the importance of the research and to increase the funding in the area of diet and cancer prevention. A proposed outline of discussion for the workshop was presented (Appendix 1).

IMPORTANT, BUT NOT WELL FUNDED, RESEARCH FIELD

Barbara Pence outlined the importance of diet and cancer prevention in public health. Cancer is the leading cause of death in adults aged 25 through 64, and about one-third of these deaths have been attributed to dietary factors. In the U.S. Department of Health and

* *Report of An AICR Workshop*

Human Services document entitled "Healthy People 2000", dietary modification is a major component of the "Cancer Risk Reduction Objectives" (Appendix 2). Much research is needed to support these recommendations. Relevant research in this area and proper dissemination of the information will empower the American public to take responsibility for their own health. In a recent article by Ames *et al.*,[1] it was stated that "Further research on the ways in which diet influences cancer risk is important because it is likely to have the greatest impact on future prevention strategies."

Michael Wargovich then discussed the funding situation based on his experience of serving on grant review panels. The general impression by him and others is that the funding for diet and cancer prevention related grants is declining. Many grants in this area have been viewed as scientifically weaker than grants in rapidly progressing fields such as molecular biology. Although many of these grant applications are indeed not rigorous scientifically, it is also possible that some of the grants were given poor priority scores by reviewers who do not appreciate the importance and complexity of the research. Moreover, the composition of study sections has changed over the years; many of the reviewers in the diet and cancer prevention area have been replaced by molecular biologists and clinicians. Nevertheless, this change may reflect the rapidly increasing research developments in molecular biology and the emphasis of extending laboratory studies to clinical investigations.

A suggestion was made for urging the creation of a "chemoprevention study section."[2] Many participants believed that this would increase the funding in this area. Others, however, cited examples to indicate that the most severe criticisms and poorest priority scores given to diet and cancer prevention applications were by reviewers in the same field, rather than by molecular biologists. Early this year, a special study section was created by the National Cancer Institute to review program project grant applications which were submitted in response to a "Request for Grant Applications" in the diet and cancer prevention area. The study section did have adequate members with expertise in this research area, but none of the nine applications were funded because the priority scores were not high enough.

It is not clear whether grants in the diet and cancer prevention area receive a lower funding rate than other areas of research, but there is consensus that more relevant research should be conducted in this important area. Whereas it is important to have qualified reviewers on study sections and review panels, it is also vital that research in the diet and cancer prevention field be strengthened.

COMMON WEAKNESSES AND PROBLEMS IN GRANT APPLICATIONS

In order to strengthen our field of research, it is useful to examine some of the common criticisms often given to grant applications in the diet and cancer prevention area. C.S. Yang presented several commonly perceived weaknesses (Appendix 1) as a starting point for discussion. The comment of "too descriptive" was discussed extensively. It was pointed out that this term may have been an expression reflecting the reviewers' dislike of the grant. Another comment was that unless the fundamental biology is known, the molecular studies may be premature.[3] Limiting grants to mechanistic studies may neglect cancers that have not been extensively studied; for example, testicular cancer, histiocytoma, and multiple myeloma which kills patients in less than a year.[4] It was pointed out that a mechanistic study may not be required in certain cases, such as in testing the chemoprotective effect of an agent on a certain cancer. However, it is also important to know whether studies in a given animal model will provide information relevant to human cancer prevention or will generate new concepts. Species differences in biological responses are common, and we have to be careful in the interpretation of results. For example, studies of an agent thatthat blocks the activation of a carcinogen which is not found in the human environment, by an enzyme which does not exist in the human target tissues, would not be as valuable as studies on an agent which can block the carcinogenic process in humans. Much mechanistic information is required in order to make such an evaluation. For human trials, we need extensive mechanistic information concerning the possible preventive or toxic effects of an agent. In the public health area, it has been suggested that as long as the epidemiological evidence is strong, we can promote a dietary modification without knowing the detailed mechanisms of the action. In this context, it is important to consider how our research projects are really contributing to the goals of the "Healthy People 2000" initiative. It is not enough to merely state that "This area of research is important to these goals; my grant application should be funded."

A comment was made that nutritional science is lagging behind molecular biology by about ten years. A counter point is that although molecular biology is at the cutting edge of technology, it does not relate directly to the goals of Healthy People 2000. Considering the spectacular and rapid advancements in molecular and cell biology, nutritional science indeed progresses slowly. Perhaps we should not consider molecular biologists as our competitors, but rather our allies and collaborators.

WAYS TO STRENGTHEN RESEARCH ON DIET AND CANCER PREVENTION

As an introduction to the discussion, C.S. Yang presented some of his views (Appendix 1). Stephen Hecht (American Health Foundation) then spoke on how to develop fundable research projects on diet and cancer. His points were as follows:

1. Focus on the part of the carcinogenic process that you understand.
2. Find an appropriate and relevant inhibitor.
3. Focus on one inhibitor in pure form.
4. Develop and test hypotheses regarding possible mechanisms of action. Ask the right questions. Most reviewers are good scientists and thus can understand a relevant question.
5. Design appropriate bioassays based on biochemical and molecular biological data.
6. Extend to humans using appropriate biomarkers.
7. Extend to mixtures/interactions based on knowledge of single component mechanisms.

Stephen Hecht also stated that he believes nutrition scientists are tough reviewers. He stressed that it is important to define the model and explain the scientific basis of the proposal to the reviewers.

Whereas studying one purified compound at a time and then combinations is an accepted prudent approach, studies with mixtures can also be very informative if the experiments are properly designed[5]. When mixtures are used, it is important to analyze the contents of the suspected active compounds in the mixture. The bioavailabilities and tissue levels of these compounds should also be studied.[5]

Stephen Barnes (University of Alabama, Birmingham) briefly described the pharmacological principles of a chemopreventive agent which include absorption, distribution, concentration at the target site, metabolism, and excretion. He also stated that it is important to know the concentrations of the test compound in specific cells of the target organs.

The topic of interdisciplinary collaboration was discussed extensively. In order to develop the most relevant studies, knowledge on etiological factors of cancer and the biological actions of possible dietary constituents are required. Knowledge of the molecular mechanisms of carcinogenesis and of how that process might be curtailed by a possible inhibitory agent will provide depth in the study designs. Applications of appropriate methods in molecular biology, bioanalytical chemistry, pharmacology, and biostatistics will enable us to reach our goals more effectively. If we do not have expertise in such areas, we need to seek collaborators and consultants in these areas. Many funding agents are encouraging multidisciplinary collaboration in biomedical research. As investigators in the diet and cancer prevention area, we should take the lead in organizing collaborative teams in attacking key problems to help achieve the goals of cancer prevention. This should be a fruitful research area for talented young scientists.

A suggestion was made to set up an electronic bulletin board to enhance interactions and collaborations among researchers in the field of diet and cancer prevention. This idea is worthy of further consideration. We also can use scientific journals and meetings more effectively for learning the varied expertise of researchers and setting up collaborations when needed.

NEW OPPORTUNITIES AND APPROACHES

Many recommendations were made in the "Surgeon General's Report on Nutrition and Health" in 1988 and the "Diet and Health: Implication for Reducing Chronic Disease Risk" report by the National Research Council in 1989 (Appendices 3 and 4). Some of these recommendations are still relevant. C.S. Yang presented some specific topics that may be promising for more in-depth research (Appendix 1).

Michael Wargovich, speaking for Gary Stoner (Ohio State University) who could not be present, mentioned a few areas of funding priorities. He remarked that it would be wise for us to initiate or be collaborators on human cancer prevention trials. Because the goal of our research is to prevent human cancers, translational research is thus of vital importance. The development of biomarkers is an extremely promising approach. It is important to validate the biomarkers for future laboratory and clinical investigations. Other suggestions included studies on the interactions between chemopreventive agents and relevant oncogenes and tumor suppressor genes. The use of agricultural biotechnology to increase cancer preventive chemicals in foods should also be of practical importance.

Bruce Dunn (British Columbia Cancer Agency) discussed the importance of research in the area of dietary regimens (as ancillary treatments) for cancer patients. He stated that cancer kills by the progression of the original tumor and metastasis. Diets which are beneficial for the prevention of the onset of the original tumor may not be the same for the prevention of metastasis or the recurrence of tumors after surgery. In reference to diet and recurrence, it is important to understand the effect of diet and dietary constituents on 1) cancer cell growth and differentiation, 2) the immune system, 3) the hormonal status, and 4) the action of chemotherapeutic agents. Research on diet and cancer recurrence has the following benefits: 1) highly relevant to patient care, 2) relatively understudied, 3) favorable for intervention trials because the subjects are motivated, 4) high event rate and rapid follow-up (i.e., 5 year disease-free survival measurement), and 5) potential for alliance with the health care system. The importance of such studies was highly appreciated by the participants. Several groups have already started working in this area, but more activity is needed. Experimental metastasis models are available for studies on dietary effects.

APPROACHES TO INCREASE FUNDING

In addition to strengthening the science in our field of research, several other approaches for increasing funding opportunities were discussed.

Justifying the Importance of the Research

Diet and cancer prevention is generally recognized as an important area, and we can cite "Healthy People 2000" to stress this point. It is up to us, however, to state how our research projects can lead to specific measures that will reduce cancer incidence or mortality. It is true that some studies yield unexpected results, and the value of some research cannot be prejudged. However, grant reviewers have to decide on the priority for funding. We should provide information to the reviewers in order to convince them that our studies will indeed provide vital information for cancer prevention.

The importance of diet and cancer prevention should be brought to the attention of the medical establishment[6] and the general public. In Texas, all of the medical schools have formed a cooperative NCI sponsored project for teaching medical students the latest information on cancer prevention, including nutritional intervention. When speaking to the public or receiving their questions, it is important to point out why research in this area is necessary and how our research results can contribute to cancer prevention. We should make a convincing argument why more research funding in this area is needed. We can also suggest promising areas of research to funding agencies such as the National Institutes of Health.[7]

Increasing Support and Fund Raising

The idea of forming advocacy or lobbying groups has been brought up; such activity is beyond the capability of our workshop. As individuals we should actively communicate with our legislators and strongly request their support for biomedical research when budget decisions are being made. We should also support the fund-raising activities of private organizations such as the AICR and the American Cancer Society. The AICR is especially focused on the area of nutrition and cancer and deserves our support. We greatly appreciate the contributions of the AICR and hope that more funds will be available to support research on diet and cancer prevention.

At a time when federal research funds are tight, it is important to seek alternative funding sources.[8] A discussion on funding from produce marketing boards and food companies ensued. Food companies do not fund projects exclusively, but there is money available to fund projects in partnership. The AICR is requesting matched funds from industry to support research projects jointly. Industry-academic partnerships may make a proposal more attractive. Many agencies, such as the Florida Citrus Fruit Department, are providing research samples for investigations. Some seed money for preliminary experiments may also be available. Preliminary studies are important for successful grant applicants. Basic food companies may also have funds available.

Strengthening the Study Sections and Review Panels

It is important to have our grants reviewed by study sections containing enough qualified reviewers who appreciate the importance and complexities of research on diet and cancer prevention. The study sections are also eager to recruit qualified members with varied expertise to cover the research areas in the grant applications. The appointment of the reviewers usually comes from recommendations, although volunteers are also considered.

We should make our talents available and serve on study sections or review panels, even though it is a time consuming task. The benefit for the hard work of reviewing grants is that it can help to sharpen our own research approaches.

Writing Better Grant Applications

There are many articles and books on grant writing; some of them are very useful. It is always helpful to discuss our research ideas and plans with colleagues who are familiar with the research field or grant. It is usually easier for people other than ourselves to see problems with the application. After writing an early draft, it is also prudent to ask colleagues to read it for clarity and general organization of the application. The assistance of a scientific writer is also helpful.

At a time when the NIH funding rates for grants are about 12 to 15%, most grant applications will not be funded. Although it is disappointing to receive notices of non-fundable priority scores, we should not be discouraged if we think we have an important project with a good system and effective approaches for study. It does not help to blame the study section or the funding system, unless one has the ability to influence them. One may politely rebut criticisms which are unjust, but it is usually helpful to take advantage of the constructive critiques and to revise the application.

Keeping the Lines of Communication Open

Most participants considered this workshop a success and suggested that position papers based on these discussions should be sent to funding agencies to point out the importance of funding in this area of research. It was also agreed that future discussions or workshops on how to strengthen research and increase funding in the area of diet and cancer prevention would be useful. A possible activity under consideration is to have a "Diet and Cancer Prevention" dinner at the 87th Annual Meeting of the American Association for Cancer Research in Washington, DC in April of 1996.

ACKNOWLEDGMENT

The chairpersons gratefully acknowledge the pre-eminence of the American Institute for Cancer Research for their generous and sustained funding of nutrition and cancer prevention.

REFERENCES

1. Ames, B.N., Gold, L.S., and Willett, W.C. (1995) The causes and prevention of cancer. *Proc. Natl. Acad. Sci.* 92:5258-5265.
2. Suggested by Hasan Mukhtar (Case Western Reserve University).
3. Comments by H. Shinozuka (University of Pittsburgh School of Medicine).
4. Comments by Arthur Furst (Neo Life Company of America). Much of the needed information may come from epidemiological and clinical studies. The development of experimental systems to study these cancers is also important.
5. Comments made by C.S. Yang during the preparation of this report. A more detailed discussion of the use of mixtures is given in the article by Yang *et al.* in this volume.
6. Comments by Floyd Dunn (Chiang Mai University).
7. Comments by Yung-Pin Liu (NCI).
8. Comments by Richard Black (Kellogg Canada Inc.).
9. Healthy People 2000 Full Report: Stock No. 017-001-00474-0 Superintendent of Documents, Government Printing Office, Washington, DC.

Appendix 1

PROPOSED OUTLINE OF DISCUSSION

1. Important Research Field, But Not Well Funded

2. Common Weaknesses and Problems

Too descriptive
Lack of mechanistic insights into the carcinogenic process and its inhibition
Lack of imaginative experimental approaches
Experimental system and preventive agents not well defined
The project will yield a lot of data, but little useful information
Lack of relevance to human cancer

3. Ways to Strengthen this Field of Research

To conduct mechanistic research for the purpose of gaining useful information for the prevention of human cancer by:

Selecting more relevant projects
Understanding the carcinogenic process
Understanding the chemical properties and biological activities of the agents
Understanding the hormonal and physiological factors involved
Utilization of newer molecular and chemical approaches
Collaboration or consultation with scientists from disciplines such as pathology, molecular biology, chemistry, epidemiology, and biostatistics

4. New Opportunities and Approaches

Molecular studies on caloric restriction
Dietary effects on cell proliferation – molecular mechanisms involved
Modulation of signal transduction pathways by dietary constituents
Protection of oxidative damage by dietary antioxidants
Dietary effects on hormonal levels which influence carcinogenesis
Nutrient – gene interactions
Development of relevant biomarkers
Extrapolation from high to low doses
Translation of animal results to human situations
Extending to human studies

5. Justify the Importance and Avoid Pitfalls

6. Possible Future Workshops and Other Activities

Appendix 2

Cancer Risk Reduction Objectives *(Healthy People 2000[9]):*

16.7 Reduce dietary fat intake to an average of 30% of calories or less and saturated fat intake to less than 10% of calories among people aged 2 and older. (Baseline: 36% of calories from total fat and 13% from saturated fat for people aged 20 through 74 in 1976-80; 36% and 13% for women aged 19 through 50 in 1985)

16.8 Increase complex carbohydrate and fiber-containing foods in the diets of adults to 5 or more servings daily for vegetables (including legumes) and fruits, and to 6 or more daily servings for grain products. (Baseline 2-1/2 servings of fruits and vegetables and 3 servings of grain products for women aged 19 through 50 in 1985)

16.10 Increase to at least 75% the proportion of primary care providers who routinely counsel patients about tobacco use cessation, diet modification, and cancer screening recommendations. (Baseline: About 52% of internists reported counseling more than 75% of their smoking patients about smoking cessation in 1986)

Appendix 3

The Surgeon General's Report on Nutrition and Health, USDHHS, PHS, USGPO, Washington, DC, 1988.

Research and surveillance issues of special priority related to the role of diet in cancer should include investigations into:

- Molecular mechanisms of carcinogenesis and the ways in which specific components of the diet may affect initiating or promoting events
- Chemoprevention and dietary clinical trials
- Effect of specific dietary components on cancer etiology
- Interactions between specific dietary factors in cancer prevention and causation
- Development of biomarkers for dietary intake
- Patterns of food intake associated with cancer prevention
- Development of national population data bases on food and nutrient consumption patterns and specific cancer rates
- Levels of carcinogenic substances in the food supply
- Dietary guidance methods to help people improve patterns of food intake
- Causes of cancer cachexia and effects of nutritional support

Appendix 4

Diet and Health: Implications for Reducing Chronic Disease Risk National Research Council, National Academy Press, Washington, DC, 1989.

Directions for Research:

- **Methodology.** Dietary assessment, biomarkers, and food databases to better establish relationships of dietary constituents to cancer.
- **Intervention Trials.** To obtain definitive information on the role of diet and cancer in humans, it would be desirable to conduct intervention trials in which diets are modified in specific ways and the subjects monitored for sufficient time to assess the impact on cancer incidence at a number of sites.
- **Genetic Determinants.** The role of genetic factors as they modify individual responses to environmental (dietary) exposures.
- **Quantitative Relationships.** Quantitative nature of the relationship between food constituents and cancer risk is poorly understood.
- **Mechanisms of Action.** Elucidation of the mechanisms of action for dietary factors that affect cancer risk are not completely understood and elucidation would help establish the causal nature of diet-cancer associations. However, this information is not essential to the formulation of dietary policy.

ABSTRACTS

INTERACTIVE EFFECT OF FAT AND VEGETABLES AND FRUIT ON 1,2-DIMETHYLHYDRAZINE INDUCED COLON CANCER IN RATS

J.M. Rijnkels, R.A. Woutersen* and G.M. Alink.
Department of Toxicology, Agricultural University, Wageningen, The Netherlands
*TNO Nutrition and Food Research Institute, Zeist, The Netherlands

High fat and low fibre diets are positively correlated with the incidence of human colon carcinogenesis. Most epidemiological and animal experimental studies focus on effects and mechanisms of one food component. However, intake of different food products may result in combined exposure to a variety of food items. It is reasonable to assume that these components interact with each other at different mechanistic levels, and that this ultimately determines the carcinogenic risk and the development of colon cancer.
In previous studies (1) the effect of heat processing and the addition of non-nutrients in vegetables and fruit in human isocaloric diets (40e% fat) was investigated in rats, pretreated with 1,2-dimethylhydrazine. It was concluded that heat processing had no effect on the incidence of colon tumors and that the addition of vegetables and fruit even may increase the colon tumor incidence. Moreover in the same study rats were fed an animal diet (20e% fat) without and with vegetables and fruit added, as control diets. Interestingly rats fed animal diets with vegetables and fruit showed a significant lower incidence of colorectal adenomas in contrast to rats fed only the animal diet. It is hypothesized that non-nutrients in vegetables and fruit exert a protective effect in diets with a low fat content. Therefore experiments were designed to further test this hypothesis.
Four week old male Wistar rats (n=120) received a semi-synthetic diet (groups A and B) or were fed a semi-synthetic diet supplemented with freezedried vegetables and fruit (groups C and D; total vegetable and fruit content was 19.5 % w/w) and were observed for 35 weeks. A and B contained 20e% fat, while groups C and D contained 40e% fat. The extra fat added in groups C and D was instead of carbohydrate, while the proteins and micronutrients for all four diets were adjusted so that the four diets only differed at the level of fat/carbohydrate content and the presence of non-nutrients in vegetables and fruit. Importantly the fatty acid composition and the choice of the vegetables and fruit approached the mean fatty acid (f.a.) composition (34.3 % saturated f.a., 42.1 % mono-unsaturated f.a., 17.2 % polyunsaturated f.a.) and the mean vegetable and fruit consumption (17 different vegetables and fruit) in the Netherlands. Three weeks after the start of the experiment the rats were given a subcutaneous injection of 50 mg 1,2-dimethylhydrazine/kg b.w. once a week for ten weeks.
The results show a higher tumor incidence in diets containing 40e% fat compared with the 20e% fat diets. Supplementation of vegetables and fruit results in a lower tumor incidence both in low and high fat diets.

[1] G.M.Alink, H.A.Kuiper, V.M.H.Hollanders, J.H.Koeman (1993). Effect of heat processing and of vegetables and fruit in human diets on 1,2-dimethylhydrazine-induced colon carcinogenesis in rats. Carcinogenesis 14(3):519-524.

QUANTITATION OF SOY ISOFLAVONOIDS IN HUMAN FLUIDS BY HPLC

Adrian Franke and Laurie Custer
Cancer Research Center of Hawaii; 1236 Lauhala St.; Honolulu, HI 96813

Increasing evidence suggests that isoflavonoids might protect against various cancers, particularly against breast and prostate cancer since Asian populations with high exposure to these agents through consumption of soy foods show a relatively low cancer rate at these sites. Additionally, soy isoflavones have been found to exhibit a variety of properties connected with cancer prevention such as antioxidant, serum cholesterol lowering, antiestrogenic, anti inflammatory and anti viral additionally to anti proliferative and anti carcinogenic activities. Epidemiologic trials concerned with the assessment of the possible role of these agents as cancer protective agents require a fast, reliable and affordable technique to measure exposure of populations to these analytes favorably through non-invasive protocols.

Therefore, we developed a simple and fast procedure to extract and hydrolyze isoflavonoids and their conjugates from human urine and from human milk. A fast and selective HPLC method is presented for baseline separations of the major isoflavonoids found in humans after soy consumption. Analytes were identified by UV absorption scans, by fluorometric and electrochemical detection and by using internal and external authentic standards. Detection limits of a 20μL injection with monitoring at the individual compound's absorption maximum were found to be 1.09, 0.53, 3.28 and 1.00 pmoles for daidzein, genistein, equol and O-desmethylangolensin, respectively. These limits could be reduced up to 1000 fold by extended concentration through partitioning with ethyl acetate, by electrochemical detection, by increased injection volumes or by a combination of these.

The proposed technique was applied to monitor isoflavone levels in human urine and in human milk after challenge with 5 to 20g roasted soybeans. Additional data are presented giving evidence about the feasibility of the proposed procedure to be used in future epidemiologic trials evaluating the cancer protective properties of soy foods and/or soy isoflavones.

FLAVONOIDS, POTENT AND SPECIFIC INHIBITORS OF THE HUMAN P FORM PHENOL SULFOTRANSFERASE.

Thomas Walle, E. Alison Eaton, & U. Kristina Walle. Department of Pharmacology, Medical University of South Carolina, Charleston, SC 29425, U.S.A.

Sulfation (sulfonation) is one of the major phase II conjugation reactions for drugs and environmental chemicals as well as for endogenous compounds such as steroids and neurotransmitters. This biotransformation reaction is usually a detoxification process, leading to greatly enhanced renal excretion of the highly charged conjugates formed. However, for an increasing number of environmental chemicals reactive intermediates have been shown to be formed, which can undergo DNA binding, leading to mutagenicity and carcinogenicity (1,2). Altered activity of the enzymes responsible, i.e. the sulfotransferases (STs), could thus have major health implications. In the present study we report for the first time on the potent inhibitory effect of quercetin and other dietary flavonoids on the human STs.

The flavonoid quercetin was a potent inhibitor of the sulfation of p-nitrophenol, the marker substrate for the human P (phenol) form phenolsulfotransferase (PST), with an IC_{50} of 0.10 ± 0.03 μM. This inhibition was highly selective, being 3 to 4 orders of magnitude more potent in inhibiting the P form PST than the other known human STs, i.e. the M (monoamine) form PST, dehydroepiandrosterone ST and estrogen ST. Surprisingly, quercetin was not a substrate for sulfation under the conditions used. Initial mechanism studies showed the inhibition to be noncompetitive, with a K_i of 0.10 μM. Among other flavonoids examined, several were found to have similar high potency as quercetin. These observations suggest the potential for important chemopreventive actions by flavonoid-containing foods and beverages.

References: 1) Miller JA, *Chem Biol Interact* **92**, 329-41 (1994). 2) Chou HC et. al, *Cancer Res* **55**, 525-9 (1995).

A TRANSDOMINANT NEGATIVE C-JUN INHIBITS TRANSFORMATION AND SENSITIZES TO CISPLATIN AND RADIATION

Olga Potapova and Dan Mercola, Sidney Kimmel Cancer Center, 3099 Science Park Road, San Diego, CA 92121 and Center for Molecular Genetics, University of California, La Jolla, CA 92093.

Recent studies in several laboratories have indicated that tumor inhibitory properties of certain retinoids may be due to interference with AP-1, a transcriptional activator complex composed of a Fos and Jun family member1. In support of this conclusion, we have previously observed that activation of c-Jun by phosphorylation at serines 63 and 73 is required for transformation of primary fibroblasts by either v-src or activated Ras[2-4]. These observations suggest that a transdominant negative inhibitor of AP-1 may be formed by a non-phosphorylatable c-Jun created by replacement of serines 63 and 73 by alanine[2-3] and that this mutant should mimic inhibition by retinoids.

We have prepared expression vectors for the mutant, c-Jun(Ala63,73) that contains the CMV promoter/enhancer which is very active in human cells. Stable clones of several human tumor lines including some which are known to be inhibited by retinoids such as MCF-7 have been prepared. These include human glioblastomas T98G and U373, breast carcinomas MCF-7 and T47-D, and prostate carcinoma PC3. Representative clones of these cell types exhibit increased expression of immunoreactive c-Jun, decreased clonogenicity, greatly reduced growth rates, and decreased saturation densities. Moreover, in several cases such as T98G and MCF-7, stable expression of the mutant c-Jun leads to much increased sensitivity to the chemotherapeutic agents cisplaltinum and radiation. Viability measured 4 days after a 1 h exposure to cisplatin (platinum chloride) is reduced by over 80%. The response is dose-dependent with an IEC_{50} of 20-30 μM cisplatin or 30 J/m^2 for UV-C irradiation.

These observations suggest that activated AP-1 may play a role in supporting the transformed phenotype of several human tumor lines consistent with the the action of certain retinoids. The c-Jun mutant may be useful for gene therapy and may potentiate the therapeutic effects of retinoids, cisplatin or radiation treatment.

References:

1. Fanjul, A., Dawson, M., Hobbs, P., Hong, L., Cameron, J. Harlev, E., Graupner, G., Lu, X.-P., and Pfahl, M. (1994) A New Class of Retinoids with Selective Inhibition of AP-1 Inhibits Proliferation. *Nature*, 372:107-111.

2. Smeal, T., Binétruy, B., Mercola, D.A., Birrer, M. and Karin, M. (1991) Phosphorylation of cJun serines 63 and 73 is required for oncogenic and transcriptional cooperation with Ha-Ras. *Nature*, 354, 494-496.

3. Smeal, T., Binétruy, B., Mercola, D., Bardwick-Gróver, A., Heidecker, G., Rapp, U. and Karin, M. (1992). Oncoprotein mediated signalling cascade stimulates cJun activity by phosphorylation of serines 63 and 73. *Molec. Cell. Biol. 12*, 3507-3513.

4. Grover-Bardwick, A., Adamson, E., and Mercola, D. (1994). Transformation-Specific Pattern of Phosphorylation of c-Jun, Jun-B, Jun-D and Egr-1 in v-*sis* Transformed Cells. *Carcinogenesis*, 15(8):1667-1674.

Carboxyalkylating Agents produced by Nitrosation of α-Amino Acids.

Xiaojie Wang[a], Sheng Chong Chen[a], Fulvio Perini[a], Barry I. Gold[a,b] and Sidney S. Mirvish[a,b,c]. ([a]Eppley Institute for Research in Cancer, [b]Department of Pharmaceutical Sciences and [c]Department of Biochemistry and Molecular Biology, University of Nebraska Medical Center, Omaha NE 68198)

W. Lutz's group found that α-amino acids react with nitrite to give agents (half-life 0.25-8.0 h) that alkylate 4-(p-nitrobenzyl)-pyridine to give purple dyes. To identify the alkylating groups, we treated α-amino acids with excess nitrite at pH 1.5-2.5, reacted the mixture with 3,4-dichlorothiophenol (DCTP) at pH 8 and made CH_2Cl_2 extracts. These were purified by TLC, recrystallized (if they were solids) and identified by ^{1}H-NMR and mass spectrometry. Glycine gave S-carboxymethyl-DCTP, m.p. 84°C (identical to the product from bromoacetic acid), glycine ethyl ester gave S-carbethoxy-DCTP (an oil) and aspartic acid gave S-1,2-dicarboxyethyl-DCTP, m.p. 155-157°C (identical to the product from bromosuccinic acid). Yields were 5, 2.8 and 13%, respectively. The alkylating species were probably diazoacetic acids (N=N=CR.COOH), because nitrosation of NH_2 group of dipeptides gives diazopeptides (N=N=CR.CONHCHR'COOH) (B.C. Challis, Cancer Surveys 8:363, 1989), which are genotoxic and carcinogenic. Diazoacetates should be more stable than simple diazoalkanes because conjugation of carboxyl group with N=CH double bond should retard electron shift into N=N to form N_2. Inductive effect of 2 carboxy groups in the diazoacetate from aspartic acid may explain why it is especially stable (half-life, 8 h cf. 15 min for glycine product).

We found free amino acid levels of 0.13, 6 and 8 mM in 3 human gastric juice samples (lit. mean, 0.37 mM) and levels of 13-49 mg/g in fish sauce, soy sauce and salted fish, which would supply free amino acids in the stomach when eaten. These foods may be involved in the etiology of gastric cancer and were obtained from high gastric cancer areas of China. We conclude that diazoacetic acids could present a carcinogenic hazard because they could form in the stomach, where both nitrite (literature mean, 2.6 μM) and amino acids occur.

In future studies, we will finish the chemical studies and determine if nitrosated amino acids can react with guanosine and DNA, and if they can alkylate gastric wall DNA when fed. Support: Grants 89B36 and 94B28 from AICR, core grant CA-36727 from NCI and core grant SIG-16 from ACS.

Effect of Phenylethylisothiocyanate (PEITC), Diallyl Sulfide (DAS) and Ellagic Acid (EA) on Methyl-n-amylnitrosamine (MNAN) Metabolism to Hydroxy-MNANs (HO-MNANS) by Rat Tissues or Tissue Slices

S.S. Mirvish[a,b,c], S.C. Chen[a], C. Hinman[a] and C. Morris[a] ([a]Eppley Inst. Res. Cancer, [b]Dept. Pharm. Sci., [c]Dept. Biochem. Molec. Biol., Univ. Nebraska Med. Center, Omaha NE 68098)

DAS, PEITC and EA all inhibited carcinogenesis by nitrosamines in the rat esophagus. Here, we examined effects of inhibitors on MNAN metabolism. Test compounds were administered to male MRC-Wistar rats under conditions reported to inhibit nitrosamine metabolism or carcinogenesis. Rats were killed and esophagi, liver slices and nasal mucosa were incubated for 2 h at 37°C with 23 μM MNAN in Eagle's medium (Huang et al., Cancer Lett. 69:107, 1993). The medium was extracted with CH_2Cl_2, which was analyzed for 2-. 3- and 4- HO-MNAN by gas chromatography-thermal energy analysis.

DAS (200 mg/2.5 ml corn oil/kg b.w.) was gavaged 18 h before death. The esophagus showed a mean 28% inhibition (3.5 ± 0.4 vs. 4.9 ± 0.4 in controls); liver, 41% inhibition (3.2 ± 0.3 vs. 5.5 ± 0.4); and nasal mucosa, 77% inhibition (12 ± 2 vs. 51 ± 6 % total HO-MNANs/100 mg tissue/2 h incubation for 6-11 incubations, mean $\pm$ SE). For the same DAS dose given 3 h before death, esophagus, liver and nasal mucosa showed 14, 35 and 76% inhibitions (4.3 ± 0.4 vs. 5.0 ± 0.5, 3.8 ± 0.3 vs. 5.9 ± 0.5, and 13 ± 2 vs. 56 ± 5 %/100 mg/2 h for 5-9 incubations). Inhibitions were less when DAS was given 42 h before death, and were significant for nasal mucosa at 18 and 3 h, and for liver slices when 3 h and 18 h results were combined ($P < 0.01$). PEITC (16.3 and 163 mg/2 ml corn oil/kg b.w.; gavaged 2 h before death) inhibited esophageal metabolism of MNAN by 42 and 72% (confirming Huang et al., Carcinogen. 14:749, 1993), but had no effect with nasal mucosa or liver. EA (400 mg/kg AIN-76 semipurified diet fed for 7 days) had little effect on MNAN metabolism by esophagus, liver and nasal mucosa.

Hence we confirmed the strong inhibition by PEITC of MNAN metabolism by rat esophagus, but saw weak inhibition of esophageal MNAN metabolism with DAS and no inhibition with EA. The strong inhibition by DAS of the nasal mucosal metabolism of MNAN was striking. Support: Grants RO1-CA-35628, P30-CA-36727 (National Cancer Institute) and SIG-16 (American Cancer Society).

Prevention of Papillomavirus Induced Tumors with Dietary Indoles

Karen Auborn[1], Allan Abramson[1], Liora Newfield[1], Dan W. Sepkowiz[2] and H. Leon Bradlow[2]
[1]Department of Otolaryngology, Long Island Jewish Medical Center, New Hyde Park, NY 11040
[2]Strang Cornell Cancer Research Center, New York, NY 10021

Our hypothesis is that increasing the consumption of indoles (e.g. indole-3-carbinol, I3C) found in cabbage, broccoli, cauliflower or brussels sprouts can prevent human papillomavirus (HPV) induced tumors in estrogen sensitive tissue. We postulated that 16α-hydroxyestrone, a product of 16α-hydroxylation of estradiol, drives HPV transformation while alternate estrogen metabolism, 2-hydroxylation induced by indoles prevents HPV transformation.

Prevention of HPV induced benign tumors with Indoles: We have previously determined that: 1) Increased 16α-hydroxylation of estradiol occurs in cells derived from HPV induced laryngeal tumors, 2) I3C abrogated the proliferative effect of estradiol on laryngeal cells, and 3) Dietary I3C prevented growth of laryngeal papillomas in a mouse model.* In patients with recurrent laryngeal papillomas, we show that the severity of disease is worse in patients showing greater 16α-hydroxylation of estradiol. In a small study, a subset of patients improved with a diet enriched with cruciferous vegetables, and improvement correlated with a change in estrogen metabolism.

Prevention of malignant transformation of HPV induced tumors by indoles: We previously showed a relationship between estrogen metabolism and HPV induced immortal and malignant genital cells.** Since 16α-hydroxyestrone potentiates estrogen action and increases unscheduled DNA synthesis, we tested the ability of the 2 and 16α- metabolites to alter anchorage independent growth (AIG) of HPV immortalized cells. This marker correlates with malignant conversion. Both estradiol and 16α-hydroxyestrone increased AIG while 2-hydroxyestrone, which is induced by I3C inhibits AIG.

These studies suggest that HPV lesions and their progression to cancer in estrogen sensitive tissues can be prevented by a diet that induces 2-hydroxylation of estradiol.

*Newfield L, Goldsmith A., Bradlow HL and Auborn KJ, Estrogen metabolism and human papillomavirus-induced tumors of the larynx: Chemo-prophylaxis with indole-3-carbinol. *Anticanc Res* 13:337-342, 1993.

**Auborn KJ, Woodworth C, DiPaolo J and Bradlow HL. The interaction between HPV infection and estrogen metabolism in cervical carcinogenesis. *Int J Cancer* 49:867-869, 1991.

Indole-3-carbinol: A Selective Chemopreventive for Human Breast Cancer. **Li G, Bradlow HL, Osborne MP, Kumar R[1] and Tiwari RK. Strang-Cornell Cancer Research Laboratory, Cornell University Medical College, New York, [1]Pennsylvania State University, College of Medicine, Hershey, PA.**

Indole-3-carbinol (I3C). an active chemopreventive, is present in cruciferous vegetables such as cabbage, broccoli, and brussels sprouts. Its anticancer properties has been attributed to its effect as an anti-initiator as determined in animal models where the compound was administered with or prior to carcinogens. Our own study shows that spontaneous mammary tumors in C3H/HeJ are significantly reduced in a dose dependent manner when I3C was incorporated in the diet. The objective of the present study is to examine the cellular and molecular mechanism of action of I3C in human breast cancer cells. I3C acts specifically on estrogen receptor (ER) positive cells and not on ER-negative cells. This sensitivity was also observed using cell clones differing in estrogen responsiveness isolated from ER-positive MCF-7 cells. I3C also selectively modulated estradiol metabolism and induction of cytochrome P-4501A1. Estradiol metabolism was shifted preferentially to C-2 hydroxylation of estrone and cytochrome P-4501A1 was inducible only in ER-positive cells. I3C does not bind with ER but affects phosphorylation of ER. Experiments are in progress to determine the role of methylation of the ER in sensitivity to I3C. These studies demonstrate the action of a dietary micronutrient on fundamental biochemical processes that control cell proliferation in human breast cancer and suggest a novel mechanism of I3C action .

(Supported by the American Institute of Cancer Research , Grant # 94B66)

EFFECTS OF INHIBITORS OF EICOSANOID BIOSYNTHESIS ON LINOLEIC ACID (LA)-STIMULATED HUMAN BREAST CANCER CELL INVASION AND METASTASIS

Jeanne M. Connolly and David P. Rose, Division of Nutrition and Endocrinology, American Health Foundation, Valhalla, NY

Previous work by us has shown that a high-fat, LA-rich, diet stimulates metastasis of MDA-MB-435 human breast cancer cells in nude mice, and the invasive capacity of these cells in an *in vitro* assay system. In the present study, we examined the influence of two pharmacological inhibitors of LA-derived eicosanoid biosynthesis on the metastatic cascade.

MDA-MB-435 cells, 10^6, were injected into the mammary fat pads of female nude mice fed 20% (wt/wt) fat diets containing 12% (wt/wt) LA. Indomethacin, principally a cyclooxygenase inhibitor, and esculetin, a selective inhibitor of lipoxygenases, were administered in the drinking water at estimated received doses of 1.0 mg and 28.0 mg/kg body weight, respectively. Eleven weeks later, necropsies were performed: these showed that the incidence of metastatic lung nodules in the controls (71%) was significantly higher than in the indomethacin (33%; $p<0.01$), or esculetin (41%; $p<0.05$) treated mice. Likewise, the estimated total metastatic volumes per mouse (mm^3) were greater in the controls (58.0 ± 15.1; $x \pm SE$) compared with the indomethacin (17.1 ± 7.2; $p<0.01$) or esculetin (36.1 ± 12.7; $p<0.05$) treatment groups.

Primary tumors from indomethacin-treated mice contained significantly lower levels of prostaglandin E_2 (PGE_2); both $PGF_{1\alpha}$ and thromboxane B_2 (TBX_2) were also reduced, but a greater change in TBX_2 resulted in a significant elevation in the $PGF_{1\alpha}/TBX_2$ ratio. Esculetin-treated mice showed reductions in tumor 12-hydroxyeicosatetraenoic acid (12-HETE) and 15-HETE; PG and TBX_2 levels were unchanged.

In vitro, LA-stimulated tumor cell invasion was blocked by esculetin; 12-HETE also stimulated invasion, and bypassed the effect of the 12-lipoxygenase inhibitor. Equimolar doses of indomethacin, or of the specific cyclooxygenase inhibitor piroxicam, had no effect on tumor cell invasion.

CONCLUSION: The stimulation of breast cancer cell metastasis in nude mice by dietary LA involves both lipoxygenase product (12-HETE)-mediated enhancement of local invasion, and cyclooxygenase-mediated events which do not appear to include regulation of local invasion.

Supported by a grant from the AICR, and PHS Grant CA 53124.

EFFECTS OF DIETARY OMEGA-6 AND SHORT AND LONG CHAIN OMEGA-3 FATTY ACIDS ON THE GROWTH OF DU145 HUMAN PROSTATE CANCER CELLS IN NUDE MICE

David P. Rose, Jeanne Connolly, and Melissa Coleman, Division of Nutrition and Endocrinology, American Health Foundation, Valhalla, NY

A menhaden oil (MO)-containing diet, rich in eicosapentaenoic acid (EPA) and docosahexaenoic acid (DHA), two long-chain omega-3 fatty acids (n-3 FAs), suppresses the growth of the DU145 human prostatic cancer cell line in nude mice. However, a recent epidemiological study found that a diet high in α-linolenic acid, a short-chain n-3 FA and the metabolic precursor of EPA, is associated with an increased risk for biologically aggressive prostate cancer.

We compared the effects of feeding high levels of corn oil (CO, linoleic acid-rich) linseed oil (LO, α-linolenic acid-rich) and MO on DU145 cell growth *in vivo*. Groups of 25 male nude mice were fed 23% (wt/wt) fat diets containing either 18% CO:5% LO, 18% LO:5% CO, or 18% MO:5% CO for 1 week. The DU145 cells, 10^6, were then injected s.c., the diets continued for 6 weeks, and the solid tumors were measured weekly. At necropsy, the tumors were weighed, and subsequently analyzed for phospholipid FAs. The tumor growth rates were similar in mice fed 18% CO or 18% LO; comparatively, there was a significant retardation in the 18% MO-fed mice ($p<0.001$ by week 4). Also, the tumor weights in the 18% MO group (0.14 ± 0.01g; $\bar{x} \pm$ SE) were significantly lower than those in the 18% CO (0.18 ± 0.02g; $p<0.05$), and the 18% LO (0.19 ± 0.01g; $p<0.01$) groups.

Linoleic acid levels were no different in the phospholipids of tumors from mice fed the 18% CO or 18% LO diets, but were significantly lower in tumors from the 18% MO group ($p<0.001$). Tumors from both the high MO and LO dietary groups contained high EPA levels, but whereas DHA was 1.6-fold higher in the 18% MO group, this n-3 FA was reduced by feeding LO compared with levels in the 18% CO group ($p<0.001$).

It is concluded that, in contrast to MO, a high α-linolenic acid diet does not suppress growth in the DU145 prostate cancer cell model; differences in effects on tumor eicosanoid biosynthesis from n-6 FAs may be involved.

Supported by a grant from the American Institute for Cancer Research.

PREDICTIVE VALUE OF PROLIFERATION, DIFFERENTIATION AND APOPTOSIS AS INTERMEDIATE BIOMARKERS FOR COLON CANCER. W-C. L. Chang, R.S. Chapkin and J.R. Lupton. Faculty of Nutrition, Texas A&M University, College Station, TX 77843.

To determine the prognostic significance of proliferation, differentiation, and apoptosis as intermediate biomarkers for colon tumor development, these indices were measured during the promotion phase of tumorigenesis. Two hundred forty weanling male Sprague Dawley rats were provided with one of 2 fats (corn oil or fish oil); 2 fibers (pectin or cellulose); plus or minus the carcinogen azoxymethane (AOM) in a 2 x 2 x 2 factorial design. AOM was injected at weeks 2 and 3 following diet administration. Eighty rats were killed at week 18; 160 at week 36. In vivo cell proliferation was measured immunohistochemically using incorporation of bromodeoxyuridine (BrdU) into DNA. Differentiation was assessed by binding of dolichos biflorus agglutinin (DBA) to colonocytes. Apoptosis was measured by immunoperoxidase detection of digoxigenin-labeled genomic DNA. At week 18, in the proximal colon, BrdU positive cells in the colonic crypt were 15% higher in corn oil + AOM animals vs the fish oil - AOM animals ($p<0.05$). Patterns of differentiation were opposite to proliferation. DBA binding for corn oil + AOM treatment was 57% lower than fish oil - AOM treatment in the bottom 1/3 of the crypt ($p<0.05$). Patterns of apoptosis and differentiation in the crypts were similar. Crypts from corn oil + AOM animals had 31% fewer apoptotic cells vs crypts from fish oil - AOM animals ($p<0.05$). Carcinoma incidence at week 36 was 79% for corn oil + AOM treatment and 53% for fish oil + AOM treatment ($p<0.001$). These data suggest that dietary corn oil resulted in enhanced tumor development compared to fish oil in part due to a coordinated increase in colonic cell proliferation and a decrease in differentiation and apoptosis. Supported in part by AICR and NIH CA 59034 and CA 61750.

Protective effects of retinoic acid against tumor progression in mouse skin: Inhibition of ras p21 processing enzyme farnesyltransferase and Ha-ras p21 membrane localization

Rajesh Agarwal, Department of Dermatology, Case Western Reserve University, Cleveland, Ohio.

In recent years, considerable efforts are directed to identify agents which may have the ability to inhibit, retard or reverse one or more stages of multistage carcinogenesis which is comprised of initiation, promotion and progression. In this context, the progression of benign tumors into malignancy is most critical because the later lesions are capable of metastatic spread leading ultimately to the death of the host. In this study, the protective effects of all *trans* retinoic acid (RA) were assessed against skin tumor progression in three different protocols of mouse skin multi-stage carcinogenesis. Under these protocols, when papilloma yield is stabilized, animals were divided into two groups and treated twice weekly either with acetone or RA (10 μg). In each case, RA showed significant protective effects against % of mice with carcinomas, number of carcinomas per mouse, and the rate of malignant conversion; the protective effects of RA were more pronounced in high-risk TPA-protocol than in low-risk. Since activation of ras oncogenes by point mutations has been implicated in the genesis of tumor induction, and since ras oncogenes encode a protein termed p21 which transforms mammalian cells only when localized at the inner side of the plasma membrane by a cytosolic enzyme known as farnesyltransferase (FTase) followed by other biochemical events, FTase activity and the levels of both membrane-bound and cytosolic Ha-ras p21 were determined in uninvolved skin of tumor-bearing mice, papillomas and carcinomas obtained from both vehicle-treated controls and RA-treated experimental groups under different treatment protocols. No matter how the data are analyzed and what comparisons are considered, in all the cases RA showed statistically significant inhibition ($p < 0.01$ to 0.001) of FTase activity. Furthermore, tissue samples from RA-treated groups in three different protocols also showed significantly diminished membrane localization of Ha-ras p21 with an increase in cytosolic Ha-ras p21. Summarizing these results together, it is logical to conclude that RA treatment results in the inhibition of FTase activity which leads to a decrease in the farnesylation of Ha-ras p21. This decrease results in a decreased membrane localization of Ha-ras p21 with a concomitant increase in cytosolic Ha-ras p21. In conclusion, a strong correlation was observed between the inhibition of ras p21 farnesylation by RA and its protective effect against malignant conversion of papillomas to carcinomas.

Synergistic Effects in the Inhibition of Proliferation of MDA-MB-435 Human Breast Cancer Cells by Tocotrienols and Flavonoids

Guthrie, N.[a], Gapor, A.[b], Chambers, A.F.[c] and Carroll K.K.[a] Departments of Biochemisty[a] & Oncology[c], University of Western Ontario, London, ON, Canada, N6A 5C1 and Palm oil Research Institute of Malaysia[b], Kuala Lumpur, Malaysia.

Palm oil is a rich source of tocotrienols, a form of Vitamin E having an unsaturated isoprene side-chain, rather than the saturated side-chain of the more common tocopherols. The tocotrienol-rich fraction (TRF) of palm oil contains α-tocopherol and a mixture of α-, γ- and δ-tocotrienols in the approximate ratio of 3:3:4.5:1. Our interest in tocotrienols was stimulated by the observation that dietary palm oil promotes chemically-induced mammary carcinogenesis only when stripped of the vitamin E fraction (Nesaretnam et al., Nutr. Res. 12:63-75, 1992). Flavonoids comprise a large number of related compounds that are widely distributed in plants. Some flavonoids, such as genistein from soybeans, have been shown to have anti-cancer properties (Messina et al., Nutr. Cancer 21:113-131, 1994).

In earlier experiments, we observed that TRF and the individual tocotrienols inhibited proliferation of MDA-MB-435 estrogen receptor-negative human breast cancer cells (IC_{50}s 30-180 μg/mL) whereas α-tocopherol was ineffective. These cells were also inhibited by flavonoids, including genistein, hesperitin, naringenin and quercetin, at IC_{50}s of 23-130 μg/mL. In the present studies, 1:1 combinations of TRF or individual tocotrienols with flavonoids inhibited these cells more effectively than any of the compounds alone. The most effective combinations were α-tocotrienol with hesperetin (IC_{50} 0.8 μg/mL) and γ-tocotrienol with naringenin (IC_{50} 1 μg/mL). This synergistic effect of tocotrienols and flavonoids suggests that they are inhibiting cell proliferation by different mechanisms.

(Supported by the Palm Oil Research & Development Board of Malaysia).

Effects of Citrus Flavonoids on Proliferation of MDA-MB-435 Human Breast Cancer Cells and on Mammary Tumorigenesis Induced by DMBA in Female Sprague-Dawley Rats

Carroll, K.K.[a], So, F.[b], Guthrie, N.[a] & Chambers, A.F.[c], Departments of Biochemistry[a], Pharmacology & Toxicology[b] and Oncology[c], The University of Western Ontario, London ON, Canada, N6A 5C1

Flavonoids are naturally occuring compounds that are widely distributed in plants. Some flavonoids, such as genistein from soybeans, have been reported to possess anti-cancer activity (Messina *et al.*, Nutr. Cancer 21: 113-131, 1994). Experiments in our laboratory have shown that naringenin and hesperetin, flavonoids present in grapefruit and oranges respectively, inhibit the proliferation of MDA-MB-435 human breast cancer cells *in vitro* more effectively than genistein.

Naringenin is found in grapefruit mainly in its glycosylated form, naringin. These compounds, as well as grapefruit and orange juice concentrates, were fed to female Sprague-Dawley rats to investigate their ability to inhibit the development of mammary tumors induced by 7,12-dimethylbenz(a)anthracene (DMBA) (Parenteau *et al.*, Nutr. Cancer 17: 235-241, 1992). The rats were fed a semipurified diet containing 5% corn oil and were given a 5 mg intragastric dose of DMBA at approximately 50 days of age while in diestrus. Groups of 21 rats were given double strength grapefruit juice or orange juice or were fed naringin or naringenin at levels comparable to that provided by the grapefruit juice. Tumor development was delayed in the rats given orange juice or the naringin-supplemented diet compared to control animals on the 5% corn oil diet and plain drinking water. Rats in the orange juice group had fewer and smaller tumors, although they gained more weight than those in any of the other groups.

These observations *in vitro* and *in vivo* indicate that citrus flavonoids merit further investigation as possible agents for breast cancer prevention and/or treatment.

(Supported by the State of Florida Department of Citrus Research)

ANTIGENOTOXIC EFFECTS OF LYCOPENE ON HUMAN COLON CELLS *IN VITRO*

Smedman, A.E.M.[1,2], Smith, C.[2], Davison, C.[2] and <u>Rowland, I.R.</u>[2]

[1] Dept. of Medical Nutrition, Karolinska Institute, Sweden.
[2] BIBRA International, Woodmansterne Rd., Carshalton SM 54 DS, UK.

Lycopene, a γ - carotenoid found at high levels in tomatoes (3.1 mg/100g) [Mangels *et* al 1993, J Am Diet Assoc 93 284 - 296], has been associated in epidemiological studies with decreased cancer incidence in man. We have studied the effects of lycopene on the induction of DNA damage in Caco-2 cells, a human adenomacarcinoma colon cell line. Caco-2 cell suspensions (2 x 10^6 cells/ml) were incubated for 30 minutes at $37^{O}C$ with 1-methyl-3-nitro-1-nitrosoguanidine (MNNG; 3 μg/ml final concentration), or hydrogen peroxide (200 μM final concentration) in the absence or presence of different concentrations of lycopene (0.1 - 500 ug/ml). After the incubation period, the cells were washed and collected by centrifugation and DNA damage was assessed using the single cell microgel electrophoresis assay ('comet assay').

Both MNNG and H_2O_2 induced DNA damage in the colon cells as defined by measuring tail moment. Lycopene, at concentrations up to 500ug/ml (the highest concentration tested) did not induce DNA damage. When lycopene was added to the incubations with MNNG or H_2O_2 , the DNA damage induced by MNNG and H_2O_2 was significally decreased . The greatest inhibitory effect (approximately 50%

of the control value) was observed at lycopene concentrations of 0.1 - 5.0 ug/ml, which are similar to serum levels reported in subjects consuming normal levels of tomatoes. At higher doses of lycopene the inhibitory effect decreased. The cell viability was over 90 % in all incubations.

Although lycopene was effective against H_2O_2-induced damage the protective effect would not appear to be due simply to the antioxidative capacity of lycopene, since genotoxicity of the alkylating agent MNNG was also inhibited.

These results provide evidence that lycopene, at concentrations found in serum, may protect against chemically-induced DNA damage in human colon cells.

A.S. is grateful for a grant from WITEC. We thank the Coca Cola Company for financial support. I.R. , C.D and C.S. are supported by a grant from MAFF, UK.

DNA DAMAGE IN FOLATE-DEFICIENT ERYTHROBLASTS: THE ROLES OF URACIL MISINCORPORATION INTO DNA AND P53 ACCUMULATION AND THE POTENTIAL FOR INCREASED LEUKEMOGENESIS.

M.J. Koury, D.W. Horne, B.C. Blount, and B.N. Ames, Vanderbilt University and V.A. Medical Centers, Nashville, TN and University of California at Berkeley, Berkeley, CA.

Deficiency of either folate or vitamin B_{12} causes the disease megaloblastic anemia which is characterized by destruction of hematopoietic cells in the bone marrow with decreased production of mature blood cells. Previous studies have demonstrated DNA breakage in megaloblastic anemia. In patients treated for vitamin B_{12} deficiency, the incidence of developing leukemia is increased several fold. Both the mechanism of hematopoietic cell destruction in megaloblastic anemia and whether the incidence of leukemia is increased in those patients treated for folate deficiency are unknown. In order to study the effects of folate deficiency on normal hematopoiesis and leukemogenesis, a murine model of folate-deficient hematopoiesis was developed using a folate-free diet. An *in vitro* system of folate-deficient erythropoiesis was established by using erythroblasts collected from mice that had received the folate-free diet and that were in the acute, non-malignant phase of the disease caused by the Friend leukemia virus. This *in vitro* system demonstrated that folate-deficient erythroblasts underwent a programmed cell death (apoptosis) that was inhibited by thymidine. In the present study, uracil misincorporation into DNA and total cellular content of p53 protein were measured in erythroblasts cultured in folate-deficient or folate-replete medium. Uracil misincorporation into DNA occurs in cells in which thymidylate/deoxy-uridylate ratio is low. The repair of misincorporated uracil residues that are closely spaced on opposite strands of DNA leads to double-stranded cleavage of DNA. DNA damage such as double-stranded breaks results in intracellular accumulation of p53 protein which appears to play a role in inducing apoptosis. The amounts of uracil in DNA were determined by gas chromatography-mass spectrometry. The amounts of wild-type p53 protein were determined by Western blotting. When cultured in folate-deficient medium, both the uracil content of DNA and the cellular content of p53 protein increased significantly in folate-deficient cells as compared to the same cells cultured in folate-replete medium. Since DNA damage and p53-associated apoptosis cause erythroblast death and result in megaloblastic anemia in folate deficiency, it can be hypothesized that an increased rate of leukemic transformation exists in those folate-deficient erythroblasts in which DNA damage is not corrected and the cell survives. To test this hypothesis, the effect of folate deficiency on the rate of transformation from the preleukemic phase to the leukemic phase was examined for the disease that occurs in mice infected with Friend leukemia virus. The transformation to the leukemic phase was measured by growth factor-independent colony growth of spleen cells. The results indicate that folate deficiency decreases hematopoietic cell numbers during the preleukemic phase, but after correction of the folate deficiency the rate of transformation to the leukemic phase is increased.

GLUTATHIONE S-TRANSFERASE GENOTYPES AS RISK FACTORS FOR HEAD AND NECK CANCER

Zoltán Trizna, MD, PhD, Gary L. Clayman, MD, DDS, Margaret R. Spitz, MD, MPH, Katrina L. Briggs, Helmuth Goepfert, MD:

The University of Texas M. D. Anderson Cancer Center, Department of Head and Neck Surgery

ABSTRACT

BACKGROUND: Several enzymatic systems, including glutathione S-transferases, are involved in the metabolism of environmental agents. The absence of glutathione S-transferases mu (GST-M1) and theta (GST-T1) results in decreased detoxification of carcinogens, e.g., chemicals in cigarette smoke. These metabolic deficiencies may predispose individuals to the development of smoking related tumors such as cancers of the lung, head and neck, and bladder.

METHODS: The glutathione S-transferase genotypes of 90 previously untreated patients with squamous cell carcinoma of the head and neck and 90 healthy controls were determined with polymerase chain reaction (PCR) methodologies. Lymphocytes separated from heparinized peripheral blood or whole blood extracts served as sources of genomic DNA. The presence or absence of the gene-specific PCR products revealed the positive or negative genotypes, respectively.

RESULTS: Odds ratios were calculated with Taylor's confidence intervals. The test showed an odds ratio of 1.25 (95% CI: 0.70-2.25) for patients with the absent GST-M1 genotype. The odds ratio was 2.47 (95% CI: 1.31-4.67) for patients in whom the GST-T1 gene was missing. The combined absence of the GST-M1 and GST-T1 genotypes conferred an odds ratio of 3.08 (95% CI: 1.38-6.87).

CONCLUSIONS: Our preliminary data suggest that genetically determined factors of carcinogen metabolism, i.e., the absence of the GST-M1 and GST-T1 genotypes may be associated with increased risk for head and neck cancer.

Differential cell proliferation responses of human Hs578T breast cancer cells to saturated and unsaturated fatty acids occur via EGFR-associated G-proteins:

Wickramasinghe NSMD, McDonald JM, Jo H, and Hardy RW

Long chain saturated fatty acids (LCSFA) have been shown to inhibit, while long chain unsaturated fatty acids stimulate breast cancer cell proliferation in animal and cell culture models. We have previously demonstrated that LCSFA may be acting via an EGFR-associated G-protein. We now show that the time course of LCSFA inhibition of breast cancer cell proliferation parallels that of the EGFR associated G-protein ADP-ribosylation. We further show that this inhibitory effect is hormone specific. In identical experiments using IGF1 in place of EGF, stearate pretreatment had a minimal effect on IGF1 stimulated cell proliferation (22% ± 10%, n=3) compared to EGF (complete inhibition). In addition we demonstrate that the inhibitory effect of stearate is not due to non- specific changes in membrane fluidity. After treatment with stearate or fatty acid free media, Hs578T cells were harvested, treated with the spin label 5-doxyl-stearate and subjected to electron paramagnetic resonance measurements at various temperatures. The changes in maximum hyperfine splitting were collected and plotted vs temperature with the resulting line being a measure of membrane fluidity. There were no significant differences in these lines and thus no difference in the membrane fluidity of cells treated with stearate compared to control cells. Interestingly in contrast to LCSFA, pretreatment with oleic acid (0.05 mM) increased basal and additively increased EGF induced cell proliferation. Also in contrast to stearate oleate increased the amount of Gi protein that associates with the EGFR in both basal and EGF treated states. Thus the differential effects of fatty acids on breast cancer cell proliferation are likely mediated via EGFR associated G-proteins.

An Abnormal Receptor for a Vitamin A Derivative in Promyelocytic Leukemia

Robert L. Redner, Elizabeth A. Rush, Susan Faas, William A. Rudert, and Seth J. Corey, University of Pittsburgh Medical Center, Children's Hospital of Pittsburgh, and the University of Pittsburgh Cancer Institute, Pittsburgh PA 15213

Vitamin A derivatives induce differentiation of malignant cells in Acute Promyelocytic Leukemia (APL). This leukemia is characterized by rearrangement of the retinoic acid receptor alpha (RAR), a retinoic acid-dependent nuclear transcription factor. In most APL cases the t(15;17) chromosomal rearrangement results in expression of a retinoic acid receptor fusion protein, called PML-RAR. The mechanism whereby expression of this aberrant protein gives rise to the APL phenotype remains unclear. In an effort to gain insight into the molecular biology of this disease, we have been studying rare patients with APL who do not show the t(15;17) abnormality. We have reported a patient with APL who manifested t(5;17). This translocation gives rise to fusion between the genes encoding nucleophosmin (NPM) and RAR. NPM is a nucleolar phosphoprotein thought to be involved in RNA processing or packaging. The NPM-RAR fusion contains the same RAR sequences as in the PML-RAR t(15;17). Two alternatively spliced forms of NPM-RAR are expressed. Both have NPM and RAR sequences in the same reading frame, and give rise to proteins of 57 kDa and 62 kDa. Both proteins are expressed in the leukemic cells. In transient transcriptional assays the two forms of NPM-RAR act as retinoic acid-dependent transcriptional factors. The NPM-RAR fusion promises to be a critical reagent that can be used to further dissect the molecular mechanism underlying APL.

The effects of a low fat high carbohydrate diet on the time to menopause.

Martin LJ, Lockwood GA, Boyd NF, Tritchler DL, Kriukov V. Division of Epidemiology and Statistics, Ontario Cancer Institute, Toronto, Ontario, Canada.

We investigated the effect of a low fat, high carbohydrate diet on the time to menopause in women at increased risk for breast cancer participating in a long term randomized controlled intervention trial. The ongoing trial is designed to test the hypothesis that a reduction in dietary fat intake (to a target of 15% of energy) will lead to a reduction in breast cancer incidence. Time of menopause was ascertained by questionnaire and was defined as the absence of menstrual periods for at least 6 months. Of the 695 women who were premenopausal at entry to the trial , 98 had become postmenopausal at the time of this investigation; 55 in the intervention group and 43 in the control group. We compared the time from randomization until menopause between the intervention and control groups by generating curves using the Kaplan-Meier method and comparing them using the log-rank test. Women who remained premenopausal at follow up were considered censored at their last clinic visit. The log-rank statistic was significant ($p=0.045$) indicating that subjects in the intervention group were experiencing menopause at an earlier time from randomization. When age and weight were controlled for using a Cox proportional hazards regression, the effect of study group remained significant ($p=0.03$). In this analysis, both age and weight significantly affected time to menopause in the directions expected. Greater age was associated with shorter time from randomization to menopause ($p=0.0001$) and greater body weight was associated with a longer time from randomization to menopause ($p=0.02$). Earlier menopause in women in the low fat, high carbohydrate group may reflect an effect of the diet on ovarian function which may lead to a reduced risk of developing breast cancer.

VITAMIN E RADICAL IS REPAIRED BY DIETARY CAROTENOIDS: AN ELECTRON TRANSFER AND CELLULAR STUDY

Fritz Böhm[1], Ruth Edge[2], Edward J Land[3], David J McGarvey[2], Jane H Tinkler[2] and T George Truscott[2]; [1] Department of Dermatology (Charité), Humbolt University, 10117 Berlin, Germany; [2] Department of Chemistry, Keele University, Keele, ST5 5BG, U.K.; [3] C.R.C. Department of Biophysical Chemistry, Paterson Institute for Cancer Research, Christie Hospital N.H.S. Trust, Manchester, M20 9BX, U.K.

Carotenoids and vitamin E are lipid soluble antioxidants present in foodstuffs. They are efficient free radical and singlet oxygen quenchers but little is known about how they interact with each other and what effect this has on their antioxidant potential. We have used the technique of pulse radiolysis to investigate the interactions between a group of carotenoids (β-carotene, lycopene, canthaxanthin, septapreno-β-carotene and 7,7'dihydro-β-carotene) and the (dl) α-tocopheroxyl radical, ($EO^{\bullet}$), which is produced during radical scavenging. The α-tocopherol radical was generated in hexane, using pulse radiolysis in the absence ane presence of carotenoid. In the presence of a carotenoid, the carotenoid radical cation was observed to grow in, with different kinetics to that in the absence of $EO^{\bullet}$, and in a larger amount. **Thus electron transfer occurs from the carotenoid to the α-tocopherol radical;**

$$EO^{\bullet} + CAR \rightarrow EO^{-} + CAR^{\bullet+}. \qquad (1)$$

The reverse reaction did not occur. From the growth kinetics of the carotenoid radical cations the rate constants for the reaction were calculated and found to be high for all five carotenoids studied, (8.8×10^{9}-$1.8\times10^{10} M^{-1}s^{-1}$). The α-tocopherol anion produced picks up a hydrogen ion reverting to its original form. In this way each of the carotenoids is regenerating the vitamin E and so maintaining its antioxidant ability. Cellular studies using human lymphocytes are consistent with the above results. Also, in separate experiments using lymphocytes we show that lycopene (the red pigment in tomatoes) protects these cells much more efficiently than β-carotene [1].

1. Böhm, F.Tinkler J.H. and Truscott T.G. *Nature Medicine* **1** (2): 98-99 (1995).

RB Protein Concentration In Nucleus vs. Cytoplasm Is Independent of Phosphorylation During the Cell Cycle or Differentiation

Andrew Yen and Susi Varvayanis, Department of Pathology, College of Veterinary Medicine, Cornell University, Ithaca, NY 14853

Unphosphorylated RB (retinoblastoma tumor suppressor) protein is known to bind isolated nuclear matrix in vitro whereas phosphorylated RB has a lower affinity, suggesting a mechanism driving differential nuclear localization, a presumed determinant of its regulatory function. This motivates interest in the in vivo localization of the endogenous RB protein during changes in its phosphorylation. It is known that in proliferating HL-60 cells all the RB protein is phosphorylated, but the extent of phosphorylation increases with progression from G1 to S to G2+M. It has also been previously shown that retinoic acid and 1,25-dihydroxy vitamin D3 shift the RB protein to the unphosphorylated state with cell differentiation (Yen, A. and Varvayanis, S. Exp. Cell Res. 214:250-257, 1994). This provides two natural instances where the phosphorylation of the RB protein changes in vivo. The dependence of RB nuclear versus cytoplasmic localization on phosphorylation state can thus be tested in these two instances. Confocal image analysis of the RB protein in vivo shows that in all cases the concentration of the RB protein in the nucleus was approximately 2 times that in the cytoplasm. The results show that contrary to expectation, the ratio of RB protein nuclear versus cytoplasmic density is independent of RB phosphorylation state. RB's enhanced nuclear concentration is thus not driven by phosphorylation regulated affinity for the nuclear matrix.

Supplemental nutrition with ornithine α-ketoglutarate in cancer - associated cachexia: surgical treatment of a tumor improves efficacy of nutritional support, Thierry Le Bricon, Luc Cynober†, Catherine J. Field, Vickie E. Baracos
Department of Agricultural, Food and Nutritional Sciences, University of Alberta, Edmonton, Alberta, Canada, T6G 2P5 and †GRENEMH, CHU Saint-Antoine, 75571 Paris Cedex 12, France

We investigated the use of ornithine α-ketoglutarate in treatment of rats bearing Morris hepatoma 7777. Rats received diets containing ornithine α-ketoglutarate, which has been used in other catabolic states (i.e. injury, sepsis), or an isonitrogenous, isocaloric diet containing glycine. Untreated tumors grew to a mass of 11 g / 100 g body weight over 3 wk after implantation, and induced progressive anorexia, negative N balance and body and tissue wasting. Compared with glycine, ornithine α-ketoglutarate had no effect on tumor growth, but also did not alter the catabolic effects of the tumor on its host. We hypothesized that capture of amino acids by the tumor limited the efficacy of supplemental nutrition here and in published reports where tumor burden comprised 4-30% of body weight. This is supported by our observation that at 3 wk of implantation the rate of protein deposition plus amino acid oxidation by the tumor was equivalent to ~70% of the host's daily protein intake. To parallel the clinical situation where tumor burden is small at diagnosis and initiation of treatment, the same diets were tested in rats treated by excision of the tumor at a limited stage of the disease. Rats received 3 d preoperative nutrition with ornithine α-ketoglutarate or glycine, and continued on the same diets for 3 or 6 d post-operatively. Compared with glycine, ornithine α-ketoglutarate - fed rats showed more positive N balance, higher concentrations of glutamine and branched chain amino acids in muscle, and accelerated protein deposition in small intestine ($P<0.05$). Our results explain the lack of success of nutritional support in untreated cancer, and underline the need for clinically relevant animal models for further studies.

Metabolism Decreases the Growth Inhibitory Effect of Genistein in Human Mammary Epithelial Cells.

G. Peterson*, M. Kirk[!], G-P. Ji*, and S. Barnes*[!]. Department of Pharmacology* and Comprehensive Cancer Center Mass Spectrometry Shared Facility[!], University of Alabama at Birmingham, Birmingham, AL 35294

Data from our laboratory show that genistein (G) inhibits the serum- and EGF-stimulated growth of normal human mammary epithelial cells (HME) with IC_{50} values 11 to 15 fold lower than transformed breast epithelial cells (MCF-7). These data suggest that G might lower breast cancer risk by slowing the growth of normal breast epithelial cells, rather than by inhibiting the growth of fully transformed cells. One possible explanation for this difference in sensitivity is that G undergoes differential metabolism by HME and MCF-7 cells. Metabolism might either increase or decrease the growth inhibitory action of G. In this study we report that MCF-7 cells metabolized G into one predominant metabolite, a mono-sulfated form of G (GS). In contrast, HME cells did not metabolize G at detectable levels. The majority of G and GS were found in the extracellular media. This suggested that metabolism of G to GS produced a less potent growth inhibitor. However, GS was an equally potent growth inhibitor in MCF-7 cells as G. Subsequently, the intracellular concentration of G was determined in HME and MCF-7 cells using 3-O-methyl glucose as a marker for intracellular volume and ^{3}H-inulin to correct for extracellular contamination. While both HME and MCF-7 cells concentrated G intracellularly , in HME cells the intracellular G concentration was 11 times higher than in MCF-7 cells. In summary, metabolism of G to GS (in MCF-7 cells) might account for the differential growth inhibition of HME and MCF-7 cells to G by producing a form of G which is excreted by MCF-7 cells. This process would lead to decreased intracellular G concentrations in MCF-7 cells. Therefore, MCF-7 cells require higher extracellular G concentrations to attain equivalent intracellular G concentrations in HME cells, resulting in higher IC_{50} values.

Supported by AICR G1B58-REV2, NCI CA-61668, and United Soybean Board

Early and late stages of colon carcinogenesis are modulated differently by a low or high fat beef tallow diet in Sprague Dawley rats. **R.P. Bird, C.K. Lasko, Kaiqi Yao and C.K. Good.** Department of Foods and Nutrition, University of Manitoba, Winnipeg, Canada, R3T 2N2 Phone:204-474-9903 Fax:204-275-5299. The main objective of the present proposal was to investigate the effect of feeding a low or high fat diet on the early and late stages of colon carcinogenesis. Sprague Dawley male rats were injected with azoxymethane (20mg/kg/wk) for two weeks. One week later they were randomly allocated to eat a low (4% beef tallow + 1% corn oil) or a high fat (18.6% beef tallow + 4.7% corn oil) diet (LF or HF). After 10 weeks of feeding, 10 animals per group were killed and their colons were evaluated for tumors. Remaining animals in each group were further divided into LF and HF groups. The four experimental groups consisted of groups receiving LF or HF diet throughput the study (LF-LF or HF-HF) and the groups fed LF or HF diet for the first 10 weeks, then assigned the alternate diet for the remainder of the duration (LF-HF or HF-LF). By week 26, the remaining animals were killed and their colons were evaluated for the number, location and size of tumors. The tumor incidence in the HF-HF and HF-LF groups were higher than the LF-LF and LF-HF groups (81.6 and 84.8 vs. 71.4 and 60.0). Tumor multiplicity ranged from 1.86 ± 0.26 to 2.54 ± 0.33 in all groups. Average size of tumors and total tumor area/rat were significantly affected by the time at which the diet was fed. Average size and total tumor area in the animals fed HF diet during early stages (HF-HF and HF-LF) were significantly higher than those fed the LF diet during the early stages. Late intervention by specific diets did not affect tumor outcome. Sequential enumeration of aberrant crypt foci of different growth features representing early preneoplastic stages corroborated the findings of the tumor outcome. It was concluded that early preneoplastic stages were more sensitive than their advanced counterparts to the dietary interventions of the present study. **Acknowledgements:** Funded by the Cancer Research Society Inc. of Canada. The NSERC operating grant to RP Bird provided a student assisstantship to Kaiqi Yao.

INHIBITION OF LIGAND-INDUCED ACTIVATION OF EPIDERMAL GROWTH FACTOR RECEPTOR TYROSINE PHOSPHORYLATION BY CURCUMIN.

L. Korutla, J. Y. Cheung, J. Mendelsohn[1] and R. Kumar*. Department of Medicine, and Department of Cellular and Molecular Physiology, Pennsylvania State University College of Medicine, P.O. Box 850, Hershey, PA 17033; [1]Memorial Sloan-Kettering Cancer Center, 1275 York Avenue, New York, NY 10021.

We explored the regulation of epidermal growth factor (EGF)-mediated activation of EGF receptors (EGF-R) phosphorylation in murine fibroblast NIH3T3 cells expressing human EGF-R by curcumin (diferuloyl-methane), a recently identified kinase inhibitor. Treatment of cells with a saturating concentration of EGF for 5-15 min induced increased EGF-R tyrosine phosphorylation by 4-11 fold, and this was inhibited in a dose- and time-dependent manner up to 90% by curcumin, which also inhibited the growth of EGF-stimulated cells. There was no effect of curcumin treatment on the amount of surface expression of labeled EGF-R, and the inhibition of EGF-mediated tyrosine phosphorylation of EGF-R by curcumin was mediated by a reversible mechanism. In addition, curcumin also inhibited the EGF-induced but not bradykinin-induced calcium release. These findings demonstrate that curcumin is a potent inhibitor of a growth stimulatory pathway, the ligand-induced activation of EGF-R, and may potentially be useful in developing antiproliferative strategies to control tumor cell growth.
(Supported by grant number 94B93 (R.K.) from the American Institute of Cancer Research)

LINOLEATE DEPENDENT CO-OXYGENATION OF t-RETINOL ACETATE BY SOYBEAN LIPOXYGENASE (SLO) AND HUMAN TERM PLACENTAL LIPOXYGENASE (HTPLO).

Arun P. Kulkarni and Kaushik Datta. Toxicology Research Program, College of Public Health, University of South Florida, Tampa, FL, USA.

Pharmacological doses of vitamin A and some of its analogs have been shown to have beneficial effects in the treatment of some neoplasms of epithelial origin as well as skin conditions such as acne and psoriasis. In addition, several retinoids are potent teratogens in a number of animal species including human. Although, the importance of metabolism in retinoid toxicity has been stressed by many, no published data exist on the biotransformation of retinoids by human conceptal tissues. Here, we report linoleic acid (LA) dependent co-oxidation of t-retinol acetate (t-RAc) by SLO and HTPLO. Under optimal assay conditions, SLO exhibited a specific activity of 844 nmoles/min/nmole of enzyme. To observe the maximum rate of co-oxidation by HTPLO (370 nmoles/min/mg protein), incubation of 100 μM t-RAc, 2 mM LA and 50 μg/ml enzyme in Tris buffer pH 9.0 was essential. Both SLO and HTPLO catalyzed reactions were significantly inhibited by NDGA, BHT, BHA, gossypol and ETI. The results suggest that t-RAc is oxidatively metabolized in human term placenta via LO pathway and this may constitute an important part of puzzle worthy of consideration in the understanding of retinoid teratogenicity.

Binding of synthetic retinoids to retinol-binding proteins and their cancer preventive activity. Brahma P. Sani, Southern Research Institute, 2000 9th Ave (South), Birmingham, AL 35255.

Certain retinyl ethers and retinamides express profound effects in the inhibition of carcinogen-induced mammary tumors. The cellular transport and biological activity of natural and synthetic retinoids are mediated by their specific retinoid-binding proteins and nuclear receptors for retinoic acid (RARs and RXRs). During our attempts to understand the action of retinyl ethers and retinamides in the prevention of cancer, we observed that retinyl methyl ether (RME) and 4-hydroxyphenylretinamide (4HPR) bind, in competition with retinol, to serum retinol-binding protein (RBP). Like retinol, RME also showed binding affinity for cellular retinol-binding protein (CRBP). 4HPR, however, lacked such binding property. RME, retinyl propynyl ether and retinyl butenyl ether expressed significant binding affinity for CRBP (60-100% inhibition). These retinoids also showed activity in the rat mammary cancer prevention model. Ethers that were biologically inactive, such as retinyl butyl ether, showed negligible binding affinity (<10%) for CRBP. In general, a correlation , with some exceptions, was observed between the binding affinities of the various retinyl ethers and their chemopreventive activity against rat mammary tumors. Eventhough RBP and CRBP facilitate, respectively, the plasma and cellular transport of retinyl ethers, we could not detect any significant binding of the ethers to recombinant RARs and RXRs. However, 4HPR showed binding affinity for the RARs equivalent to 15% of that of retinoic acid. Whether the retinyl ethers and retinamides, as such, function as activators/antagonists in the RAR/RXR-mediated transcriptional activation or whether they require metabolic activation to interact more efficiently with the nuclear receptors are yet to be determined (Supported by Grants AICR 93B39 and NIH 34968).

1,25-DIHYDROXYVITAMIN D_3 PROTECTS HEMATOPOIETIC STEM CELLS BUT NOT BREAST CANCER CELLS AGAINST APOPTOSIS

GEORGE P. STUDZINSKI, HALINA ORGACKA, TIM KANTER AND PETER DAMBROWSKI

DEPARTMENT OF MOLECULAR AND DIAGNOSTIC PATHOLOGY
UMD - NEW JERSEY MEDICAL SCHOOL
NEWARK, N.J. 07103

Brief treatment of HL60 cells with differentiation-inducing concentrations of 1,25-dihydroxyvitamin D_3 (1,25D_3) has been previously shown to make these cells resistant to cell death by apoptosis (Xu et al, Exp Cell Res 209,367, 1993). Apoptosis was detected by the characteristic morphology under light microscopic examination, presence of DNA "ladders" on agarose gel electrophoresis, DNA fragmentation by filter elution assay, and the "apoptotic index" obtained by comparison of damage to mitochondrial and nuclear gene DNA on Southern blots. The protective effect of 1,25D_3 treatment was apparent before phenotypic evidence of differentiation and before altered traverse of the cell cycle could be detected. We extended these studies to a model chemotherapy system by treating with 1,25D_3 and its analogs ("deltanoids") the hematopoietic stem cells from human umbilical cord blood (HUCB) obtained at delivery, and cultured human breast cancer cells MCF 7. Subsequent exposure of the cells to the chemotherapeutic drugs doxorubicin (DOX) or arabinocytosine (araC) showed that HUCB cells have an increased resistance to these drugs when pretreated with deltanoids. Conversely, MCF7 cells have an increased susceptibility to DOX when pretreated with deltanoids. This suggests that deltanoids have the potential to increase the therapeutic index of drugs used to treat systemic breast cancer. (Supported by Susan G. Komen Foundation).

PHENOLIC ANTIOXIDANTS PRODUCE REACTIVE PHENOXYL RADICALS THAT CAN DAMAGE CRITICAL THIOLS AND DNA

Valerian E. Kagan, Jack C. Yalowich, H. Gregg Claycamp, Billy W. Day, Rado Goldman, Detcho A. Stoyanovsky, Vladimir B. Ritov

Departments of Environmental and Occupational Health and Pharmacology
University of Pittsburgh, Pittsburgh, PA 15238

Phenolic compounds act as radical scavengers due to their ability to donate a mobile hydrogen to peroxyl radicals producing a phenoxyl radical. If the phenoxyl radical formed in the radical scavenging reaction efficiently interacts with vitally important biomolecules, this interaction may result in cytotoxic effects rather than in antioxidant protection. Using two model compounds - phenol and a phenolic antitumor drug, VP-16, we demonstrated that their phenoxyl radicals readily oxidize ascorbate, low molecular weight thiols (glutathione, dihydrolipoate) and protein sulfhydryls in model systems, cells and tissues. This results in depletion of water-soluble antioxidants and oxidative modification of proteins (e.g., thioredoxin, Ca^{2+}-ATPAse, Na^{+},K^{+}-ATPase, NADPH-cytochrome P450 reductase) due to the formation of disulfides of mixed disulfides (reversible modification) or sulfoxygenated species (irreversible modification). Using a newly developed HPLC technique to detect ESR-silent spin adducts of glutathionyl radicals with a spin trap, DMPO, we found that myeloperoxidase-driven redox-cycling of phenoxyl radicals yields GSH-thiyl radicals in HL-60 cells. Phenoxyl radical-initiated generation of thiyl radicals (in isolated thiols) and disulfide-anion radicals (in isolated and vicinal thiols) may be accompanied by generation of reactive oxygen species and subsequent oxidation of DNA bases (e.g., accumulation of 8-OHdG). The above reactivity of phenoxyl radicals was not characteristic for vitamin E and its water-soluble homologues (Trolox or 2,2,5,7,8-pentamethyl-6-hydroxychromane). The reactivity of phenoxyl radicals towards critical biomolecules should be carefully considered in the design and development of biomedical natural and synthetic phenolic antioxidants (e.g., flavonoids).

Dietary Black Tea Extracts Inhibit 7,12-dimethylbenz(a)anthracene Binding to Mammary Cell DNA.

K. Sakamoto[1] Y. Hara[2]. J.A. Milner[1].

[1]Department of Nutrition, Penn State University, University Park, PA and [2]Food Research Laboratories, Mitsui Norin, Fujieda, Japan.

This study examined the influence of dietary black tea extract powder obtained from Mitsui Norin, Japan on the binding of the 7,12-dimethylbenz(a)anthracene (DMBA) metabolites to mammary cell DNA. This black tea powder was incorporated into a 15% casein based diet at 0, 1, or 2%. Female Sprague Dawley rats, 41 days of age, were fed these diets for 14d. Food intake was not influenced by dietary black tea supplementation. Weight gains also were not influenced by black tea consumption. Rats were sacrificed 24h after DMBA treatment and DNA was extracted from the mammary tissue. Total DMBA-DNA adducts were depressed (ca. 45%) by black tea (1%) consumption. Providing more than 1% black tea did not result in a further reduction in DMBA-DNA binding. A depression in anti-dihydrodiolepoxide adducts accounted for most of the reduction in total binding. The activity of hepatic glutathione-S-transferase (GST) was increased by 18.8 and 35% in the rats consuming 1 and 2% black tea, respectively.

These studies demonstrate that black tea extract is effective in altering DMBA bioactivation and presumably the initiation phase of chemical carcinogenesis. Enhanced GST activity may partially account for this protection.

Supported by American Institute for Cancer Research Grant 92B65.

Curcumin irreversibly inactivates protein kinase C activity and phorbol ester binding: its possible role in cancer chemoprevention

Rayudu Gopalakrishna, Usha Gundimeda, and Zhen-Hai Chen

Department of Cell and Neurobiology, School of Medicine,
University of Southern California, Los Angeles, CA 90033

Curcumin is the major pigment of turmeric, a product isolated from plant *Curcumin longa Linn,* possesses antitumor promoter, anti-inflammatory, and antioxidant properties. However, its mechanism of action is not clearly known. Since protein kinase C (PKC) serves as a receptor for not only phorbol esters but also for other tumor promoters such as oxidants, it is logical that curcumin may counteract at a certain stage in the metabolic events triggered by tumor promoters. However, the early studies revealed either a lack of or a weak inhibition of PKC by curcumin. Since we have been interested in the oxidative regulation of PKC in tumor promotion, curcumin effects on PKC studied taking into consideration of its phenolic group, ferric binding properties, and ability to react with mercapto agents. Curcumin inhibited purified PKC (Ca^{2+}-dependent isoenzymes) in the absence of thiol agents (IC_{50} = 15 μM). However, in the presence of thiol agents such as dithiothreitol (1 mM) the inhibition was weaker (IC_{50} = 105 μM). Nonetheless, when curcumin was used in the form of ferric complex, it inhibited PKC strongly (IC_{50} = 3 μM) even in the presence of thiol agents. Furthermore, curcumin inhibited PKC activity reversibly, whereas curcumin-ferric complex inhibited irreversibly. Other metals tested were found to be ineffective in promoting the action of curcumin. Detailed studies carried out with purified PKC suggested that curcumin-ferric complex permanently modified PKC resulting in the loss of both kinase activity and phorbol ester binding. The phenolic group present in curcumin was found to be necessary for mediating this effect.

Curcumin also irreversibly inactivated PKC in intact cells and the sensitivity of various cell types varied. Since curcumin alone inhibits PKC in a reversible way, the observed irreversible inactivation of PKC intact cells as well as the difference in susceptibility of various cell types suggest that other cofactors may be necessary to mediate this effect. It is possible that such an irreversible effect of curcumin on PKC may be mediated by an intracellular formation of a curcumin-ferric complex. The difference in sensitivity of various cell types to curcumin-mediated inactivation of PKC may be related to differences in the intracellular accessible iron pool. Since iron plays an important role in oxidations, carcinogenesis, and cell growth, a chelation of ferric by curcumin may facilitate its beneficial effects. Taken together these results suggest that curcumin may mediate its cancer chemopreventive effects, at least in part, by chelating iron and inducing an irreversible inactivation of PKC.

Supported by a grant 93B43 from the American Institute for Cancer Research.

INSIGHT INTO THE MECHANISM OF EARLY TUMOR ANOREXIA

Michael M. Meguid, MD, PhD; Alessandro Laviano, MD; and Zhong-jin Yang, MD. Surgical Metabolism and Nutritoin Laboratory, Department of Surgery, University Hospital, Syracuse, NY

An increase in the brain neurotransmitter serotonin (5-HT) is known to cause anorexia while an increase in its plasma pre- cursor free tryptophan (FTRP) vs other large neutral amino acids (FTRP/LNAA) occurs in cancer patients. Whether in tumor-bearing rats, a temporal relationship exists between an increase in plasma FTRP, an increase in brain 5-HT and onset of anorexia [reduction in food intake (FI)] was studied. **Methods**: 112 rats were assigned to 3 groups: Tumor-Bearing (TB); TB-Pair Fed (PF); and Controls. FI was recorded daily. In TB rats anorexia developed on day 18 and thereafter FI decreased progressively until end of study. After tumor inoculation, tumor became palpable on day 10 and continued to grow exponentially until end of study. Rats were killed on days 6, 10, 16, 18, 22, and 26 to determine plasma FTRP, FTRP/LNAA and brain 5-HT,

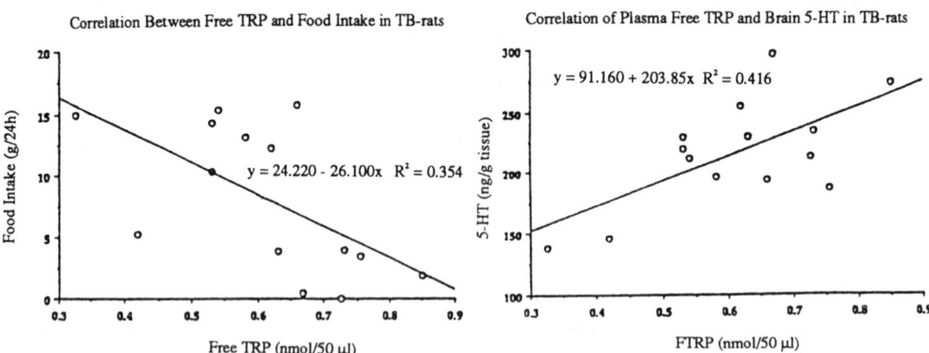

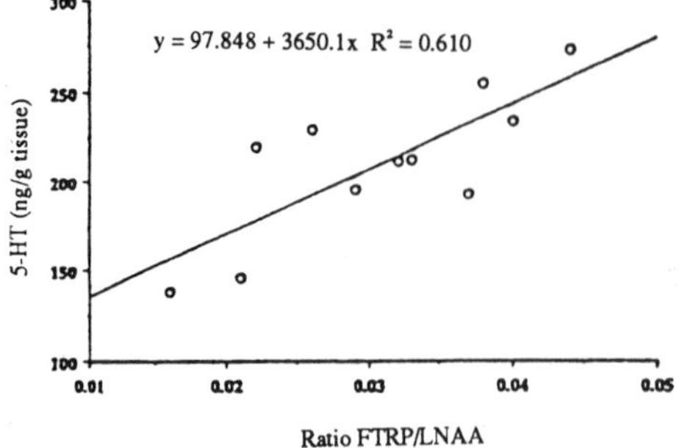

and compared to TB-PF and Controls using ANOVA. **Results**: On day 6, before tumor became detectable in TB rats vs PF and Controls, FTRP and FTRP/LNAA were increased ($p=0.05$). Both continued to increase so that by day 18 when FI had started to decrease $p(<0.05)$, brain 5-HT increased and correlated to onset of anorexia ($R^2=0.6$, $p<0.05$).

Conclusion: Increases in plasma FTRP the precursor to brain 5-HT occurred in TB rats before physical appearance of tumor and increased until an increase in brain 5-HT occurred, leading to anorexia. The reduction in FI associated with early cancer anorexia is related to a temporal increase in plasma FTRP and an increase in brain 5-HT.

IRON STATUS, CANCER INCIDENCE, HISTOPATHOLOGY, AND CELL CYCLE DISTRIBUTION IN DMBA-TREATED RATS.

Sherman, A.R., Hrabinski, D., Costa, A., Millian, B. Department of Nutritional Sciences, Rutgers, The State University, New Brunswick, NJ.

Previous research from our laboratory and others has suggested a relationship between iron and cancer. The present study examined the effects of dietary iron on the tumor incidence, pathology of DMBA-induced mammary tumors, and progression of tumor cells through the cell cycle. 21d old rats were fed AIN 76 diets containing: 5 (Severe), 10, 15 (Moderate), 35, PF, 50 (Control), 500 and 1000 (High) mg Fe/kg diet. To account for reduced food intake by iron-deficient rats, the PF group was fed the 35 mg Fe/kg diet in the amount consumed by the 5 group. Desired iron status was reached after feeding diets for 6 wk. DMBA was given i.g. (5 mg/100 g body weight) at 63d of age. Rats were palpated weekly and detailed anatomical records of tumors were kept. Necropsy was performed at 20 wk post-DMBA. The 5 group had a significantly lower tumor incidence (42%) than controls (69%). Moderately iron-deficient rats and high-iron rats had a significantly higher tumor incidence than controls (77% and 94%). Tumor burden and weight were lowest, and tumor latency was longest in the 5 group. Malignancy rate and anatomical locations of tumors were comparable across treatment groups. Histopathology of tumors was quantified according to 4 parameters: epithelial content, connective tissue content, inflammation, and pleomorphism. No significant differences were found between dietary iron level and tumor histologic score. Preliminary analysis indicates that neither severe nor moderate iron deficiency significantly altered the progression of tumor cells through the cell cycle. While severe iron deficiency protects against mammary cancer, moderate deficiencies and iron excess exacerbate cancer incidence. Neither cell cycle analysis nor histopathology of tumors provide explanations for these findings.
This research was supported by the American Institute of Cancer Research.

PHYTOCHEMICALS FROM TWO CHINESE HERBAL TEA MODULATE RAT HEPATIC S9-DEPENDENT MUTAGENESIS, DNA BINDING, AND METABOLISM OF AFB_1 AND BaP.

Brian Wong[a], Benjamin Lau and Robert Teel[b]. [a]Caribbean Union Col., P.O. Box 175, Port of Spain, Trinidad,[b]Loma Linda U., Sch. of Med., CA 92350, USA.

Herbal tea *Oldenlandia diffusa* (OD) and *Scutellaria barbata* (SB) have been used in traditional Chinese medicine for treating liver, lung and rectal tumors. In this study, we determined the effects of aqueous extracts of these two herbal tea on AFB_1, BaP, BaP 7,8-DHD and BPDE-induced mutagenesis, mediated by enzymes in rat hepatic 900 xg (S9), in *Salmonella typhimurium* TA100. We also studied the effects of these two herbal tea on AFB_1 and BaP binding to calf thymus DNA and their effects BaP 7,8-DHD and BPDE binding to calf thymus DNA. After incubation with rat hepatic S9, organosoluble metabolites of AFB_1, BaP, and BaP 7,8-DHD were analyzed by HPLC, and water-soluble conjugates of BaP, BaP 7,8-DHD and BPDE were analyzed by alumina column LC. Water-soluble conjugates of AFB_1 were determined by passing through C_{18}-Sep-pak cartridge and analyzed by HPLC. Mutagenesis assays showed that aqueous extracts of the herbal tea produced a concentration-dependent inhibition of histidine-independent revertant (His^+) colonies induced by AFB1, BaP, BaP 7,8-DHD, and BPDE. OD and SB inhibited AFB_1 metabolite binding to calf thymus DNA, formation of AFB_1-DNA adducts and metabolism of AFB_1. They also inhibited the binding of BaP metabolites to calf thymus DNA. SB decreased BaP-DNA adduct formation and inhibited the metabolism of BaP. OD and SB significantly inhibited the metabolism of BaP 7,8-DHD and the formation of water-soluble conjugates of BaP 7,8-DHD and BPDE. Both tea modified the distribution of conjugates within the water-soluble fraction. SB inhibited BaP 7,8-DHD and BPDE binding to DNA and decreased the formation of adducts. Our results therefore suggest that the Chinese herbal tea OD and SB possess antimutagenic and anticarcinogenic phytochemicals against the common environmental AFB_1 and BaP. They may possess these properties because they are able to scavenge the reactive metabolites and thus prevent their interaction with DNA.

SULFATION OF ENDOGENOUS, THERAPEUTIC AND ENVIRONMENTAL ESTROGENS BY EXPRESSED HUMAN CYTOSOLIC SULFOTRANSFERASES.

Richard J. Dudley, Victor N. Krasnykh, Josie L. Falany and Charles N. Falany.
Department of Pharmacology and Toxicology, University of Alabama at Birmingham, Birmingham, AL 35294.

Conjugation of steroids, drugs and xenobiotics with a sulfate moiety is an important metabolic process in humans. Addition of the charged sulfate group greatly decreses the biological activity and increases the excretion of most compounds. Steroid sulfates are biologically inactive in that they do not bind their receptors. Four cytosolic sulfotransferases (STs) have been identified in human tissues. Three of the STs are capable of conjugating steroids and related compounds. To analyze the steroid sulfation and kinetic properties of the individual human STs, the four STs were cloned and expressed in *E. coli*: the phenol-sulfating form of phenol ST (P-PST), the monoamine-sulfating form of PST (M-PST), estrogen-ST (EST) and dehydroepiandrosterone ST (DHEA-ST). For expression, the human ST cDNAs were subcloned into the expression vector pKK233-2 and expressed in *E. coli* XL-1 Blue cells. The cytosolic fraction was isolated and the expressed human STs were partially purified by anion-exchange chromatography prior to kinetic characterization. Expressed P-PST sulfated small phenols and estrogens at micromolar (μM) concentrations but was not active towards hydroxysteroids. DHEA-ST conjugated hydroxysteroids such as DHEA and pregnenolone, as well as testosterone and estrogens at μM concentrations. EST sulfated both ß-estradiol and estrone at nanomolar concentrations with maximal activity at 15-20 nM. EST also sulfated 3-hydroxysteroids at μM concentrations but did not conjugate testosterone. M-PST did not sulfate any of the steroids tested. Also none of the STs conjugated glucocorticoids. The phytoestrogens genestein and daidzein were sulfated by P-PST, DHEA-ST and EST; however, naringenin was not sulfated by any of the STs. Diethylstilbestrol and the equine estrogen equalenin were also sulfated by P-PST, DHEA-ST and EST but not M-PST. Expression of the human STs in *E. coli* provides a convenient method for characterization of the individual STs without contamination from other ST activities. [Supported by NIH grant GM38953].

Inhibitory effects of *d*-limonene on the development of colonic aberrant crypt foci induced by azoxymethane in F344 rats.

Toshihiko Kawamori, Takuji Tanaka and Hideki Mori
First Department of Pathology, Gifu University School of Medicine, 40 Tsukasa-machi, Gifu City 500 Japan.

The modifying effect of the monoterpenoid *d*-limonene in drinking water on the development of azoxymethane (AOM)-induced colonic aberrant crypt foci (ACF) was investigated in male F344 rats. The effects of *d*-limonene intake on ornithine decarboxylase (ODC) activity and on the silver-stained nucleolar organizer regions protein (AgNOR) count in the colonic mucosa were also estimated. Animals were given 3 weekly s. c. injections of AOM (15 mg/kg body weight) to induce ACF. These rats were treated with 0.5% *d*-limonene or without *d*-limonene in the drinking water, starting one week before the first dosing of AOM. All rats were killed 2 weeks after the last AOM injection, to measure the number of ACF, ODC activity, and AgNOR count per nucleus in the colon. In rats given AOM and *d*-limonene, the frequencies of ACF and aberrant crypts per colon, and aberrant crypts per focus were significantly decreased compared with those of rats given AOM alone ($P<0.01$, $P<0.001$ and $P<0.001$, respectively). ODC activity in the colon of rats treated with AOM and *d*-limonene was lower than that of rats treated with AOM alone, but the difference was not significant. Number of AgNOR count per nucleus of rats treated with AOM and *d*-limonene was significantly smaller than that of rats treated with AOM alone ($P<0.001$). These results suggest that the monoterpenoid *d*-limonene might be a chemopreventive agent for intestinal tumorigenesis in rats.

Compositional and structural changes in exfoliated leukemia membranes induced by docosahexaenoic acid. Laura J. Jenski, Lian Zhang, La-Dawn Caldwell, A. C. Dumaual, and William Stillwell, Department of Biology, Indiana University - Purdue University at Indianapolis, IN 46202-5132.

Membrane exfoliation occurs commonly in various cell types; in cancer cells, exfoliation may alter plasma membrane enzymatic activity, antigenicity, and structural integrity, ultimately affecting cell survival. Docosahexaenoic acid (DHA), the longest and most unsaturated omega-3 fatty acid with purported anti-cancer properties, is readily incorporated into cancer cell plasma membranes where it apparently alters membrane permeability, bilayer structure, and antigen expression. Because DHA-containing phospholipids interact poorly with the major membrane lipid cholesterol, we predict that DHA enrichment of membranes may foster formation of DHA-rich, cholesterol-poor and DHA-poor, cholesterol-rich membrane domains, with some segregation of membrane proteins into their preferred domains. In the current project we test this hypothesis by exploring DHA's effect on the composition and structure of exfoliated murine leukemia (T27A) membranes (putative membrane domains). Ascites T27A cells were DHA enriched with dietary fish oil, and cultured T27A cells were incubated with fatty acid-supplemented medium; all cells were then allowed to exfoliate membrane vesicles into DHA-free medium. Exfoliated and parent plasma membranes were isolated by differential centrifugation, and compared with respect to fatty acid and protein composition, cholesterol content, and membrane structure. Exfoliated membranes from fish oil-fed ascites or DHA-supplemented cultures were rich in DHA compared to parent plasma membrane and control membranes, and consequently had a high double bond index (unsaturation indicator). Consistent with the compositional measurements, the membrane structure probe pyrene reported fish oil-fed exfoliated membrane to support enhanced lateral mobility. Material exfoliated from DHA-enriched cells had a relative cholesterol content lower than controls, although this may be due to dilution of cholesterol-containing vesicles by concurrently exfoliated DHA-rich vesicles. SDS-PAGE of the various membranes suggested subtle differences in protein segregation induced by DHA. Overall these results are compatible with DHA-induced formation and subsequent exfoliation of select membrane domains. Our investigation is now directed toward understanding the functional changes in cancer cells resulting from DHA-induced exfoliation.

THE EFFECT OF DIETARY SOY ON HUMAN BREAST CANCER METASTASIS IN NUDE MICE

Xin-Hua Liu, Pramod Upadhyaya, Karam El-Bayoumy and David P. Rose
Divisions of Nutrition and Endocrinology, and Chemical Carcinogenesis, American Health Foundation, Valhalla, NY

Soy is a dietary source of several isoflavones, including genistein, a demonstrated antiangiogenic compound. Neovascularization is a key element in tumor progression, and in the present study we examined the effect of a soy-rich diet on the growth and metastasis of the MDA-MB-435 human breast cancer cell line in female nude mice.

The diets all contained 20% (wt/wt) total fat with 8% linoleic acid. Toasted soy chips were substituted for carbohydrate and protein in equal proportions to provide 5, 10 and 20% soy diets. The soy-containing and 8% linoleic acid control diets were fed to groups of 20 mice for 7 days before the mammary fat pad injection of 10^6 MDA-MB-435 cells, and continued for a further 11 weeks. At necropsy, body and primary tumor weights were no different between the 4 dietary groups. However, the incidence of metastatic lung nodules was reduced in both the 10% and 20% soy-fed groups compared with that in the controls ($p<0.05$). When present, the extent of grossly visible metastatic involvement was reduced in the 10% soy-fed mice ($p=0.01$); the difference in the 20% soy compared with the control group just failed to achieve statistical significance ($p=0.064$). The lungs of mice without macroscopic nodules were examined for microscopic metastases; small clusters of tumor cells were detected more frequently in the 10% soy ($p=0.01$) and 20% soy ($p<0.001$) fed mice compared with the control group. HPLC analysis after β-glucuronidase hydrolysis of urine that had been collected immediately prior to the termination of the experiment demonstrated, based on chromatography with synthetic standard, the presence of high levels of both genistein and daidzein in the urine of mice fed the 20% soy diet. Structural confirmation is in progress.

It is concluded that dietary soy, most likely because it is a source of antiangiogenic phytochemicals, suppresses the proliferation rate of micrometastases in a breast cancer nude mouse model; however, in this pilot study, there was no effect on the growth rate of the primary tumor.

Supported by grants 94A12 from the American Institute for Cancer Research and CA46589 awarded by the National Cancer Institute.

Can palmitic acid, a significant antimutagen in yogurt, explain yogurt's putative anticarcinogenicity?

Sudarshan R. Nadathur [a], John R. Carney [b], Steven J. Gould [b], and Alan T. Bakalinsky [a]

[a] *Department of Food Science and Technology, Wiegand Hall, Oregon State University, Corvallis, OR 97331-6602.*

[b] *Department of Chemistry, Gilbert Hall, Oregon State University, Corvallis, OR 97331-4003.*

A number of studies have examined the antimutagenic, anticarcinogenic, and anti-cancer activities of yogurt, other fermented milks, and the dairy lactic acid bacteria used in their production. Yogurt and extracts thereof have been shown to be antimutagenic against a range of mutagens and promutagens in microbial and mammalian cell systems. Although certain epidemiological studies have suggested that consumption of yogurt and fermented milk may reduce the incidence of colon or breast cancer, others have been equivocal.

We recently used the Ames test (*Salmonella typhimurium* TA 100) to direct fractionation of antimutagenic activity in yogurt and identified palmitic acid as a significant anti-*N*-methyl-*N*'-nitro-*N*-nitrosoguanidine-active compound (Nadathur et al., 1996, Mut. Res. 359:179-189). The possibility that antimutagenic activity arises in yogurt due to partial lipolysis of milk triacylglycerols by the bacteria used in yogurt manufacture is under investigation in this laboratory.

Can palmitic acid contribute to the putative anticarcinogenicity of yogurt? Consumption of nuts, oils, and meats that contain higher levels of palmitic acid has not been associated with a reduced incidence of colon cancer. Because triacylglycerides are hydrolyzed in the small intestine, yogurt ought to provide no more free palmitic acid than the milk from which it is made. Free fatty acids are absorbed in the small intestine, not the colon, suggesting an indirect role if any, in inhibiting colon carcinogenesis. Salerno and Smith (Anticancer Res. 1991, 11:209-216) reported that palmitic and lauric acids among others, inhibited the growth of HT-29 cells, a malignant human colon cell line. We are currently attempting to determine the relevance of palmitic acid's antimutagenicity to anticarcinogenesis in light of these considerations.

Effects of estrogen-mimicking or antiestrogenic polychlorinated aromatic hydrocarbons on rat mammary tumors depend on the initiating carcinogen: preliminary results. B. A. Diwan, L. E. Beebe, E. G. Snyderwine, L. M. Anderson. BCDP, SAIC Frederick and LCC, NCI-FCRDC, Frederick, MD 21702; and LEC, NCI, Bethesda, MD 20892.

Blood or breast tissue levels of polychlorinated biphenyls (PCBs) and dichlorodiphenyltrichloro-ethane (DDT), which are weakly estrogenic, have been associated with human mammary cancer in some studies, whereas 2,3,7,8-tetrachlorodibenzo-p-dioxin (TCDD) is antiestrogenic in human mammary cancer cells in culture, and at a high dose suppressed the development of rat mammary tumors arising spontaneously or initiated by 7,12-dimethylbenz[a]anthracene (DMBA). TCDD, however, is a promoter of liver, lung and skin tumors in rodents and had not been adequately tested in the mammary tumor model at a nontoxic dose. We are investigating the effects of these compounds on the development of rat mammary tumors initiated by DMBA (female F344 rats, 5 mg/kg, 50 days old) or the food mutagen/carcinogen 2-amino-1-methyl-6-phenylimidazo[4,5-b]pyridine (PhIP) (75 mg/kg, 10X daily from 47 days). Eight days later, rats received TCDD (1 μg/kg, 1X), Aroclor 1254 (250 mg/kg, 1X), or a diet containing 500 ppm DDT. Mammary tumors started to appear 5-10 weeks after initiation and have an incidence of 47% for both initiators thus far (35 weeks). The low TCDD dose completely prevented the appearance of DMBA-initiated tumors for 20 weeks, until about 85% of the TCDD dose would have been cleared. Remarkably, this suppression was not seen for the PhIP-initiated tumors. Aroclor also inhibited the DMBA-initiated tumors, and DDT had a similar though lesser effect on incidence, but accelerated tumor growth, resulting in a significant increase in morbidity ($p=0.04$). By contrast, the PhIP-initiated tumors showed a slight (nonsignificant) increase in incidence following DDT treatment but no change in morbidity. They were unaffected or slightly inhibited by Aroclor treatment. Thus, both enhancing and suppressive effects are occurring in this model so far, and they are dependent on the initiating chemical, suggesting major phenotypic differences in these tumors. These findings have implications not only with regard to the contributions of bioretained chlorinated aromatic hydrocarbons to breast cancer risk, but also for natural dietary estrogenic or antiestrogenic compounds, and for interactions between the latter and the former.

PRETREATMENT OF COLON CELLS WITH SODIUM BUTYRATE BUT NOT *ISO*-BUTYRATE PROTECTS THEM FROM DNA DAMAGE INDUCED BY HYDROGEN PEROXIDE

B. L. Pool-Zobel, S.L. Abrahamse, and G. Rechkemmer, Institute for Nutritional Physiology, Federal Research Centre for Nutrition, Karlsruhe Germany

Foods with high starch or fibre content are connected with a lower risk for bowel diseases including colon cancer, maybe due to the formation of butyrate and other short chain fatty acids. One protective mechanism shown for butyrate is to counteract cell proliferation which is a step of cancer progression (1,2). Next to proliferation, genotoxic events leading to activation of proto-oncogenes, inactivation of tumor suppressor genes, or to various types of cell death are also important early events of carcinogenesis. However, neither the individual carcinogens that may induce these lesions, nor the role of butyrate in preventing these events are known. Suspected aetiological factors however include lipidperoxidation yielding peroxides or other reactive oxygen species. Therefore, we studied sodium butyrate for its ability to prevent cytotoxicity and genotoxicity induced by hydrogen peroxide (H_2O_2) in rat colon epithelium cells. We also studied *i*-butyrate, which in contrast to butyrate, is not effectively utilised as an energy source by colon cells (3). The parameters measured were membrane damage by trypan blue exclusion or by determination of intracellular calcium levels using FURA-2/AM (4), and genotoxicity. The studies were performed by first establishing the effective concentration ranges in which H_2O_2 could induce cytotoxic and genotoxic effects. The most sensitive of these parameters was genotoxicity detected with the technique of single cell microgel-electrophoresis, also called "Comet Assay" (5). With this technique damaged DNA is visualized in single cells embedded in gels on microscopical slides after their electrophoresis and staining with ethidium bromide. The formation and extension of comets from individual cells is quantified and the increase in median image length (med IL) of all evaluated cells is a reflection of induced genotoxicity. The susceptibilities of cells treated with butyrate-isomers were compared to untreated control cells. It was shown that colon cells pretreated with 6.25 mM butyrate15 minutes prior to treatment with 200 µM H_2O_2 were significantly less sensitive to the DNA damaging effects of H_2O_2 (med IL 64±7) than the control cells (med IL 46±14; n=4; $p<0.05$). In contrast, pretreatment of cells with *i*-butyrate did not alter their susceptibility to H_2O_2 in comparison to control cells (med IL 72±3 and 79±13, respectively). These results are the first demonstration of butyrate's ability to prevent the DNA damage induced by a carcinogen. The effect may be due to an increased energy turnover, since *i*-butyrate was ineffective. The elucidation of mechanisms of protective effects is under current investigation.

References: **1**. BUGAUT, M. ANDBENTéJAC, M. Ann Rev Nutr, 1993, 13, 217; **2**. SCHEPPACH, W. GUT, 1994, supplement 1, S35.; **3**. CLAUSEN, M.R. AND MORTENSEN, P.B. Gastroenterology, 1994, 106, 423. **4**. BORLE, A.B. Environ Health Perspect, 1990, 84, 45. **5**. POOL-ZOBEL, B.L., LOTZMANN, N., KNOLL, M., et al. Environmental and Molecular Mutagenesis, 1994, 24, 23.

Supported by the EC programmes ECAIR -2-CT94-0933 and CIPA -CT94-0129 and by the EDEN-Stiftung, Bad Soden, Germany.

NUTRIENT INTAKE OF CANCER PATIENTS AND NEED FOR DIETARY COUNSELING

Faye B. Bass, Ph.D. candidate
Department of Human Nutrition and Foods
Virginia Polytechnic Institute and State University
Blacksburg, VA 24061

One hundred six cancer patients, recruited from three cancer clinics and a private oncology center in southwest Virginia, completed a questionnaire and were asked to complete a 3-day food diary. The study objectives were to determine the extent of dietary counseling, side effects of cancer treatment, types of cancer, and the nutrient intake of patients both from food and dietary supplements. Sixty-five women (mean age of 58) and 41 men (mean age of 62) completed the 3-day food record. Mean nutrient intakes of cancer patients were compared with the Recommended Daily Allowances (RDAs) and with intakes of the general population, as reported in three national surveys. Supplement usage by the cancer patients was compared with results of a Food and Drug Administration (FDA) Survey.

Significantly more female (38 of 65) than male (12 of 41) cancer patients reported taste aversions. Dietary intake of nearly all nutrients was significantly lower for the cancer patients than the general population. Dietary analysis of female cancer patients for all B-vitamins, except folate, showed significantly lower intakes compared to national surveys. Male cancer patients' intakes of B-vitamins were also significantly lower than the general population, except for thiamin. Both female and male cancer patients had significantly lower intakes of vitamin E and significantly higher intakes of beta-carotene than national surveys. Protein and energy consumption were significantly lower among male cancer patients, while fiber intake was significantly lower for both sexes, compared to national surveys.

Cancer patients were also deficient in several nutrients, as compared to the RDAs, especially among the men. Supplement usage among the cancer patients was significantly lower for several vitamins and minerals, compared to the FDA survey. Only 31% of the patients had received dietary counseling from a member of the health care team. whereas, a majority reported they were interested in dietary counseling. Nutritional professionals should be aware that many cancer patients both need and desire dietary counseling.

Antitumor Promoting Activities of Glycyrrhetic Acid Monoglucuronide Derived from Glycyrrhizin and Dihydroflavonoids from Kohki Tea

Akifumi Higurashi[1], Kenji Mizutani[1], Osamu Tanaka[1], Harukuni Tokuda[2], Mutsuo Kozuka[3], Hoyoku Nishino[4]

Maruzen Pharmaceuticals Co., Ltd.[1], Omomichi, 722, Japan, Kyoto Prefectural University of Medicine[2], Kyoto 602, Japan, Kyoto Pharmaceutical University[3], Kyoto 607, Japan, and Cancer Prevention Division, National Cancer Research Institute[4], Tokyo 104, Japan.

Glycyrrhetic acid monoglucuronide (MGGR) derived from glycyrrhizin of licorice by enzymic hydrolysis, is 941 times sweeter than cane sugar and used as a sweetening and flavoring agent in Japan. The leaves of Kohki (Huang-Qi in Chinese, *Engelhardtia chrysolepis* Hance, Juglandaceae) has been used as a folk medicine and a healthgiving tea in the southern region of China, and recently popularized as Kohki tea in Japan. Kohki tea contains 4-6% astilbin, a dihydroflavonol glycoside which shows several activities such as antioxidation, antiallergy, antiinflammation, *etc.* In the course of studies on foods and food ingredients for health supplements, MGGR, astilbin and (+)-taxifolin (an aglycone of astilbin) were found to indicate potent antitumor promoting activities by a primary screening test using a short-term *in vitro* assay of Epstein-Barr virus early antigen activation induced by the tumor promoter 12-O-tetradecanoylphorbol-13-acetate (TPA). In the two-stage carcinogenic experiments of 7,12-dimethylbenz[*a*]anthracene (DMBA)/TPA induced mouse skin tumor, the topical application of MGGR inhibited tumor formation more effectively than that of glycyrrhizin or glycyrrhetic acid. Further, on 4-nitroquinoline-*N*-oxide (4NQO)/glycerol induced mouse pulmonary carcinogenesis and DMBA/ultraviolet-rays induced skin carcinogenesis, the oral administration of MGGR was significantly effective to prevent tumor promotions. The antitumor promoting activities of astilbin, (+)-taxifolin and Kohki-extracts were also examined *in vivo*, and promising results as a tea for cancer prevention were obtained. The details of the antitumor promoting activities will be presented.

ANTI-INVASIVE ACTIVITY OF DIETARY BIOPOLYPHENOLICS AGAINST SOLID TUMOURS

VS Parmar [a*], KS Bisht [a], Suman Gupta [a], Amitabh Jha [a], Rajni Jain [a], SK Sharma [a], PM Boll[b], MM Mareel [c] and ME Bracke [c]

[a] *Department of Chemistry, University of Delhi, Delhi-110 007 (India)*
[b] *Department of Chemistry, Odense University, DK-5230 Odense M, Denmark*
[c] *Laboratory of Experimental Cancerology, Department of Radiotherapy and Nuclear Medicine, University Hospital, De Pintelaan 185, B-9000 Gent (Belgium)*

Invasion is the hallmark of malignant tumours and generally leads to metastasis, which is the major cause of death of cancer patients.[1] Anti-invasive agents are being studied for both development of new therapeutic rationales in cancer treatment and for the analysis of tumour invasion mechanisms.[2] Controlling the tumours before the metastasis takes place is a sort of prevention and is of immense importance until an ultimate cure is available.

With an aim to find anti-invasive phytochemicals or their analogs which are not cytotoxic and could hence be used as tools for studying targets implicated in invasion, we have been studying their effect on the invasion of MCF 7/6 human mammary carcinoma cells into embryonic chick heart fragments in organ culture.[3] In this contribution, we wish to present the results of our investigations on the effect of fifty additional phytochemicals of dietary origin, most of which happen to be polyphenolics against invasion in the above assay at concentrations ranging from 1-100 μM.

Among the natural products tested, cerasinone (**1**, isolated from the sour cherry tree, *Prunus cerasus*[4]), 5,6,7-trimethoxyflavone (**2**, isolated from *Physalis minima*,[5,6] the leaves and fruits of which are edible), piperine (**3**, a common constituent of *Piper* species, a few of which are dietary), crotepoxide (**4**) and 2-(3-methylbut-2-enyl)-3,4,5-trimethoxyphenol (both isolated from *Piper clarkii*[7,8]) have shown significant anti-invasive activity with varying degree of cytotoxicity.

The anti-invasive activity of 3,7-dimethoxyflavone (**5**), the dimethyl ether of naturally occurring 7-hydroxyflavonol (**6** in *Platymiscium praecox*[9]) was found to be the best. Even at a concentration of as high as 100 μM, no cytotoxic effect could be detected. The anti-invasive effect was reversible upon omission of the compound from the medium and it did not inhibit the growth of MCF 7/6 cell aggregates of heart fragments kept in suspension culture.

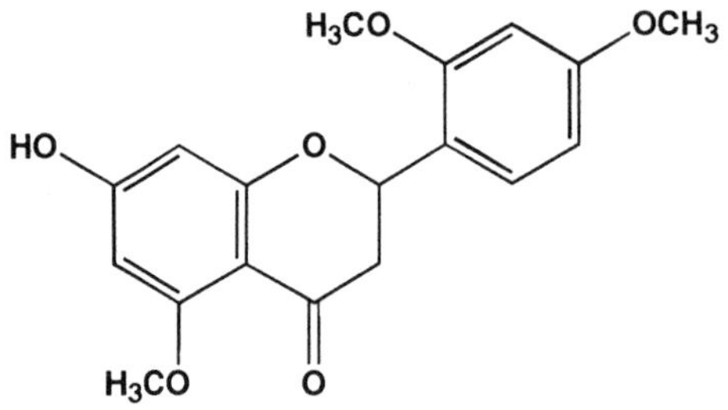

1

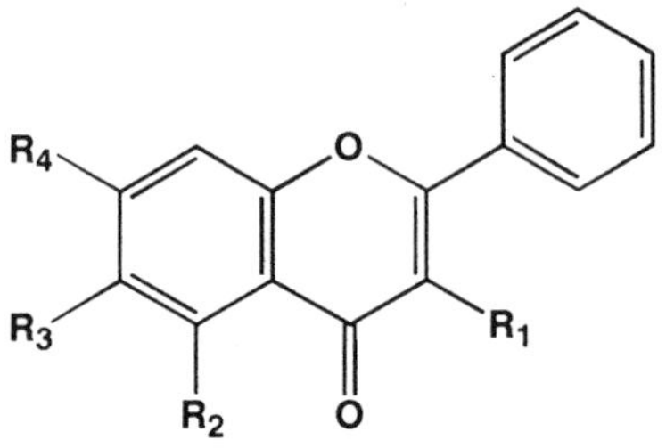

2 R_1=H, R_2=R_3=R_4=OCH_3
5 R_1=R_4=OCH_3, R_2=R_3=H
6 R_1=R_4=OH, R_2=R_3=H

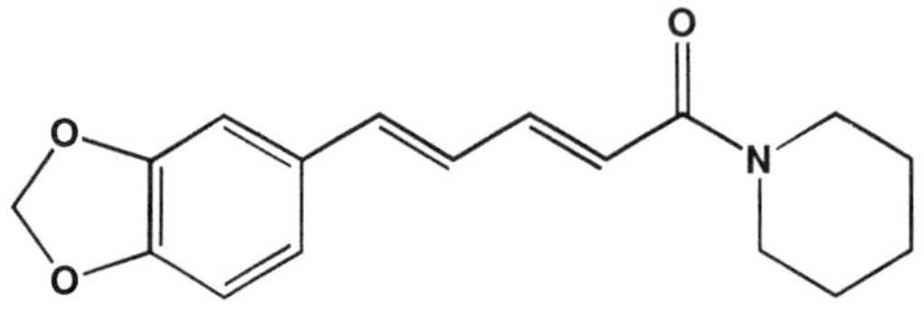

3

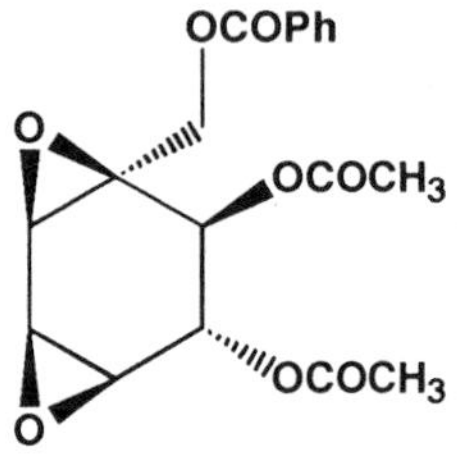

4

Acknowledgement. We thank DANIDA (Danish International Development Agency) for financial assistance.

References

1. *Mechanisms of Invasion and Metastasis*; MM Mareel, P De Baetselier and FM Van Roy, Eds.; CRC Press : Boca Raton, Florida (USA), pp. 7 (1991).
2. MM Mareel and M De Mets, *CRC Rev. Hematol. Oncol.* **9**, 263 (1989).
3. VS Parmar, R Jain, SK Sharma, A Vardhan, A Jha, P Taneja, S Singh, BM Vyncke, ME Bracke and MM Mareel, *J. Pharm. Sci.* **83**, 1217 (1994).
4. GR Nagarajan and VS Parmar, *Phytochemistry* **16**, 1317 (1977).
5. VS Parmar, S Gupta, R Sinha and SK Sharma, *Indian J. Chem.* **32B**, 244 (1993).
6. NA Ser, *Phytochemistry* **27**, 3708 (1988).
7. PM Boll, M Hald, VS Parmar, OD Tyagi, KS Bisht, NK Sharma and S Hansen, *Phytochemistry* **31**, 1035 (1992).
8. S Jensen, CE Olsen, OD Tyagi, PM Boll, FA Hussaini, S Gupta, KS Bisht and VS Parmar, *Phytochemistry* **36**, 789 (1994).
9. AB de Oliveira, LGF e Silva and OR Gottlieb, *Phytochemistry* **11,** 3515 (1972).

CHEMOPREVENTION OF CARCINOGEN (AFLATOXIN B_1)-DNA BINDING : THE RELATIVE ACTIVITY OF ACETOXY, HYDROXY AND METHOXY SUBSTITUENTS ON COUMARINS OF EDIBLE PLANTS ORIGIN IN THE INHIBITION OF LIVER MICROSOME CATALYSED AFB_1-DNA BINDING *IN VITRO*.

Virinder S Parmar*, HG Raj[a], Sangita Gupta, Suddham Singh and Kirpal S Bisht
Department of Chemistry ([a]Department of Biochemistry, V. Patel Chest Institute), University of Delhi, Delhi-110 007 (India)

Cytochrome P-450 dependant activation of carcinogens is known to be inhibited by several polyphenolic compounds, widely distributed in the plant kingdom. Earlier studies have indicated that hydroxylated flavonoids inhibit carcinogen activation, while those with methoxy groups tend to stimulate.[1] We have examined this postulate with another class of widely occurring polyphenolics in nature, i.e. coumarins. We have chosen, in particular 4-methylcoumarins **1-19**, a few of which occur in the edible plant fenugreek (*Trigonella foenumgraecum*).[2-5] It is reported that 4-methylcoumarins show enhanced biological activities than their analogs lacking the C-4 methyl substituent.[6,7]

In our studies for evaluating the inhibitory activity of 4-methylcoumarins towards aflatoxin-DNA binding, ^{3}H-AFB_1 was incubated at 37°C for 30 min with calf thymus DNA in a reaction mixture principally including rat liver microsomes and NADPH. The test compounds were added before addition of microsomes. The DNA was isolated free of proteins and RNA, and the amount of ^{3}H-AFB_1 bound to DNA was measured according to our earlier published procedure.[8]

We have examined a number of coumarins bearing a combination of three functional groups, i.e. methoxy, hydroxy and acetoxy and an aliphatic side chain of 0-5 carbon atoms at the C-3 position on the 4-methylcoumarin nucleus for their ability to inhibit rat liver microsome-mediated AFB_1 adduction to DNA *in vitro*. The number of carbon atoms in the aliphatic side chain at the C-3 position does not seem to make any difference in the activity of coumarins in inhibiting the AFB_1-DNA binding. Coumarins substituted with one or more methoxy group(s) appear to be least effective in causing inhibition of AFB_1-DNA binding *in vitro*. It was observed that presence of hydroxyl group(s) on the coumarin nucleus results in higher inhibition of AFB_1-DNA binding *in vitro* as compared to those having methoxy group(s) at the same positions. The number and position of hydroxyl group(s) appear to have no significant effect. Further it was observed that acetylation of the hydroxyl group(s) results in still enhanced inhibition. In addition, it was noticed that coumarins carrying two acetoxy groups possess much higher

1 $R=CH_2CH_2COOC_2H_5$, $R_1=R_2=R_4=H$, $R_3=OCH_3$

2 $R=R_2=R_4=H$, $R_1=R_3=OCH_3$

3 $R=R_1=R_4=H$, $R_2=R_3=OCH_3$

4 $R=CH_2COOC_2H_5$, $R_1=R_2=H$, $R_3=R_4=OCH_3$

5 $R=CH_2COOC_2H_5$, $R_1=R_4=H$, $R_2=R_3=OCH_3$

6 $R=CH_2COOC_2H_5$, $R_1=R_2=R_4=H$, $R_3=OH$

7 $R=R_1=R_2=H$, $R_3=R_4=OH$

8 $R=CH_2CH_2COOC_2H_5$, $R_1=R_4=H$, $R_2=R_3=OH$

9 $R=R_2=R_4=H$; $R_1=R_3=OH$

10 $R=CH_2CH_2COOC_2H_5$, $R_2=R_4=H$, $R_1=R_3=OH$

11 $R=R_4=H$, $R_1=R_3=OCH_3$, $R_2=OAc$

12 $R=R_1=R_2=R_4=H$, $R_3=OAc$

13 $R=CH_2COOC_2H_5$, $R_1=R_2=R_4=H$, $R_3=OAc$

14 $R=R_1=R_2=H$, $R_3=R_4=OAc$

15 $R=CH_2CH_2COOC_2H_5$, $R_1=R_2=H$, $R_3=R_4=OAc$

16 $R=R_1=R_4=H$, $R_2=R_3=OAc$

17 $R=R_2=R_4=H$, $R_1=R_3=OAc$

18 $R=CH_2COOC_2H_5$, $R_1=R_3=OAc$, $R_2=R_4=H$

19 $R=CH_2CH_2COOC_2H_5$, $R_1=R_3=OAc$, $R_2=R_4=H$

inhibitory activity as compared to compounds bearing one acetoxy group. It was further found that coumarins containing two acetoxy groups at the C-5, C-7 or C-6, C-7 positions were significantly more active as compared to those having acetoxy groups at the C-7, C-8 positions. The aliphatic side chain at the C-3 position was found to have minor effect with the combination of functional groups discussed above. Thus compound **18** having the two acetoxy groups at C-5 and C-7 position had the best activity, it inhibited the AFB_1-DNA binding to the extent of 37%, 58% and 84% at 10, 50 and 100 μM concentration, respectively.

We can conclude from the above observations that the contribution of functionality on coumarin nucleus towards the inhibition of AFB_1-DNA binding *in vitro* is in the order : acetoxy > hydroxy > methoxy. These studies have a bearing on AFB_1 activation and can prove useful in the designing of useful chemoprevention agents, effective against AFB_1 carcinogenesis.

References

1. AH Conney, *Cancer Res.* **42**, 4875 (1982).
2. VS Parmar, HN Jha, SK Sanduja and R Sanduja, *Z. Naturforsch.* **37B**, 521 (1982).
3. SK Khurana, V Krishnamoorthy, VS Parmar, R Sanduja and HL Chawla, *Phytochemistry* **21**, 2145 (1982).
4. VS Parmar, S Singh and JS Rathore, *J. Chem. Res.(S)*, 378 (1984).
5. VS Parmar, JS Rathore, S Singh, AK Jain and SR Gupta, *Phytochemistry* **24**, 871 (1985).
6. A Tyagi, VP Dixit and BC Joshi, *Naturwissenschaften* **67,** 104 (1980).
7. S Takeda and M Aburada, *J. Pharmacobio-Dynam.* **4**, 724 (1981).
8. A Allameh, M Saxena and HG Raj, *Cancer Lett.* **40**, 49 (1988).

BUTYRATE-INDUCED DIFFERENTIATION OF HT-29 COLON CARCINOMA CELLS IS ASSOCIATED WITH SPECIFIC ALTERATIONS IN GENE EXPRESSION. Winesett, M.P. and Barnard, J.A. Vanderbilt University School of Medicine, Nashville, TN 37232-2576.

Butyrate is a short chain fatty acid derived from colonic fermentation of dietary fiber. Butyrate and other short chain fatty acids are the most abundant intraluminal solutes in the colon and are purported to function in the regulation of colonocyte proliferation. We have previously shown that butyrate rapidly induces differentiation of the HT-29 colon carcinoma cell line (*Cell Growth and Differentiation* 4:495, 1993). Butyrate-induced differentiation is associated with growth arrest early in G_1 and down-regulation of c-*myc* expression. Expression of c-*jun*, *nup*/475, *zif*/268, and c-*fos* is unchanged. Additional studies were designed to determine the mechanism by which butyrate regulates c-*myc* expression. Nuclear run-on transcription assays showed that butyrate decreases c-*myc* transcription. c-*myc* mRNA decay curves in HT-29 cells treated with actinomycin D with or without 5 mM butyrate were identical, suggesting that butyrate does not influence mRNA half-life. Conversely, concurrent treatment of HT-29 cells with 5 mM butyrate and the protein synthesis inhibitor cycloheximide resulted in complete abrogation of butyrate-mediated down-regulation of c-*myc*. Western blot analyses revealed that a decrease in c-Myc was first detectable 8 hours after butyrate treatment. This time course is consistent with the emergence of alkaline phosphatase (AP) mRNA 12 hr following treatment. Treatment of HT-29 cells with other short chain fatty acids (acetate and propionate) did not result in growth inhibition, down-regulation of c-*myc*, or appearance of AP mRNA.

In separate studies performed very recently, we have used the technique of "differential display" to isolate genes regulated by butyrate treatment in HT-29 cells. Three butyrate-inhibited genes were cloned from HT-29 cells treated for 48 hr with 5 mM butyrate. These were identical to human nucleosome assembly protein 1 (NAP-1), human interferon-inducible protein 9-27, and the yeast gene snf-3. Differential display appears to be a valuable strategy for analysis of the molecular effects of butyrate on cells of colonic origin. Differential display of genes expressed within minutes or hours of butyrate treatment may provide further information on the mechanism by which butyrate induces differentiation in colonocytes. Collectively, these observations indicate that butyrate-induced growth inhibition and differentiation occurs in association with specific down-regulation of c-*myc* mRNA and protein expression in HT-29 cells by a complex mechanism involving both transcriptional and post-transcriptional pathways. Further study of the regulation of c-*myc* and additional genes associated with differentiation should contribute to an expanded understanding of colonocyte growth control in neoplastic and nonneoplastic disease of the colon.

PEPTIDE YY AND CLENBUTEROL REDUCE GUT ATROPHY AND CACHEXIA DURING PARENTERAL NUTRITION OF TUMOR-BEARING RATS. W.T. Chance, A. Balasubramaniam, H. Thompson, L. Zuo and J.E. Fischer. Dept. Surgery, Univ. Cincinnati Med. Ctr. and VA Med. Ctr., Cincinnati, OH.

Malnutrition contributes to morbidity and mortality of cancer patients. However, nutritional support employing total parenteral nutrition (TPN) has been associated with increased infection rate which may be related to TPN-induced gut atrophy. Our preliminary studies of this phenomenon demonstrated that co-infusing peptide YY (PYY) with TPN resulted in significant savings of gut mass and protein. In the present experiment we sought to extend these observations to tumor-bearing (TB) organisms and investigate whether treating the rats with the anabolic beta-2 adrenergic agent, clenbuterol (CLE), would amplify the effect of PYY on gut. These hypotheses were investigated in Fischer 344 rats bearing methylcholanthrene sarcomas. Two weeks after tumor inoculation TB and control (C) rats received jugular vein catheterizations or sham operations. Beginning 3 days later TPN (isocaloric and isonitrogenous to rat chow) was begun for 8 days, with groups receiving CLE injections (2 mg/kg/day) or CLE plus PYY co-infusions (1 nmol/kg/hr). After euthanization, protein levels were determined in 20 cm segments of jejunum and ileum as well as in gastrocmenius muscle.

Group	N	Jejunum (mg/g/20 cm)	Ileum (mg/g/20 cm)	Gastrocnemius (mg/muscle)
C-CHOW	16	177 + 13	130 + 7	290 + 11
TB-CHOW	16	156 + 10	118 + 4	227 + 18*
TB-TPN-SAL	14	91 + 4*	76 + 3*	256 + 10**
TB-TPN-CLE	6	82 + 6*	91 + 8*	287 + 7
TB-TPN-CLE-PYY	8	128 + 4+	109 + 9+	282 + 10

* $p < 0.01$, ** $p < 0.05$ vs C-CHOW; + $p < 0.01$ vs TB-TPN-SAL

These results suggest that the combination of PYY and CLE results in significant reduction of gut atrophy and savings of skeletal muscle mass in TB rats maintained on TPN. Therefore, gut atrophy and cachexia may be preventable in cancer patients maintained on TPN.

Preliminary results from the deoxyribose assay suggest that some tannins are OH radical scavengers. However, some types of tannins behave as prooxidants in the deoxyribose assay. We have also noted that some tannins irreversibly damage proteins when conditions favor tannin oxidation.

Our results show that each of the different structural types of tannins has unique redox reactivity which cannot be predicted from studies of simpler phenolics. In order to better understand the potential role of dietary tannins in cancer prevention, more detailed studies of the redox reactivities of tannin are needed.

This work was supported by the Research Corporation, by the Miami University Undergraduate Summer Scholars Program, and by the Howard Hughes Medical Institute.

PHYTOESTROGENS, A PLANT-BASED DIET, AND BREAST CANCER RECURRENCE

John P. Pierce, Cedric Garland., Vicky Newman. The Womens Healthy Eating and Living Study (WHEL) is a multicenter randomized controlled trial of the effect of a high vegetable-low fat diet on the incidence of secondary events for women previously diagnosed with early stage breast cancer. This study aims to enroll 2,400 women over a four year period and follow them for an average of 6 years. The rationale for such a study is that chemoprevention achieved from a plant-based diet may represent the cumulative effect of multiple weak anticarcinogens. Thus, a whole-diet approach is proposed, with components relating to several proposed preventive mechanisms. The consumption of plant estrogens and estrogenic products from fermentable fiber is thought to interfere with the stimulatory activity of endogenous estrogen. To elucidate the role of this mechanism, a portion of the WHEL women will provide urine samples, which will be analyzed using HPLC-mass spectrometry for phytoestrogens, enterolactone, enterodiol and related compounds, including aglucones and glucuronides. The results will be analyzed in relation to dietary assignment, recurrence, and other measures of macronutrient and micronutrient intake. Data will be presented from the feasibility study on enrollment, reported dietary changes and validation of dietary change from serum nutrient concentrations.

Gene Expression Induced by Chemoprotective Agents

T. Primiano, J.A. Gastel, T.W. Kensler, and T.R. Sutter

The chemoprotective actions of dithiolethiones have been correlated with induction of a battery of carcinogen metabolizing enzymes which, through their increased activity prevent formation of certain cancers. Evidence exists that a number of unknown proteins are induced by this class of compounds, and in order to isolate candidate genes expressed in response to dithiolethiones, a differential hybridization screening method was employed. A dithiolethione-induced rat liver cDNA library consisting of 30,000 clones was hybridized using a subtracted probe resulting in 670 potential positive clones. Additional screening and Northern blot analysis in concert with cross-hybridization analysis revealed that 12 of these cDNA clones represented inducible genes. Subsequent DNA sequence database searches indicated that clones for the known carcinogen metabolizing enzymes epoxide hydrolase, aflatoxin aldehyde reductase, quinone reductase and multiple subunits of glutathione S-transferase were among those isolated. Additionally, cDNAs for ferritin heavy and light chains, ribosomal proteins L18a and S16, and two novel cDNAs were isolated. Levels of RNA recognized by each clone were increased from 2- to 31-fold 24 h after treatment. The time-courses for steady-state RNA induction varied so that relatively early or late responding genes were identified. These elevations in RNA were correlated to increased gene transcription with the exception of epoxide hydrolase, suggesting that multiple regulatory pathways are activated by dithiolethiones. Immunoblot analysis indicated that levels of the ferritin L , which participates in sequestration of iron, were increased 3.7-fold. Inhibition of iron-mediated oxidative stress through induction of ferritin may represent an important additional component of the chemoprotective actions of dithiolethiones and related agents. (Supported by grants CA 39416, ES 07141, and ES 03819.)

CROSS RESISTANCE TO SELENITE ACCOMPANIES CISPLATIN RESISTANCE IN HUMAN OVARIAN TUMOR CELLS.

P. B. Caffrey (SPON. G. D. Frenkel). Dept. of Biological Sciences, Rutgers University, Newark, NJ 07102.

Our previous studies have shown drug resistant tumor cells with high glutathione (GSH) levels to be sensitive to selenite. This sensitivity has been associated with the formation of selenodiglutathione (SDG), a product of the metabolic reaction of GSH and selenite. We have now compared the selenite sensitivity of a line (A2780) of human ovarian tumor cells with that of two sublines: one (A2780-CP) selected for cisplatin (CP) resistance and another (A2780-ME) selected for resistance to melphalan. Both sublines contain higher GSH levels than the A2780 cells. The A2780-ME cells were 4-fold more sensitive than the A2780 cells to exposure to selenite as determined by trypan blue exclusion. In contrast, the A2780-CP cells were 2-fold *resistant* compared to the A2780 cells. This suggests that despite having high GSH levels the A2780-CP cells accumulate less SDG than the other cell lines. All of the cell lines were several-fold more sensitive to exposure to SDG than to selenite. However, the A2780-CP cells were less sensitive to SDG than were the other cell lines. One explanation is that A2780-CP cells may efflux SDG by a glutathione S-conjugate export pump (Ishikawa and Ali-Osman., JBC, 268:20116-20125, 1993) associated with the development of cisplatin resistance.

TANNINS IN BIOLOGICAL REDOX REACTIONS

Ann E. Hagerman, Nicole T. Ritchard, G. Alexander Jones and Thomas L. Riechel

Department of Chemistry, Miami University, Oxford OH 45056

Tannins, which are widespread in fruits, grains, and beverages, are polymers of simple phenolics such as ellagic acid and/or gallic acid (hydrolyzable tannins), epicatechin (condensed tannins), or phloroglucinol (phlorotannins). The common characteristics of these chemically distinct tannins are their high molecular weights and their ability precipitate proteins. Although simple phenolics are redox active, oxidation reactions of tannins have not been examined. Since dietary antioxidants may play an important role in cancer prevention, we have started to examine the redox properties of tannins.

We have evaluated the redox reactivity of purified tannins between pH 3 and 8 using cyclic voltametry. For all phenolics, the oxidation potentials were more negative at higher pH values, consistent with easier oxidation at higher pH. In general, the redox potentials of the polymers were 0.05-0.10 V more positive than the redox potentials of the corresponding simple phenolics, indicating that the polymers are more difficult to oxidize than the monomers. Oxidation of the polymers rapidly fouled the glassy carbon working electrode, but electrodes were not fouled during oxidation of the simple phenolics. This suggests that complex reactions, perhaps involving polymerization, occur subsequent to oxidation of tannins. At pH 6, the redox potentials of the various tannins ranged from 0.35 to 0.95 V (vs. Ag/AgCl).

VITAMIN D AND INTRACELLULAR CALCIUM REGULATION IN COLONIC CRYPTS

Bruce M. Brenner*, Nils Russell*, Svenja Albrecht*, Richard J. Davies*#. Departments of Surgery, UMD-New Jersey Medical School*, Newark, N.J., and Hackensack University Medical Center#, Hackensack, N.J.

Low dietary and serum Vitamin D have been implicated as risk factors in the development of colorectal cancer. Although Vitamin D is known to stimulate intestinal calcium absorption, it is unclear what effects it may have on intracellular calcium (Ca^{2+}_i) at the level of the colonic crypt. Ca^{2+}_i plays an important role in intracellular signaling and control of the cell cycle. Differentiation and proliferation of skin keratinocytes are highly dependent on calcium, with differentiation being induced by an increase in calcium as keratinocytes migrate from the basal to the suprabasal zone of the skin. A similar calcium gradient may occur in colonic mucosa and be involved in the regulation of proliferation, differentiation and apoptosis along the crypt axis. The crypt Ca^{2+}_i profile and its dependence on Vitamin D were examined in this study. Forty female CF_1 mice were fed a low (0.25 IU/g) or a supplemented (2.5 IU/g) Vitamin D diet, both with a calcium level of 0.15%. The mice were maintained on this diet for six weeks and then sacrificed, the distal colon excised and colonic crypts isolated. Ca^{2+}_i was measured at five levels along the crypt using the calcium sensitive ratiometric fluorescent dye FURA-2 on a digital imaging system. Statistical analysis was performed using either the paired Student's t-test or Wilcoxon signed rank test. In mice fed a high Vitamin D diet a gradient in Ca^{2+}_i was identified along the length of the crypt ranging from 70 ± 10 nM in the suprabasal region to 157 ± 34 nM at the mouth ($p < 0.05$). There was also an elevation of Ca^{2+}_i in the most basal portion of the crypt (p N.S.). In mice fed a low vitamin D diet this gradient was lost, with a Ca^{2+}_i of 167 ± 52 nM at the base and 153 ± 35 nM at the mouth of the crypt (p N.S.). This data suggests that Vitamin D deficiency may function in the development of colorectal cancer by contributing to a loss of the normal Ca^{2+}_i crypt gradient resulting in a dysregulation of proliferation, differentiation and apoptosis.

(Supported in part by American Institute for Cancer Research Grant #92A37)

EFFECT OF SOYBEAN SAPONINS ON ABERRANT CRYPT FORMATION IN MICE.

Koratkar R., Sung M-K., Torzsas T.L., and Rao A.V.

Saponins are naturally occurring amphiphilic compounds present in many plants. Recently, they have been reported to have anticarcinogenic properties. However, most of the studies have used saponins from non-dietary sources. Thus, the objective of this study was to examine the effect of soybean saponin on the incidence of chemically induced colonic preneoplasia in mice.

CF1 male mice were divided into three groups. Group 1 and 2 were administered Azoxymethane (AOM) to induced aberrant crypt foci (ACF). AOM was administered intra peritoneally at a concentration of 5 mg/kg b.w./week for 4 weeks. One week after the last AOM injections, group 2 was placed on a 3% soybean saponin supplemented diet. Mice in group 3 were on 3% soybean saponin diet, without AOM initiation. The mice were monitored weekly for body weights and food intake. At the end of 14 weeks (post-initiation), animals were sacrificed and colons processed to score for ACF formation.

Results indicated that food consumption and body weight gain were not significantly different between the groups. However, soybean saponin fed animals had significantly lower number of both ACF and aberrant crypt per focus. Mice fed soybean saponins alone did not induce any change in colonic morphology as observed under light microscope. These results imply that soybean saponins, an important dietary source of saponins, play a significant role as chemotherapeutic compounds in the management of colon carcinogenesis.

Effect of tannic acid on benzo[a]pyrene-DNA adduct formation in mouse epidermis. Comparison with synthetic gallic acid esters.

W. Baer-Dubowska., J.Gnojkowski. and W. Fenrych.

K. Marcinkowski`s University Medical School, Dept. of Biochemistry, Poznañ, Poland

Tannic acid (TA) is a naturally occurring plant phenol, which has been found to inhibit xenobiotic metabolism and skin tumorigenesis induced by polyaromatic hydrocarbons in SENCAR mice. In this study the effect of topical application of TA on epidermal aryl hydrocarbon hydroxylase (AHH), gluthatione-S-transferase (GST) and binding of benzo[a]pyrene(B[a]P to epidermal DNA was compared with the synthetic gallic acid esters. Single topical application of TA, octyl (OG) or dodecyl (DG) gallate at a dose of 400 μmol/kg had no effect on AHH, but propyl gallate (PG) applied under the same conditions increased the enzyme activity. Elevated GST activity was observed after treatment with all tested compounds. The application of the same dose of TA to mouse skin 1 hour before the application of 0.2 or 1 μmol of B[a]P resulted in 65% inhibition of covalent B[a]P binding to epidermal DNA. Gallic acid esters were much less effective B[a]P binding inhibitors especially when the higher dose of B[a]P was used. These results indicate that TA is more potent inhibitor of carcinogen binding than synthetic antioxidants, gallic acid esters and suggest that the altering the induction of GST may be partly responsible for this effect.

Effect of Selenite on Fibronectin-Stimulated Cell Spreading and Migration

Lin Yan, John A. Yee, and Michael H. McGuire
Departments of Biomedical Sciences and Surgery, Creighton University, Omaha, NE 68178

The role of integrins in malignant invasion is documented. Integrins are a family of receptors that mediate the attachment of cells to the extracellular matrix (ECM). In addition, ligand binding to some integrins is known to activate intracellular signal transduction pathways that lead to altered cell function. Fibronectin (Fn) is an ECM molecule that enhances cell attachment, spreading, and migration. These effects may be mediated by protein kinase C (PKC). The purpose of the present study was to determine if selenite (Se) affects Fn-stimulated spreading and migration of Chinese hamster ovary (CHO) cells, and if so, whether this is due to an effect on the PKC pathway. Cell spreading was determined by measuring changes in cell size using an image analysis system, and migration of CHO cells through 12 μm pores of polycarbonate filters was monitored in Boyden chambers. Cells were plated into Fn-coated or uncoated plastic culture wells and cell size was measured every 20 minuts for two hours. The rate of cell spreading was significantly greater on Fn-coated compared to uncoated surfaces. The addition of 50 nM 12-myristate 13-acetate (PMA), a stimulator of PKC, further increased the rate of spreading on Fn, but had no effect on cells plated on plastic. Treatment of cells before plating with 10 to 50 nM calphostin C, an inhibitor of PKC, or 1-5 μM Se significantly inhibited spreading of CHO cells on Fn in both the absence and presence of PMA. Adding Fn to the lower compartment of Boyden chambers stimulated the migration of CHO cells. This chemotactic effect was enhanced by PMA, and inhibited by both calphostin C and Se. Based on these results it is concluded that Fn-stimulated spreading and migration of CHO cells is mediated by the ECM/integrin/PKC signal transduction pathway. Furthermore, the inhibition caused by Se may be due to an effect on this pathway. (Supported by the Creighton University Health Future Foundation).

INHIBITORY EFFECTS OF DIALLYL SULFIDE ON CHEMICALLY INDUCED TUMORIGENESIS AND MUTAGENESIS

Mark Shlyankevich[1], Kwang-Kyun Park[2] and Young-Joon Surh[1*]

[1]Department of Epidemiology and Public Health, Yale University School of Medicine, New Haven, CT 06520-8034 and [2]College of Medicine, Yonsei University, Seoul, Korea

Introduction: Diallyl sulfide (DAS), one of the major volatile organosulfur components present in garlic (*Allium sativum*), has been shown to be an efficient chemopreventive agent in various animal tumor models (1-3). One of the most likely mechanisms of chemopreventive action of DAS is considered to involve interference with microsomal carcinogen metabolizing enzymes, particularly the cytochrome P-450 2E1 (CYP2E1) isoform (4,5). In the present study, we investigated the effects of DAS on mutagenicity and/or carcinogenicity of vinyl carbamate (VC) and *N*-nitrosodimethylamine (NDMA) that are known to be preferentially activated by CYP2E1.

Materials and Methods: Groups of 30 female ICR mice (6- to 7-week-old) were given DAS (8 μmol) by gavage 10 min prior to a single topical dosage of VC (5.8 μmol) in 0.2 ml acetone. In another experiment, DAS (0.2 mg/10 μl trioctanoin/g body wt.) or the equivalent volume of vehicle alone was administered to female CD-1 mice by gastric intubation 1 day and 2 h before topical application of 0 μmol or 11.5 μmol of VC in 0.2 ml acetone. One week after initiation, animals were treated topically with 12-*O*-tetradecanoylphorbol-13-acetate (2.5 μg) twice weekly. Bacterial mutagenicity of VC and NDMA was determined by the Ames-*Salmonella* assay.

Results: DAS protected against mutagenesis induced by VC or NDMA. Thus, mutagenicity of VC (8.3 mM) in *S. typhimurium* TA100 was suppressed about 50% in the presence of 1.67 mM DAS. Bacterial mutagenicity of NDMA (80 mM) was also inhibited 30% and 74% by DAS at concentrations of 0.67mM and 1.0 mM, respectively. Oral administration of DAS (8 μmol) to female ICR mice 10 min prior to topical application of VC (5.8 μmol) lowered the average number of skin tumors by 58% at 22 weeks after promotion with 12-*O*-tetradecanoylphorbol-13-acetate. Incidence of VC-induced papillomas in mouse skin was also reduced in the animals given DAS orally prior to carcinogen challenge (70%, control vs. 57%, DAS-pretreated). Likewise, pretreatment of female CD-1 mice with DAS by gavage resulted in significant attenuation of skin tumor formation by this carcinogen:

Pretreatment	Initiator	Tumor response at 16 weeks	
		% Mice with papillomas	Average number of tumors/mouse
Trioctanoin	Acetone	0	0
DAS	Acetone	0	0
Trioctanoin	VC	58	1.0 ± 1.6
DAS	VC	28	0.4 ± 0.9

Conclusion: DAS exhibited the protective effects against bacterial mutagenicity of VC and NDMA mediated by rat hepatic S9 and NADPH. Oral administration of DAS to female ICR or CD-1 mice significantly lowered the VC-initiated skin tumorigenesis. The antimutagenic and antitumorigenic effects of DAS against VC or NDMA appear to be associated with its inhibition of CYP2E1 that is responsible for metabolic activation of these carcinogens

Acknowledgements: This work was supported in part by the Bristol-Myers Squipp Laboratory Science Training Program Fund and the American Cancer Society Grant IN31-34 awarded to Y.-J. Surh and the Yonsei Cancer Center Grant awarded to K.-K. Park.

References

1. Wargovich, M.J., Imada, O. and Stephens, L.C. (1992) *Cancer Lett.*, **64**: 39-42.
2. Wattenberg, L.W., Sparnins, V.L. and Barany, G. (1989) *Cancer Res.*, **49**: 2689-2692.
3. Hayes, M.A., Rushmore, T.H. and Goldberg, M.T. (1987) *Carcinogenesis*, **8**: 1155-1157.
4. Brady, J.F., Ishizaki, H., Fukuto, J.M., Lin, M.C., Fadel, A., Gape, J.M. and Yang, C.S. (1991) *Chem. Res. Toxicol.*, **4**: 642-647.
5. Brady, J.F., Li, D., Ishizaki, H. and Yang, C.S. (1988) *Cancer Res.*, **48**: 5937-5940.

ASPARAGINE STIMULATION OF HUMAN COLON CANCER CELLS IS BLOCKED BY VALINE.

Nancy J. Emenaker, PhD, Marc D. Basson, MD, PhD.

Dept of Surgery, Yale University School of Medicine, New Haven, CT 06520-8062.

Asparagine, an amino acid endogenously synthesized, has been reported to stimulate the growth of transplantable colonic tumors in animal models. Since asparagine (Asn) has been observed to induce ornithine decarboxylase activity in some cell lines and valine (Val) blocks this increase in ornithine decarboxylase activity, we hypothesized that Val supplementation would reverse the stimulation of tumor cell proliferation by Asn. We therefore studied the effects of 10 mM Asn and 10.8 mM Val supplementation on the proliferation, motility and differentiation of the human Caco-2 colon cancer cell line which is normally cultured without Asn and in the presence of 0.8 mM Val. Cell doubling times were calculated by logarithmic transformation of serial cell counts and cell motility by monolayer expansion across a type I collagen substrate. Cell differentiation was assessed by the specific activity of the brush border enzyme alkaline phosphatase. 10 mM Asn significantly stimulated proliferation, shortening the duration of the cell cycle to 76.3 $\pm$ 2.4% of control (n=6, p<.001). By contrast, 10.8 mM Val lengthened the doubling time to 184.7$\pm$ 26.2% of control (n=6, p<.01). In combined Val and Asn supplementation, Val overcame the Asn induced mitogenic effects and slowed proliferation. The combination increased the doubling time to 174.4$\pm$9.4% compared to control (n=3, p<.001), not significantly different from the effect of Val alone. Asn also stimulated cell motility to 117.5$\pm$3.5% of control (n=6, p<.005) and inhibited alkaline phosphatase activity to 88.3$\pm$5.7% of control (n=6, p<.05). As was observed for proliferation, Val exhibited the opposite effects to Asn, and furthermore reversed the effects of Asn on motility and alkaline phosphatase. Thus, the promotion of tumor growth by Asn *in vivo* can be modelled in cell culture. Asn stimulates proliferation and motility and inhibits differentiation in culture, effects which would be expected to result in more malignant biology *in vivo*. Furthermore, the effects of Asn can be blocked by Val supplementation. Although the mechanism of these effects and their relationship to ornithine decarboxylase activity await further exploration, these results raise the possibility that pharmacological Val supplementation can reverse the deleterious effects of exogenous or endogenous Asn on tumor growth.

REGULATION OF COLONIC EPITHELIAL DIFFERENTIATION BY TYROSINE PHOSPHORYLATION. Z Rashid, MBBS, FRCS, MD Basson, MD, PhD.

Departments of Surgery, Yale University and University of Connecticut

Colonic epithelial differentiation is closely regulated during normal renewal and maturation, and deregulated in malignant transformation. Since tyrosine phosphorylation influences differentiation in other cell types, we studied the role of tyrosine phosphorylation in colonocyte differentiation, using human colonic Caco-2 cells as a model and the brush border enzymes alkaline phosphatase (AKP) and dipeptidyl peptidase (DPDD) as differentiation markers. We studied three tyrosine kinase inhibitors with different modes of action and specificities, genistein, erbstatin analog (EA), and tyrphostin, as well as the tyrosine phosphatase inhibitor orthovanadate. AKP and DPDD specific activity were assayed in protein-matched cell lysates by synthetic substrate digestion and correlated with tyrosine phosphorylation of intracellular phosphoproteins by Western blotting. Genistein (5-75 μg/ml) dose-dependently stimulated AKP and DPDD with maximal effect to 158.6±17.51 and 228.6±37.1% of control respectively (n=12,p<0.001) Tyrphostin (25 μM) similarly stimulated AKP and DPDD by 138±6.6 and 131.8±1.5 % of control (n=12, p<0.001). Unexpectedly, EA (0.1-10μM) had the opposite effect, dose-dependently inhibiting AKP and DPDD to 85.2±7.3% and 74.6±4.6% of control respectively (n=12,each p<0.001). Orthovanadate had a discordant effect on enzyme activity, dose-dependently increasing AKP maximally to 233±16.10% of basal but decreasing DPDD to 47.25±3.86% (n=9, p< 0.001 each). The effects of each agent were preserved when proliferation was blocked with mitomycin C, suggesting that the modulation of phenotype by these agents was independent of proliferation. Three phosphoprotein bands (87, 116 and 194 kD) were affected discordantly by these agents. EA and orthovanadate increased the tyrosine phosphorylation of these bands, while genistein and tyrphostin decreased their phosphorylation. The different effects of these modulators of tyrosine phosphorylation suggest that at least two independent tyrosine phosphorylation events modulate intestinal epithelial differentiation. Tyrosine phosphorylation of the identified phosphoproteins may regulate colonocytic differentiation.

The *Real World* of Phytochemicals (I): A Geographic Survey of Nutraceutical/Phytonutrients that America Eats

An Approach to Standardization

Carl M. Ruyter, Ron Carstens
BeWell Inc., 12420 Evergreen Drive, Mukilteo WA 98275

The well publicized recommendations of the National Cancer Institute and the Food Guide Pyramid are 3 - 5 servings of vegetables and 2 - 4 of fruits per day respectively to reduce your risks of heart disease and cancer [1]. Numerous studies have shown mortality and incident reduction when sufficient quantities of vegetables and fruits were consumed. A recent study showed a 70% reduction in strokes among men who consumed at least eight 1/2 cup servings of fruits and vegetables per day[2].

The on-going research on the molecular working mechanisms of different phytonutrients is mainly based on animal studies and *in-vitro* techniques where pharmacokinetic problems like bioavailability, absorption, distribution and elimination are not included in the experimental design. Therefore recommendations for the daily dosage of certain phytonutrients like Sulforaphane, Lyopene, or decomposition products of Glucobrassicin are still lacking a real scientific basis.

A "serving" size would seem to be a reasonable beginning standard for the nutraceutical/phytonutrient quantity which is a RDM - "recommended daily minimum". However, serving size varies and as is well known, phytochemical content varies in plants from region to region, season to season and is influenced by horticultural treatment.

In an effort to discern typical geographic differences in the U.S., peak season samples of 4 different vegetables (carrots, tomatoes, cabbage, broccoli) were randomly obtained from local supermarkets and analyzed for the levels of

Lycopene, Betacarotene, the glucosinolate Glucobrassicin[3] (a precursor of Indole-3-carbinol) and the isothiocyanate Sulforaphane[4] (aglycone of Glucoraphanin).

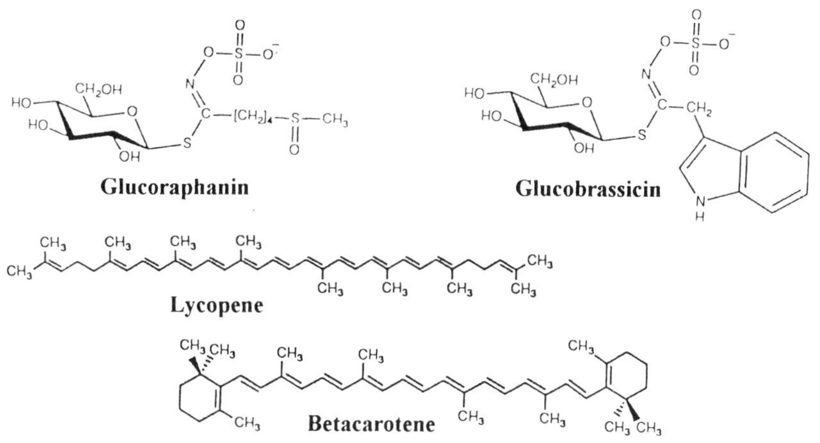

Glucoraphanin **Glucobrassicin**

Lycopene

Betacarotene

HPLC determination supported by UV/DAD-detection was utilized throughout the study. Results from six regions (CA, GA, IL, NJ, TX, WA) indicate a wide variation in phytonutrient content for all the vegetables sampled. However, an overall average content with these chemopreventive compounds was applied to typical servings and frequency of consumption per serving to arrive at a tentative daily phytonutrient consumption for American people.

---

1 U.S. Department of Agriculture and U.S.Department of Health and Human Services.

2 Gilman, M.W., Cupples, L.A., Wolf, P.A. (1995) Protective Effect of Fruits and Vegetables on Development of Stroke in Men. JAMA **273**, 1113-1117.

3 Betz, J.M., Fox, W.D. (1994) High Performance Liquid Chromatographic Determination of Glucosinolates in Brassica Vegetables. In: Food Phytochemicals for Cancer Prevention I: Fruits and Vegetables. Huang M.T., Osawa, T., Ho, C.T., Rosen R.T. (eds.) ACS Symposium Series No. 546, p.181-196.

4 Zhang, Y., Talalay, P., Cho, C.-G., Posner, G.H. (1992) A major inducer of anticarcinogenic protective enzymes from broccoli: Isolation and elucidation of structure. Proc. Natl. Acad. Sci. USA **89**, 2399-2403.

The *Real World* of Phytochemicals (II): Loss of Phytonutrient Content in Juicing Economical Calculations for the Health Conscious Kitchen

Ute F. Fischer, Ron Carstens
BeWell Inc., 12420 Evergreen Drive, Mukilteo WA 98275

Juicing has been and continues to be a popular way to "eat your veggies" and fruits in the 90's. Juice bars are proliferating and juicing machine sales continue to thrive. In addition, the sale of refrigerated ready-to-consume juices through grocery outlets is rapidly growing.

This study examines the fate of 3 important phytonutrients - Betacarotene, Lycopene, Sulforaphane - through the juicing process of carrots, tomatoes and. broccolirespectively. Since numerous health benefits are touted for this food preparation method; it is important to know if juicing delivers the significant quantities of chemoprotective phytonutrients that are recommended (e.g. 5 servings/day) for good health.

These results vary of course, but in general they indicate that juicing delivers about one-half (30-60%) of the important phytonutrients present in the original food. The rest are to be found in the discarded pulp. Instant analyses were necessary to prevent phytonutrient degradation by the endogeous enzymes. Material balances on the juicing yields together with sophisticated HPLC analyses confirm a high degree of accurracy. The peak identity was confirmed by rentention time of reference compounds and DAD/UV spectral analysis.

These losses are not surprising when one considers that some phytonutrients resides in or near the cell membranes and would only be incidentally carried through filters in the juicer into the food juice. If one is to obtain the full measure of phytonutrients available in vegetables by juicing, then it would be advisable to double the juice input and therefore output to be assured of delivery.

EVALUATION OF ENDOGENOUS RETINOIC ACID RECEPTOR FUNCTION IN CERVICAL TUMOR CELLS.

Doris M. Benbrook, Kevin W. Brewer, Taylor L. Jordan, Coy Heldermon, and Kent Robinson. Section of Gynecological Oncology, Department of Obstetrics and Gynecology, University of Oklahoma Health Sciences Center, P.O. Box 26901, Oklahoma City, OK 73190.

The use of natural and synthetic retinoids in the prevention and treatment of cervical cancer requires detailed knowledge of retinoic acid receptor function in cervical tumor cells. There are two classes of nuclear receptors (RARs and RXRs) which form homodimers and heterodimers. These dimers bind to and transactivate retinoic acid response elements (RAREs). RXREs are RAREs which are RXR/RXR-specific. Several orphan receptors have been identified which dimerize with RXR's and bind to these DNA elements.

The cervical tumor cell lines (HeLa, SiHa and CC-1) evaluated did not differ from normal keratinocytes in that the RAR/RXR-mediated pathway dominated over that mediated by RXR/RXR, as demonstrated by the retinoid-induced transactivation of the RARE from the RARβ gene but not the RXRE from the CRBPII gene. The RARE was transactivated to a greater extent by 9 cis retinoic acid than by all trans retinoic acid, but not transactivated by beta carotene. The degree of transactivation however was 4 to 5 fold less than that reported for normal keratinocytes.

A quantitative EMSA (electrophoretic mobility shift assay) evaluation of the endogenous DNA binding specificity in nuclear extracts of these lines demonstrated that the strength and specificity of binding was greater for the RXRE than for the RARE. The RXRE competed binding to the RARE with an inhibition constant (Ki) of 0.5 micromolar, while the RARE competed the RXRE with a 10 fold higher Ki of 5 micromolar. Both the RARE and RXRE competed binding to self with Ki's of 0.5 and 0.3 micromolar, respectively. Expression of RARβ in HeLa which normally does not express this receptor did not alter the Ki's, but slightly increased RARE transactivation. This lack of correlation of the strength and specificitv of binding with transactivation is not surprising, given the number of potential dimers that bind to these DNA elements.

Enhancement of 4-(Methylnitrosamino)-1-(3-pyridyl)-1-butanone (NNK)-Induced Lung Tumorigenesis in Mice by Oral Administration of Fresh Garlic Homogenate. Jun-Yan Hong, Theresa J. Smith, Wei-qun Huang, Yongyu Wang and Chung S. Yang *Laboratory for Cancer Research, College of Pharmacy, Rutgers University, Piscataway, NJ 08855-0789*

Garlic is widely used in culinary practice and sometimes as a popular folk medicine. Recent epidemiological studies indicated that frequent consumption of garlic and other allium vegetables was associated with a reduction in risk of certain human cancers. In animal studies, garlic-related organosulfur compounds were shown to inhibit tumorigenesis induced by different chemical carcinogens. Our previous work demonstrated that diallyl sulfide, a component of garlic oil, and its metabolite, diallyl sulfone inhibit the metabolic activation and lung tumorigenicity of the tobacco-specific nitrosamine 4-(methylnitrosamino)-1-(3-pyridyl)-1-butanone (NNK) in mice. The present study investigated the effect of fresh garlic preparations on NNK-induced lung tumorigenesis in female A/J mice. In two separate experiments, a 3-day treatment with fresh garlic homogenate (FGH, 5 g garlic/kg/day, p.o.) prior to a single administration of NNK (2 mg/mouse, i.p.) significantly increased (by 88 and 52%) the tumor multiplicity (tumors per mouse). Neither a similar 3-day treatment with fresh garlic juice (5 g garlic/kg/day, p.o.) nor a 7-day FGH treatment (5 g garlic/kg/day, p.o.) started one week after the NNK administration showed any enhancing effects on the lung tumorigenesis. There were also no changes in lung tumor multiplicity in mice receiving a lower daily dose of FGH (1 g garlic/kg, given either prior to or after NNK), or 3% FGH in their drinking water for a total of 9 weeks started 1 week prior to NNK. These results suggest that FGH, but not fresh garlic juice, contains compound(s) which enhances the lung tumorigenicity of NNK in the early initiation stage. To understand the relationship between the enhancing effect of FGH on tumor formation and the metabolic activation of NNK, microsomal activities in activating NNK were determined in the lungs and livers of mice pretreated with FGH (1 and 5 g garlic/kg/day, p.o. for 3 days). In contrast to its enhancing effect on tumorigenesis, FGH treatment led to a dose-dependent inhibition in the rates of formation of keto aldehyde and keto alcohol (products of NNK α–oxidation) in both lung and liver microsomes. In mice receiving the high dose of FGH (5 g garlic/kg/day for 3 days with the last dose 2 hr prior to NNK), a significant reduction in NNK-induced DNA methylation was observed in the livers, which appears to be consistent with the inhibitory effects of FGH on NNK α-hydroxylation. In the lungs of these FGH-treated mice, however, the level of NNK-induced DNA methylation was not altered at early time points and a 30% increase was observed at 48 hr after NNK administration. Our results clearly demonstrate the enhancing effect of high dose FGH on NNK-induced lung tumor formation in mice. Before the mechanisms of the presently observed enhancing effect are elucidated, caution should be applied in the use of garlic products in large quantities by humans (Supported by New Jersey State Cancer Research Commission Grant 693-016 and NIH Grant ES03938).

GARLIC PREVENTS ENHANCEMENT OF ONCOGENIC CELL TRANSFORMATION DUE TO CO-EXPOSURE OF BPL WITH NON-TRANSFORMING DOSES OF THE PROMOTER TPA. D. Green[1], S. Green[2], N. Baturay[1]. (College of Pharmacy, St. John's University[1], New York; Bronx Lebanon Hospital Center[2], New york)

3T3 mouse embryo cells cultures, subcloned in this laboratory were used throughout these experiments. Colonies were chosen on the basis of their sensitivity, both to topoinhibition and contact-inhibition of division.

β-Propiolactone (BPL), a combustion by-product is a defined direct acting chemical carcinogen. We exposed log phase cells grown in 100 mm petri dishes to increasing doses of BPL (0, 0.1, 1.0, 10.0 ng/mL), alone or in concert with increasing doses of garlic.

0-13-tetradecanoyl phorbol myristate acetate (TPA), an oil extracted from the roots of the *Croton tiglium* plant, is a tumor promoter. Promotion was determined by repeated exposures of these cells to TPA, following the initial low dose of the carcinogen.

Cells co-exposed to both BPL and TPA further highlighting the concept of the importance in carcinogenesis of multiple compound exposures.

Our results indicate that the addition of an optimal amount of garlic and/or its active metabolites to cell cultures with induced neoplastic potential significantly decreases the number of transformed foci/dish.

NUMBER OF TRANSFORMED COLONIES/DISH*

BPL	WITHOUT GARLIC	WITH GARLIC (0.5 mg/mL)
0	1.33 ± 0.57	1.00 ± 1.00
0.1	15.00 ± 4.58	3.66 ± 1.15
1.0	25.33 ± 1.53	5.66 ± 0.58
10.0	29.00 ± 2.00	8.33 ± 1.15

* In the presence of 0.1 μg/mL TPA

Each experiment was performed in triplicate with the appropriate controls. Cytotoxicity determinations were performed simultaneously on replicate cultures.

In addition, when the cocarcinogen catechol was added simultaneously with low doses of BPL, the same protective effects were observed.

Phorbol ester and 12(S)-HETE modulate in vitro keratinocyte differentiation

R.A. Hagerman, S. M. Fischer and M.F. Locniskar

The effect of the tumor promoter 12-0-tetradecanoylphorbol-13-acetate (TPA) on cellular functions such as adhesion is known to be mimicked by the lipoxygenase metabolite of arachidonic acid, 12(S)-hydroxyeicosatetraenoic acid (HETE). Experiments were designed to investigate the hypothesis that TPA and 12(S)-HETE have similar effects on keratinocyte differentiation. In primary cultures of mouse keratinocytes, differentiation can be induced by increasing medium calcium concentration from 0.05 mM to 0.12 mM. This treatment induces the expression of keratin 1 (K1) mRNA and protein at 24 hours. Treatment with 50 nM TPA or 10 nM 12(S)-HETE (but not 12(R)-HETE) simultaneously with increased calcium inhibits the expression of K1 mRNA and protein at 24 hours. Additional experiments were designed to investigate the role of the cellular receptor for TPA, protein kinase C (PKC), in this process. Pre-treatment of cultures with 60 nM bryostatin, which tightly binds PKC, inhibited the effect of TPA and 12(S)-HETE on K1 mRNA expression. However, direct assay of PKC activity does not support a role for PKC in this process. In control low calcium cultures (0.05 mM), PKC activity is distributed 80:20 $\pm$ 3 (cytosolic:particulate); in control high calcium cultures (0.12 mM), the distribution is 76:24 $\pm$ 8. After 15 minutes of 50 nM TPA in high calcium medium, PKC distribution is 77:23 $\pm$ 3, and after 10 nM 12(S)-HETE, 80:20 $\pm$ 4. However, after 15 minutes of 500 nM TPA in high calcium medium, PKC distribution shifts to 38:62 $\pm$ 3. These results are reflected in Western blots of PKC α, the primary isoform of PKC present in these cultures. Thus, TPA and 12(S)-HETE modulate keratinocyte differentiation similarly, but this effect does not appear to occur by direct activation of PKC.

Supported by AICR and NIH CA46886.

Effects of Saponin, Curcumin and Chlorophyllin on Colonic Cell Proliferation and Multiple Immune Functions in Rats

B.A. Magnuson, J.H. Exon, E.H. South and K, Hendrix. Department of Food Science and Toxicology, University of Idaho, Moscow, ID 83844-2201

The objective of this study was to investigate the effects of various non-nutritive food compounds on colonic cell proliferation indices and immune function as a means of understanding the possible anticarcinogenic effects of these compounds. Sprague Dawley rats were fed palatable forms of either vehicle, saponin (40 mg/kg), curcumin (40 mg/kg) or chlorophyllin (40 mg/kg) 5 days a week for 6 weeks. The crypt height, labeling index, proliferating index, and labeled cell distributions in both the ascending and descending colon were determined using proliferating nuclear cell antigen (PCNA) immunohistochemistry. Immune parameters measured were IgG response to keyhole limpet hemocyanin (KLH) immunization; delayed-type hypersensitivity; and natural killer cell activity. In ascending colon, animals fed the chlorophyllin diet had higher a labeling index ($p<0.05$) and higher proliferative index ($p<0.005$) compared with other diet groups. Similar trends were observed in the descending colon, but differences were not statistically significant. Labeled cells in the colons of animals fed the saponin diet were shifted upwards in the colonic crypt from the lower third to the middle third of the crypt. This was observed in both the ascending and descending colon. Serum IgG levels to KLH were increased approximately 2-fold in saponin-fed rats ($p<0.005$) and 1.5-fold in curcumin-fed rats ($p<0.01$) compared to vehicle-fed rats. A slight enhancement of other immune functions were also observed in animals fed saponin and curcumin. There were no significant differences in body weights during or at the completion of the experiment. Our findings suggest that saponin and curcumin may exert a portion of their anticarcinogenic activity through enhancement of immune function. The elevation of colonic cell proliferation indices by chlorophyllin suggests a possible promotional rather than anticarcinogenic effect of this compound in the colon.

INHIBITION OF CELL GROWTH IN AN ANDROGEN-INDEPENDENT RAT PROSTATIC CELL LINE (AIT) BY PHYTOCHEMICALS. **Shuk-mei Ho and Jeremy Chun, Department of Biology, Tufts University, Massachusetts.**

Carcinoma of the prostate is now the most common cause of cancer in males and the second most common cause of cancer deaths among men (Silverberg E and Lubera JA, CA 39:3-20, 1989). The magnitude of the problem is expected to increase mainly due to an increase in life expectancy and early detection by the prostate-specific antigen test. Since the early work of Huggins (Huggins CE and Hodges CV, Can Res 1:293-297, 1941), carcinoma of the prostate has been recognized as an androgen-sensitive tumor. Thus the mainstay treatment for the disease has been androgen ablation or blockade of androgen action. Although androgen-deprivation therapies are fairly effective initially, their effectiveness are unfortunately temporary. Patients treated with one of these therapies will inevitably encounter a relapse within five years, at which juncture the disease will journey down a terminal course that is typified by uncontrolled growth of an androgen-independent tumor.

As androgen-independent cells are unaffected by androgen levels, they continue to proliferate after androgen-deprivation and pose a major challenge to therapeutic control of the disease. The long-term goal of our laboratory is to develop effective dietary intervention for late stage or androgen-independent prostate cancer growth. To this end we have recently established a rat prostatic cancer cell line, the AIT, that grows in the absence of androgen. Growth kinetic studies demonstrated that quercetin, luteolin and genistein are effective in inhibiting the growth of this cell line. Genistein is the most effective growth inhibitor; at concentrations of 10^{-5} M - 10^{-7} M it reduced AIT growth to 20-30% of that of controls. Both luteolin and quercetin were only effective at 10^{-5} M, achieving a 70-80% reduction in cell number. Current investigations now focus on revealing the mechanisms by which these and other phytochemicals act as antiproliferative agents for prostatic cancer cells.

(supported in part by grant numbers 60923, 15776 and 62269 awarded by NCI)

Glyceride forms of butyric acid induce apoptosis in colon cancer cells, but differ in their effects on nuclear proteins, including the histones.

Janet G. Smith and J. Bruce German Department of Food Science and Technology, University of California, Davis CA 95616.

Butyric acid (BuA) is the 4-carbon fatty acid present in the diet directly from bovine milk consumption or indirectly from bacterial fermentation of dietary fiber. The latter releases milimolar concentrations of BuA into the lower bowel. *In vitro* experiments have demonstrated that physiological levels of BuA induce apoptosis in colon cancer cell lines but not in healthy cells. The mechanism for this is unknown, but has been associated with BuA-induced hyperacetylation of the histone proteins, and its parallel induction of differentiation. Thus, BuA is being suggested as the fiber product at least partly responsible for the protective effects of a high fiber diet against colon cancer. We have investigated the effects of glyceride forms of BuA on nuclear proteins, and their abilities to induce apoptosis in the human colon cancer cell line HT-29. Comparisons were made between sodium butyrate, tributyrin (3 BuA per triglyceride [TG]), dibutyryl stearin (2 BuA and 1 stearate per TG), BuA rich milkfat (1 BuA per TG on the sn-3, plus 2 longer chain fatty acids), tricaproin (3 caproic acids [6 carbons] per TG) and tricaprylin (3 caprylic acids [8 carbons] per TG). Examination of the treated cells by DNA agarose electrophoresis, flow cytometry, SDS-PAGE and Triton acid urea electrophoresis revealed that each of the treatments had an effect on the HT-29 cells. However, hyperacetylation did not necessarily correlate with apoptosis induction. These glyceride forms of fatty acids including butyric acid may target different genes/mechanisms on HT-29 cells leading to differential effects on chromatin structure and apoptosis.

Comparison of Pure IP_6 and High-Bran Diet in the Prevention of DMBA-Induced Mammary Carcinogenesis

Ivana Vucenik*†, Guang-yu Yang†, Abulkalam M. Shamsuddin†
*Department of Medical and Research Technology, †Department of Pathology
University of Maryland School of Medicine, Baltimore, MD

While most studies of diet and breast cancer are focused on the role of fat, very few have adressed the effect of fiber. Inositol hexaphosphate (IP_6) is abundant in cereals and legumes, particularly the bran part of mature seeds. The aims of this study were: a) to investigate whether dietary fiber containing high IP_6 shows a dose-response inhibition of DMBA-induced rat mammary carcinogenesis, and b) to test the hypothesis that IP_6 is the active component of fiber associated with low cancer risk. Starting at 2 weeks before DMBA initiation, rats were divided into 5 groups fed with AIN76A diet only, AIN76A containing 5%, 10% or 20% Kellogg's All Bran. The fifth group received 0.4% IP_6 (an amount equivalent to the IP_6 content in 20% bran) in drinking water. Following carcinogen administration, the rats remained on these regimens for 29 weeks. The results at completion are as follows:

Diet Group	Number of Rats/Group	Tumor Incidence	Number of Tumor/Group	≥ 3 Tumors
DMBA only	38	79.0 %	84	47 %
5 % Bran	38	65.8 %	73	56 %
10% Bran	40	67.5 %	74	44 %
20% Bran	40	70.0 %	87	36 %
IP_6 0.4%	38	52.6 %*	43**	15 %**

Compared to DMBA-control, significant at $p<0.02$* or $p<0.03$** (χ^2 test)

Compared to the carcinogen control, at 29 week rats fed 5%, 10% and 20% bran had 16.7%, 14.6% and 11.4% reduction of tumor incidence respectively (not significant). However, rats given 0.4% IP_6 in drinking water, equivalent to that in 20% bran, had a 33.5% reduction in mammary tumor incidence ($p<0.02$) and 48.8% less tumors ($p<0.03$).

These data show that supplemental dietary fiber in the form of bran exhibited a negligible inhibitory effect. In contrast, animals given IP_6 showed significant reduction in tumor number, incidence and multiplicity, suggesting that pure IP_6 in drink is definitively more effective than 20% bran in diet. The data support our hypothesis that IP_6 is the active substance responsible for cereal's beneficial anti-cancer effect.

[Supported by a Matching Grant MG92B02 from American Institute for Cancer Research and Kellogg Co (I.V.)]

The Chemopreventive Agent Apigenin Blocks G_2/M Cell Cycle Progression by Inhibiting p34^{cdc2} Kinase. Lepley, Denise M.[1], Birt, Diane F.[1], and Pelling, Jill C.[2]
[1]University of Nebraska Medical Center, Eppley Institute for Cancer Research, Omaha, NE 68198
[2]University of Kansas Medical Center, Department of Pathology and Laboratory Medicine, Kansas City, KS 66160

Apigenin is a nontoxic and nonmutagenic plant flavonoid which has been shown to significantly inhibit mouse skin tumorigenesis when applied topically. Since apigenin is an effective chemopreventive agent in mice, it may represent an alternative sunscreen agent in humans. We have investigated the molecular mechanism(s) by which apigenin inhibits skin tumorigenesis. Initial studies examined the effects of apigenin on the cell cycle. DNA flow cytometric analysis indicated that a 24 hour culture with apigenin caused a G_2/M arrest in human HL-60 cells, as well as two mouse skin derived cell lines, C50 and 308, that represent normal epidermis and papilloma cells, respectively. The G_2/M arrest was fully reversible after an additional 24 hours in medium without apigenin. In view of the fact that p34^{cdc2} is the key enzyme regulating G_2/M progression, we investigated the effects of apigenin on p34^{cdc2} kinase. Western blot and immune complex kinase assays using whole cell lysates from 308 and C50 cells treated with increasing doses of apigenin for 24 hours demonstrated that apigenin-treatment did not change the steady-state level of p34^{cdc2}, but did inhibit p34^{cdc2} H1 kinase activity. In view of the fact that cyclin B1 must accumulate above a threshold level for cell cycle progression to occur, we measured the level of cyclin B1 protein in apigenin-treated keratinocytes. Western blot analysis showed that apigenin-treatment significantly inhibited the accumulation of cyclin B1 during G_2. The reduced amount of p34^{cdc2}/cyclin B1 complex would explain our observation of cell cycle arrest at the G_2/M checkpoint, and may also account for the observed decrease in p34^{cdc2} kinase activity. We next employed immufluorescent staining with anti-cyclin B1 antibody to investigate the translocation of p34^{cdc2}/cyclin B1 complexes from the cytoplasm to the nucleus. Immunofluorescence experiments illustrated that cyclin B1 was efficiently translocated from the cytoplasm to the nucleus in apigenin-treated cells. In conclusion, we hypothesize that apigenin exerts its chemopreventive effect, in part, by inhibiting the mitotic kinase activity of p34^{cdc2} , cyclin B1 accumulation, and, ultimately, G_2 to M phase progression.

The Soy Isoflavone Genistein Induces Micronucleus Formation in Mouse Splenocytes *in vitro* but not *in vivo*.

Ian R Record, Mary Jannes, Ivor E Dreosti and Roger A King

CSIRO Division of Human Nutrition, Adelaide, S Australia 5000

The effects of genistein, one of the major soybean isoflavones, genistin the glucosylated form of genistein and etoposide, a topoisomerase 11 inhibitor have been studied in mouse splenocytes in culture. Genistin (25 µmol/L), genistein (25 µmol/L) and etoposide (0.1 µmol/L) all induced the production of large numbers of micronuclei, however genistein at 12.5 µmol/L or 2.5 µmol/L had no clastogenic effect. In a second study mice were gavaged with 20 mg genistein/d for five days (approximately equivalent to a 70 Kg human consuming 2.8 Kg soybeans) and the micronucleus frequency determined. There was no observable increase in the micronucleus frequency even though the plasma genistein concentrations in the treated animals were found to be 9.2 ± 2.0 µmol/L compared with the control animals 0.1 ± 0.0004 µmol/L. The results show that even though genistein is capable of inducing micronucleus formation, an event associated with genetic damage, plasma levels are unlikely to be sufficiently elevated to produce such an effect.

Genistein Inhibits Growth of B16 Melanoma Cells *In Vivo* and *In Vitro*

Ian R Record, Jessica L Broadbent', Roger A King, Ivor E Dreosti', Richard J Head' and Anne L Tonkin

CSIRO Division of Human Nutrition, Adelaide, S Australia 5000 and theDepartment of Clinical and Experimental Pharmacology, University of Adelaide, S Australia 5000

Consumption of soy products has been linked to a reduction in mortality and morbidity of a number of cancers. Genistein, one of the principal soy isoflavones, has been shown to inhibit the growth of a number of tumour cell lines *in vitro*, however a role of genistein in retarding tumour growth *in vivo* is less well documented. In this study, in addition to examining the effects of genistein *in vitro* on the growth of murine B16 melanoma cells and of non-transformed fibroblasts, we have also examined the effects of feeding a genistein-rich diet on subcutaneous growth of the tumour cells in mice. *In vitro*, the melanoma cells became increasingly sensitive to genistein with increasing time of exposure, culminating in a stable IC_{50} of 12.5 µmol/L after five days. Growth of Balb/c 3T3 fibroblasts was inhibited with an IC_{50} of 10 µmol/L after five days. Growth of the solid tumour was inhibited by 50% when mice were fed genistein (0.33 mg per kg diet) for one week before and for one week after inoculation with B16 melanoma cells. Plasma genistein levels at the time of tumour removal were 1.1 µmol/L, which is similar to levels reported in humans consuming diets high in soybeans or soybean products while the control animals had no detectable genistein in plasma. This study provides additional *in vivo* evidence suggesting that the isoflavonoid genistein could be the significant anticarcinogenic component of soyfoods.

INDEX